THE FINITE
ELEMENT
METHOD IN
Thermomechanics

Titles of related interest

Analytical and numerical methods in engineering rock mechanics
E. T. BROWN (ed.)

Boundary element methods in solid mechanics
S. L. CROUCH & A. M. STARFIELD

The boundary integral equation method for porous media flow
J. A. LIGGETT & P. L-F. LIU

Computers in construction planning and control
M. J. JACKSON

Earth structures engineering
R. J. MITCHELL

Floods and drainage
E. C. PENNING-ROWSELL *et al.*

Geology for civil engineers
A. C. McLEAN & C. D. GRIBBLE

Hemispherical projection methods in rock mechanics
S. D. PRIEST

Hydraulics in civil engineering
A. CHADWICK & J. MORFETT

Introducing groundwater
M. PRICE

Numerical methods in engineering and science
G. de VAHL DAVIS

Plastic design
P. ZEMAN & H. M. IRVINE

Rock mechanics
B. H. G. BRADY & E. T. BROWN

*Theory of vibration, with applications**
W. T. THOMSON

* Originally published by, and available in North America from, Prentice-Hall Inc.

THE FINITE ELEMENT METHOD IN
Thermomechanics

Tai-Ran Hsu

Department of Mechanical Engineering,
University of Manitoba,
Winnipeg, Manitoba, Canada

Boston
ALLEN & UNWIN
London   Sydney

Allen & Unwin, Inc.,
8 Winchester Place, Winchester, Mass. 01890, USA

Allen & Unwin (Publishers) Ltd,
40 Museum Street, London WC1A 1LU, UK

Allen & Unwin (Publishers) Ltd,
Park Lane, Hemel Hempstead, Herts HP2 4TE, UK

Allen & Unwin (Australia) Ltd,
8 Napier Street, North Sydney, NSW 2060, Australia

First published in 1986

Library of Congress Cataloging in Publication Data

Hsu, Tai-Ran
   The Finite Element Method in Thermomechanics.
Bibliography: P.
Includes Index.
1.   Thermal Stresses−−Data Processing.   2. Finite
Element Method−−Data Processing.   3.   Continuum
Mechanics−−Data Processing.   4. TEPSAC (Computer Program).
5. Thermoelasticity−−Data Processing. I. Title.
II. Title: Thermomechanics.
TA418.58.H78   1986   620.1'121   86−7884
ISBN 0−04−620013−4 (alk. paper)

British Library Cataloguing in Publication Data

Hsu, Tai-Ran
   The finite element method in
   thermomechanics.
1. Thermomechanics————Mathematics   2. Finite
I. Title
621.402'0151'5353   QC318.F5
ISBN 0−04−620013−4

Set in 10 on 12 point Times by
Blackpool Typesetting Services Ltd, Blackpool
and printed in Great Britain by
Anchor Brendon Ltd, Tiptree, Essex

To my wife Su-Yong
and Jean, Euginette, Leigh

PREFACE

The rapid advances in the nuclear and aerospace technologies in the past two decades compounded with the increasing demands for high performance, energy-efficient power plant components and engines have made reliable thermal stress analysis a critical factor in the design and operation of such equipment. Recently, and as experienced by the author, the need for sophisticated analyses has been extended to the energy resource industry such as *in-situ* coal gasification and *in-situ* oil recovery from oil sands and shales.

The analyses in the above applications are of a multidisciplinary nature, and some involve the additional complexity of multiphase and phase change phenomena. These extremely complicated factors preclude the use of classical methods, and numerical techniques such as the finite element method appear to be the most viable alternative solution.

The development of this technique so far appears to have concentrated in two extremes; one being overly concerned with the accuracy of results and tending to place all effort in the implementation of special purpose element concepts and computational algorithms, the other being for commercial purposes with the ability of solving a wide range of engineering problems. However, to be versatile, users require substantial training and experience in order to use these codes effectively. Above all, no provision for any modification of these codes by users is possible, as all these codes are proprietary and access to the code is limited only to the owners.

The present approach attempts to strike a compromise between the above two extremes. The theoretical derivations presented in the proposed book, in particular Chapters 2–4, will serve as the basis for a computer code TEPSAC (*T*hermo*e*lastic-*p*lastic *S*tress *A*nalysis with *C*reep). The code, although limited to two-dimensional plane, or three-dimensional axisymmetric structures, can handle large classes of thermomechanical problems using a simplex element algorithm. The thermal and mechanical analyses are quasi-coupled by automatic transferring of data back and forth between these two major components. The full listing of the basic TEPSAC code will be included in Appendix 5 so that the readers may relate the corresponding theory in the text to the program in the Appendix, in addition to being able to use the code for their own research and development work.

In addition to the transient (or steady state) thermal–elastic–plastic-creep capability that has been included in the basic TEPSAC code, additional topics such as Fourier series approximation for nonaxisymmetric loading, elastodynamic stress analysis, thermofracture mechanics, finite strain theory, coupled thermoelastic-plasticity and program organization will also be included. These topics were carefully selected on the basis of industrial needs.

PREFACE

As the intention of this book is to embrace as wide a range of topics in the nonlinear thermomechanical analysis as possible, it is necessary to cut back on some of the fundamental principles of the finite element method. Topics such as element variety, in-depth study of the variational principle, integration schemes, convergence criteria, etc. will be omitted in the text. However, these are adequately discussed in many published books on the finite element method.

Chapter 1 presents only the bare fundamental principle of the finite element method. The essence of the discretization concept described in this chapter will enable the readers to appreciate the potential of the finite element method and to acquire a correct perception of its nature. The steps of the general finite element method will be presented in a way that includes the formation of some key equations. At the end of Chapter 1, readers are expected to have a firm grasp of this technique and its application in various branches of the engineering science.

Chapter 2 is devoted to the derivation of the finite element formulation of heat conduction analysis. A review of basic mathematical expressions for heat conduction in solids and convection in fluids as well as appropriate initial and boundary conditions will be presented first. It will then be followed by the discretization of axisymmetrical solids, the derivations of the element's thermal equilibrium equations and the conductivity and heat capacity matrices. An alternative method of implementing boundary conditions to the discretized models will also be included. Axisymmetric solid geometry was chosen, as planar structures can often be treated as degenerated axisymmetric geometry with a large artificial radius of curvature.

Chapter 3 deals with the finite element formulation of thermoelastic-plastic analysis, and it plays a very important part in the TEPSAC code. A brief review of relevant theories of elasticity and plasticity will be presented first. A comprehensive derivation of the element equations in both elastic and elastic–plastic cases is provided, including both the effects of thermal and the associated material properties variation. Chapter 3 also presents a detailed formulation for both isotropic and kinematic hardening behavior of the structural material. A clear distinction is made between these hardening rules and their respective applications. The solution procedure offers the reader a clear idea of how the respective mathematical expressions are implemented in the TEPSAC code.

Chapter 4 deals with the derivation of the finite element algorithm for structures subject to creep. It will be demonstrated in this chapter that the only major required modification for the creep effect on the main thermo-elastic–plastic formulation is the force matrix. The reader will follow the logical sequence of a solution presented in the book and, it is hoped, will appreciate the relative simplicity of this important aspect of nonlinear thermomechanical analysis contained in the basic TEPSAC code.

Many engineering applications may involve axisymmetric structures subject to nonaxisymmetric loadings. Typical are large horizontal pressure vessels or pipelines containing heavy fluids. The analysis is clearly of a three-dimensional nature. However, such analysis in the thermoelastic–plastic regime would require an enormous amount of computer time and storage. An approximation, by means of Fourier series, for the third directional variation of the physical quantities in load description and displacement field in a two-dimensional analysis code, may prove to be a great deal more economical and hence more practical. Chapter 5 presents the theory and algorithm used in conjunction with such a technique. Although this method has been proposed in the past for elastic stress analysis, the present approach extends the use of this technique into the elastic–plastic region. A numerical example on the sag of a heavy pipe illustrates the merit of this method.

Chapter 6 presents the finite element equations for stress wave propagation through structures, caused by dynamic loads such as impact or explosions. Again, the dynamic effect could be treated as a special module to the base TEPSAC code for the cases involving strong thermal effect. A numerical example is included to illustrate the physical significance of the dynamic effect.

The importance of fracture mechanics analysis on large structures has been well recognized by engineers and researchers. While the theory of linear fracture mechanics for brittle materials has been well established, the treatment of initially stable, but eventually unstable, cracks commonly occurring in ductile materials is less complete. In view of the many such recent catastrophic structural failures, emphasis on the fracture analysis of ductile materials is warranted.

Chapter 7 presents specifically some of the unique techniques that could be used to predict the growth of cracks in structures under combined thermomechanical loads. A review of the literature on linear elastic and elastic–plastic fracture mechanics is included with the presentation of brief relevant mathematical expressions, as well as the application of the finite element method. Detailed description is, however, given to the use of the "breakable" element concept which was originally developed at the author's research laboratory and has been incorporated into a special version of the TEPSAC code. This unique technique can provide users with detailed information associated with the growing crack, which is missing from many other existing methods. The subject of crack growth due to creep is also included. The existing C^*-integral method of solution is thoroughly described. Readers will find that the "breakable" element concept can be readily extended to handle creep crack growth, with some unique advantages over the C^*-integral approach.

In Chapter 8 is presented the theory and the corresponding finite element formulation on finite strain plasticity. The theoretical formulation of the problem is again made adaptable to the TEPSAC code as an optional module. It will be demonstrated that finite element analysis for finite strain plasticity can be achieved by including two additional element stiffness matrices, K_2 for the nonlinear terms in the strain–displacement relations and K_3 for the necessary correction of loading due to excessive rotational displacement components. Numerical examples on the simple elongation of a prismatic bar serve to illustrate the merit of this approach, as well as to indicate the inadequacy of the usual small strain approach for the instability of metal deformation.

Chapter 9 deals with the very important subject of coupled thermomechanics. The effect of thermomechanical coupling can be observed in many metal forming and thermal shock processes. This topic, in the author's view, has received far less attention than it deserves, presumably because of its complex nature and of the difficulty of its solution by classical methods. This chapter, however, provides the reader with a physical sense of the mathematical formulation, which starts with a review of the first and second law of thermodynamics, and the energy equations which link the thermal and mechanical effects for a deforming solid. The finite element discretization of the coupled thermoelastic–plastic equations is carried out, and shown to be related to the existing quasi-coupled approach. Again, a numerical example on the simple tension–compression of prismatic bars illustrates the physical significance of this effect, as well as providing direction for further research on this subject.

The remaining chapter of the book, Chapter 10, is devoted to a description of the basic TEPSAC code. It provides the reader with case illustrations on the use of the code. Most of the cases included are excerpts from published work.

The task of preparing a book of this breadth certainly cannot possibly be accomplished by a single person's effort. The author has the fortune and great pleasure to have had many young and dedicated research associates and students working for him during the past dozen years. The following are just a few names identifiable with specific chapters in the book: A. W. M. Bertels for his early, but very important, assistance in shaping up the TEPSA code (TEPSAC version without creep), in particular, the theories described in Chapter 3 and the concept of breakable element in Chapter 7; W. K. Tam for his work in improving the kinematic hardening theory, Y. J. Liu for his contribution in creep analysis and the formulation of the generalized creep-fracture theory in Chapter 7; R. Y. Wu for Chapter 5; D. A. Scarth for Chapter 6; Y. J. Kim for the excellent work in the improvement of the breakable element algorithm and significant contribution in Chapter 7; S. Y. Cheng for a very major contribution in formulating the finite strain theory given in Chapter 8; A. Banas and N. S. Sun for their significant work in the coupled thermoelastic–plastic theory in Chapter 9; and, more recently, the valuable assistance by G. G. Chen and Z. Wozniak for the solutions of several important numerical examples in the book. Assistance by other technical and research staff and graduate students at the author's laboratory is also gratefully acknowledged. Among these individuals are D. Fedorowich, D. Kuss, H. A. Ashour, Z. L. Gong, J. R. Yu, Z. H. Zhai, E. A. Abdel-Hadi and K. S. Bhatia.

A special acknowledgement is due to the author's senior research associate, G. S. Pizey, for his invaluable contribution in housekeeping the TEPSAC code and in continuously upgrading it.

The development of the TEPSAC code and other related topics covered in this book obviously was not a trivial task. It would not have been possible without generous support by several institutions such as the Whiteshell Nuclear Research Establishment of the Atomic Energy of Canada Ltd and the Natural Sciences and Engineering Research Council of Canada. Support of computing facility and time allocation by the author's university was essential to the success of this development.

Finally, the author is indebted for the excellent cooperation received from his colleagues and the support staff in the Department of Mechanical Engineering. He also gratefully acknowledges the assistance given by the book's reviewers, Professor R. W. Lewis, Dr C. Patterson and Dr P. Roberts. He is particularly grateful to Mrs Violet Lee for her extraordinary patience in typing a substantial portion of the manuscript.

Tai-Ran Hsu
Winnipeg, Manitoba,
Canada

ACKNOWLEDGEMENTS

R. Y. Wu, D. A. Scarth, Y. J. Kim and A. Banas kindly allowed the use of ideas and extracts from their theses. Extracts from Volume 2 of the *Proceedings* of the 4th International Conference on Pressure Vessel Technology, held in London on May 19–23, 1980, are reproduced by permission of the Council of the Institution of Mechanical Engineers. Extracts from "A numerical analysis of stable crack growth under increasing load" (*Journal of Fracture* **20**, 17–32) are reproduced by permission of Martinus Nijhoff. Extracts from the *Journal of Engineering Fracture Mechanics* are reproduced by permission of Pergamon Press. Parts of Chapter 10 are reproduced by permission of the American Society of Mechanical Engineers.

CONTENTS

LIST OF TABLES

FUNDAMENTALS OF THE FINITE ELEMENT METHOD

1.1 Introduction

The rapid advances in the nuclear and aerospace technologies in the past two decades compounded with the increasing demands for high performance, energy-efficient power plant components and engines have made reliable thermal stress analysis a critical factor in the design and operation of such equipment. In recent years, the need for sophisticated analyses has been extended to the traditionally empirical resource and mine industries, as has been realized by the author and several other researchers.

The required analyses for the above applications clearly are of a multidisciplinary nature, and some involve the additional complexity of multiphase and phase change phenomena. These extremely complicated factors preclude the use of classical methods, and numerical techniques such as the finite element method appear to be the most viable alternative solutions.

The versatility of the finite element method for the solution of practical problems involving complex geometries and loading/boundary conditions has been recognized by researchers since 1956[1]. Several excellent books describing the basic principles of this subject have been published in the past two decades[2-11]. Books that are devoted to special purposes have also been made available in recent years[12-14]. There have also been, of course, many worthwhile conference proceedings and monographs published dealing with the theory and application of this technique.

Generally speaking, there are two ways of developing a finite element analysis. The first approach is to place emphasis on the accuracy of the result. This approach requires the development of sophisticated algorithms, such as high order and special elements. Most of such work is tailored to some special purpose. The second approach is to develop general purpose programs and commercial codes such as ANSYS, NASTRAN, MARC analysis, etc. Special features of these codes can be found in a recent survey of finite element codes[15]. Another commercial code which fits into this category is the BERSAFE code[16]. These codes are versatile enough to handle just about any engineering problem but require large mainframe computers. Further, engineers require substantial training in order to be proficient in using these codes. Finally, the proprietary nature of these commercial packages allow the users to use them as "black boxes" only, and often the important physical sense of this powerful tool is lost.

As the objective of this book is to strike a compromise between the above two extremes and to embrace as wide a range of topics in nonlinear thermo-mechanical analysis as possible, it is necessary to cut back on some of the fundamental principles of the finite element method. Topics such as element library, in-depth study of the variational principle, integration schemes, convergence criteria, etc. will be omitted in the text. However, these topics are adequately covered in the texts cited in references[2–14].

This chapter therefore presents only the bare fundamental principle of the finite element method. The essence of the discretization concept described in the following section will enable the reader to appreciate the potential of the finite element method and hence to acquire a correct perception of its nature. The steps of the general finite element analysis will be presented in such a way as to include the formulation of some key equations. At the end of this chapter, readers are expected to have a firm grasp of this technique and its application in various branches of the engineering discipline.

1.2 The concept of discretization

Many engineering analyses involve the solutions of physical quantities such as the following: stress (or strain) *everywhere* in a solid caused by applied forces (or pressure); temperature at every *point* in a solid caused by heat sources (or sinks); velocities at *various locations* in a fluid due to difference in heads or pressure; amplitudes of vibration at *different parts* of a solid caused by sources of excitation, etc. One should realize that the substances (either solids or fluids) are considered to be *continuous*; such substances are known as **continua**. It is not hard to imagine that the number of elements in these continua implied by the terms "everywhere", "point", "various locations" and "different parts" is infinite. Therefore, we are really asked to solve problems involving an infinite number of points, which implies infinitely many degrees of freedom[†]. Common sense suggests to us that the degree of difficulty of the solution to a problem depends on the number of degrees of freedom involved. Consequently, many engineering problems which involve a large number of degrees of freedom are difficult to solve, if not impossible. The finite element method (FEM), which is based on the idea of **discretization**, is obviously a vital alternative way of solving these problems.

The use of the discretization concept in science and technology is not new. One such common practice was to measure the areas enclosed by closed

---

[†] A degree of freedom, in a commonly used terminology, can be regarded as one desired unknown quantity.

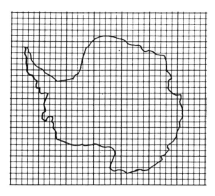

Figure 1.1 Discretized map of Antarctica. One grid square = 200×200 km = $40\,000$ km^2.

curves, before the invention of the planimeter. As illustrated in Figure 1.1, the area of Antarctica may be determined approximately by summing up the areas of the small square meshes which subdivide the entire domain. Each of the approximately 404.5 meshes represents $40\,000$ km^2, giving an approximate total area of 16 180 million km^2, which is about 13.6% in excess of the actual size. Another example is to estimate the work produced by an engine by measuring the area under the pressure–volume diagram from an indicator. Many readers may not be aware of the fact that the discretization concept was indeed the essence of the integration in calculus, in which the entirety of a physical quantity is determined by summing up all incremental values determined at infinitesimally small increments. The finite difference operators originally formulated by Newton in the 17th century constituted the foundation of a powerful numerical tool known as the finite difference method (FDM) which has been widely used in engineering analyses. An upsurge of the use of this method became even more evident with the rapid development of digital computers. The FDM, which is built exclusively on the basis of discretization of continua, has been used in many engineering disciplines for modern industrial applications. Many worthwhile monographs on this subject have been published over the past three decades[17-23].

Discretization procedures in finite element analysis are similar to those used in the FDM, except for the terminologies. It starts by the subdivision of a continuum into a finite number of subdomains called **elements**. These elements are connected at the corners, or in some cases at selected points on the edges, called **nodes**, as shown in Figure 1.2. In this illustration, the solid has a finite shape which can be described by the boundary Ω, and is supported at points A and B. The solid is under a set of specified **actions** (e.g. forces, heat sources etc.) which can be described by:

$$\{P\} = P_1, P_2, P_3, \ldots$$

By virtue of these actions, there exist the required **reactions** (e.g. deflections, strains, stresses, temperatures, etc.) which can be described by:

$$\{\phi\} = \phi_1, \phi_2, \phi_3, \ldots$$

3

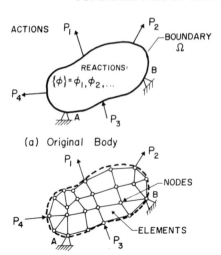

(a) Original Body

(b) Discretized Body

Figure 1.2 Discretization of a solid.

For most engineering problems, there exist conditions that a functional $\chi(\phi)$, which can be defined by:

$$\chi(\phi) = \int_v f\left(\{\phi\}, \frac{\partial\{\phi\}}{\partial \mathbf{r}}, \ldots\right) dv + \int_s g\left(\{\phi\}, \frac{\partial\{\phi\}}{\partial \mathbf{r}}, \ldots\right) ds \qquad (1.1)$$

in which $\{\phi\}$ are the unknown reactions in the continuum and $\mathbf{r}$ are the position vectors specified by the spatial coordinates.

The minimization of the functional $\chi(\phi)$ is satisfied for the equilibrium state, or

$$\frac{\partial\chi(\phi)}{\partial\{\phi\}} = \left\{ \begin{array}{c} \dfrac{\partial\chi}{\partial\phi_1} \\[2ex] \dfrac{\partial\chi}{\partial\phi_2} \\[2ex] \vdots \end{array} \right\} = 0 \qquad (1.2)$$

Since there are *infinitely* many points in the solid bounded by Ω and each point is associated with a set of values $\{\phi\}$, there are obviously infinitely many ϕ values to be determined.

Suppose the original solid in Figure 1.2a is subdivided into a finite number of subdomains (elements) with certain shapes (triangular and quadrilateral plates in this case) and these elements are interconnected at the nodes. After such discretization, the original continuum is no longer a continuous body but contains a number of "broken" pieces connected at corners or/and edges.

4

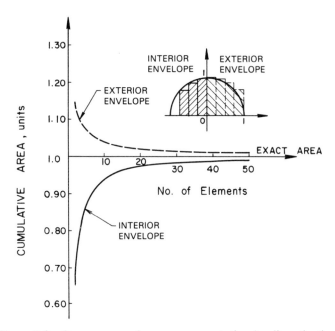

Figure 1.3 Convergence of an area computation by discretization.

Another noticeable change is the originally continuous *curved* boundary Ω has now become segments of *straight* lines[†]. It is obvious that the discretized body is geometrically *similar* but not identical to the original solid. The degree of "similarity" of the two bodies, of course, depends on the number of elements used in the discretized body (model). The more elements used in the model, the closer it gets to the original solution as illustrated by a simple example given in Figure 1.3. The reader will find that the area of a semi-circle was approximated by summing up the areas of the dividing rectangles, one set was entirely enclosed within the circular boundary and another set approached from the exterior boundary. Both sets have been shown to asymptotically converge to the exact area with increasing number of the elements used in the discretized models.

The unique advantage of discretization is that analysis needs only to be applied to the *individual elements* of certain simple geometries (e.g. triangular plates) rather than the entire solid of complex geometry.

The physical quantities $\{\phi\}$ to be determined in the element can be obtained from these values at the nodes through certain *shape* or *interpolation* functions, or:

$$\{\phi\}^e = [N]\{\phi\}^n \tag{1.3}$$

[†] We will learn later that curved "edges" are allowed for certain special elements. In that case, the curved surface of the original solid may be preserved after the discretization.

5

where $\{\phi\}^e$, $\{\phi\}^n$ = respective physical quantities in element and nodes, and $[N]$ = shape or interpolation function.

Since the solutions are required only in the individual element or at the nodes by (1.3), and there are now only *finitely* many of these elements and nodes in the discretized model, the original problem involving an *infinite* number of degrees of freedom has thus been reduced to a finite number of degrees of freedom by means of the discretization process. Solutions of $\{\phi\}$ in the discretized body are hence substantially more attainable.

The discretization by (1.3) suggests, although intuitively at this moment, that the functional $\chi(\phi)$ for the entire continuum is equal to the sum of the contributions made by all the elements, that is

$$\chi(\phi) = \sum \chi^e(\phi)$$

Then the minimization scheme in (1.2) simply becomes

$$\frac{\partial \chi(\phi)}{\partial \{\phi\}} = \sum \frac{\partial \chi^e(\phi)}{\partial \{\phi\}^e} = 0 \tag{1.4}$$

with a finite number of equations involved in (1.4).

1.3   Steps in the finite element method

The FEM is now used in a wide cross-section of engineering analysis. It is used to calculate stress and temperature distributions in solid structures of complex geometries subject to complicated loadings. It can also be used to predict the pattern of a fluid flowing in a channel of complicated geometry. The engineering behavior of substances with unusual thermophysical and mechanical properties can also be readily assessed by the FEM. The principles of the application of the FEM in many other engineering disciplines have been comprehensively illustrated by Desai[9]. It is not possible to establish a set of standard procedures for all the computations for the problems described above. However, as a general guideline, most finite element analyses follow eight steps, as will be described below.

Step 1: discretization of the real structure

Depending on the nature of the structure, there are various types of elements available. Figure 1.4 shows the typical and most commonly used elements. Bar elements are frequently used in modelling trusses and frames, and plate elements are suitable for plate type structures, or when the variation of the physical quantities is limited to a plane (i.e. plane stress case). For structures

6

SHAPES OF SIMPLEX ELEMENTS

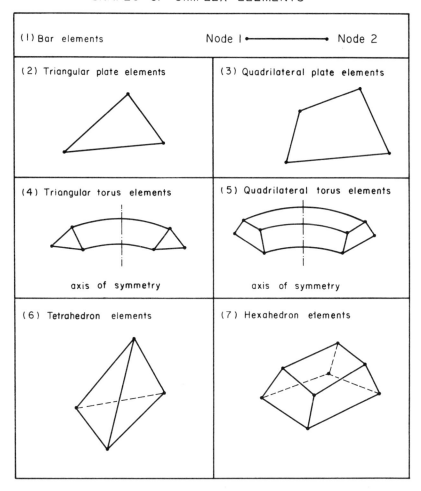

Figure 1.4 Typical element geometries.

with an axis of symmetry, e.g. cylinders and pressure vessels, torus elements can be used. Finally, for structures of extremely complicated geometry, when a three-dimensional model becomes a necessity, tetrahedron or hexahedron elements are commonly used. There are many other types of elements developed for special applications, e.g. *beam* and *shell* elements. Some elements have additional nodes on the sides which can be either curved or straight. Many *isoparametric* elements are of this type. Modern finite element computer programs allow various combinations of different element types and hence make finite elements modelling highly versatile.

Step 2: identification of the primary unknown quantities and an appropriate interpolation (shape, or trial) function

Again, depending on the nature of the problem, the *primary unknown* quantity varies from case to case. One may choose the *displacement* component as the primary unknown quantity in a stress analysis problem, or *temperature* for heat conduction analysis, *velocity* for fluid flow.

Once the primary unknown quantity, which is to be represented by a vector quantity $\{\phi\}$ in the analysis, is identified, discretization following (1.3) begins. A proper form of the interpolation (or shape, or trial) function must then be chosen. This function, according to (1.3), relates the primary unknown quantity in the elements to that of the associated nodes.

Two requirements exist for the selection of the interpolation function:

(a) The resulting functional $\chi(\phi)$ in (1.4) must be continuous at the element interfaces. This condition can be satisfied if the integrands in (1.1), i.e. the functions f and g, possess finite derivatives of one order less than the highest order derivative appearing in the functional.

(b) It must lead to convergent results in $\{\phi\}$. This requires that the functional χ approaches a limiting value as the element volume v approaches zero.

There are several forms of interpolation functions which one may use in finite element analysis. Two commonly used types are presented below.

(1) LAGRANGE POLYNOMIAL[5]

Since the primary purpose of the interpolation function $[N]$ is to interpolate the nodal values to those in the element as expressed in (1.3), suitable interpolating polynomials will result in a both efficient and accurate computational algorithm in the finite element analysis. Lagrange polynomials have been widely adopted for this purpose due to the fact that they are more convenient to use than many other popular interpolating polynomials. They can be used with nonuniform spacing of the abscissa values, as illustrated in Figure 1.5, and the discrete values $\phi_1, \phi_2, \ldots, \phi_n$ need not correspond to ascending or descending x values.

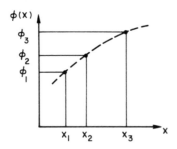

Figure 1.5 Interpolation of a function.

The polynomial function which fits the values ϕ_1, ϕ_2, ϕ_3, ... at the co-ordinates x_1, x_2, x_3, ..., in Figure 1.5, according to Lagrange, has the form:

$$\phi(x) = N_1(x)\phi_1(x_1) + N_2(x)\phi_2(x_2)$$

$$+ N_3(x)\phi_3(x_3) + \cdots \tag{1.5}$$

where

$$N_1(x) = \frac{(x - x_2)(x - x_3) \cdots (x - x_n)}{(x_1 - x_2)(x_1 - x_3) \cdots (x_1 - x_n)}$$

$$N_2(x) = \frac{(x - x_1)(x - x_3) \cdots (x - x_n)}{(x_2 - x_1)(x_2 - x_3) \cdots (x_2 - x_n)}$$

$$N_3(x) = \frac{(x - x_1)(x - x_2) \cdots (x - x_n)}{(x_3 - x_1)(x_3 - x_2) \cdots (x_3 - x_n)}$$

or to express $N_1(x)$, $N_2(x)$, ... in a general form as follows:

$$N_r(x) = \frac{(x - x_1)(x - x_2) \cdots (x - x_{r-1})(x - x_{r+1}) \cdots (x - x_n)}{(x_r - x_1)(x_r - x_2) \cdots (x_r - x_{r-1})(x_r - x_{r+1}) \cdots (x_r - x_n)} \tag{1.6}$$

with

$$N_r(x) = 1 \quad \text{at } x = x_r$$

$$= 0 \quad \text{elsewhere}$$

Example 1.1. For a bar element with the function values ϕ_1 at the node located at $x_1 = 0$ and ϕ_2 at $x_2 = L$, then according to the Lagrange interpolating polynomial given by (1.5),

$$\phi(x) = \sum_{i=1}^{2} N_i\phi_i(x) = N_1(x)\phi_1 + N_2(x)\phi_2$$

$$N_1(x) = \frac{x - x_2}{x_1 - x_2} = \frac{x - L}{0 - L} = 1 - \frac{x}{L}$$

$$N_2(x) = \frac{x - x_1}{x_2 - x_1} = \frac{x - 0}{L - 0} = \frac{x}{L}$$

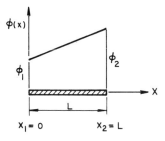

Figure 1.6 Linear interpolation function of a bar element.

9

Hence

$$\phi(x) = \left(1 - \frac{x}{L}\right)\phi_1 + \frac{x}{L}\phi_2 = \left\{1 - \frac{x}{L} \quad \frac{x}{L}\right\}\begin{Bmatrix}\phi_1 \\ \phi_2\end{Bmatrix}$$

The interpolation function is

$$[N] = \left[\left(1 - \frac{x}{L}\right) \quad \frac{x}{L}\right] \tag{1.7}$$

The Lagrangian interpolating polynomial can be expanded to a two-dimensional domain. The mathematical expression for this case takes the following form:

$$\phi(x, y) = \sum_{i=1}^{m} \sum_{j=1}^{n} N_i(x)N_j(y)\phi_{ij} \tag{1.8}$$

where m = total number of points in the x-axis, n = total number of points in the y-axis, and ϕ_{ij} = value of $\phi(x, y)$ at x_i and y_j.

Example 1.2. The value of a function $\phi(x, y)$ which represents three specified values ϕ_1, ϕ_2 and ϕ_3 located at coordinates (x_1, y_1), (x_2, y_2), (x_3, y_3), as illustrated in Figure 1.7, can be determined by using (1.8) with $m = n = 3$:

$$\phi(x, y) = N_1(x)N_1(y)\phi_1 + N_2(x)N_2(y)\phi_2 + N_3(x)N_3(y)\phi_3$$

or in matrix form:

$$\phi(x, y) = \{N_1(x)N_1(y) \quad N_2(x)N_2(y) \quad N_3(x)N_3(y)\}\begin{Bmatrix}\phi_1 \\ \phi_3 \\ \phi_3\end{Bmatrix} \tag{1.9}$$

The interpolation functions $N_i(x)$, $N_i(y)$, $i = 1, 2, 3$ can be evaluated using (1.6) to give

$$N_1(x) = \frac{(x - x_2)(x - x_3)}{(x_1 - x_2)(x_1 - x_3)}; \qquad N_1(y) = \frac{(y - y_2)(y - y_3)}{(y_1 - y_2)(y_1 - y_3)}$$

$$N_2(x) = \frac{(x - x_1)(x - x_3)}{(x_2 - x_1)(x_2 - x_3)}; \qquad N_2(y) = \frac{(y - y_1)(y - y_3)}{(y_2 - y_1)(y_2 - y_3)}$$

$$N_3(x) = \frac{(x - x_1)(x - x_2)}{(x_3 - x_1)(x_3 - x_2)}; \qquad N_3(y) = \frac{(y - y_1)(y - y_2)}{(y_3 - y_1)(y_3 - y_2)}$$

The reader should realize that the interpolation functions given in (1.9) are of fourth order, which makes subsequent formulation very tedious. A much less sophisticated but much simpler linear function can be used, as will now be described.

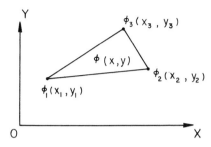

Figure 1.7 Interpolation function in a two-dimensional domain.

(2) LINEAR POLYNOMIAL

Following the configuration illustrated in Figure 1.7 the function $\phi(x, y)$ can be assumed to vary within the specified domain according to a linear function:

$$\phi(x, y) = a_1 + a_2 x + a_3 y$$

$$= \{1 \quad x \quad y\}\begin{Bmatrix} a_1 \\ a_2 \\ a_3 \end{Bmatrix} = \{R\}^T\{a\} \tag{1.10}$$

The coefficients a_1, a_2 and a_3 are constants which can be determined by substituting the coordinates of the specified nodal values ϕ_1, ϕ_2 and ϕ_3, to give

$$\phi_1 = a_1 + a_2 x_1 + a_3 y_1$$

$$\phi_2 = a_1 + a_2 x_2 + a_3 y_2$$

$$\phi_3 = a_1 + a_2 x_3 + a_3 y_3$$

or in a matrix form:

$$\{\phi\} = [A]\{a\} \tag{1.11}$$

and

$$\{a\} = [A]^{-1}\{\phi\} = [h]\{\phi\} \tag{1.12}$$

The matrix $[A]$ in the above expressions contains the coordinates of ϕ_1, ϕ_2 and ϕ_3 as

$$[A] = \begin{bmatrix} 1 & x_1 & y_1 \\ 1 & x_2 & y_2 \\ 1 & x_3 & y_3 \end{bmatrix}$$

The inversion of the matrix $[A]^{-1} = [h]$ can be readily performed to give:

11

$$[h] = \frac{1}{|A|} \begin{bmatrix} x_2y_3 - x_3y_2 & x_3y_1 - x_1y_3 & x_1y_2 - x_2y_1 \\ y_2 - y_3 & y_3 - y_1 & y_1 - y_2 \\ x_3 - x_2 & x_1 - x_3 & x_2 - x_1 \end{bmatrix} \quad (1.13)$$

where

$|A|$ = the determinant of the elements in $[A]$

$= (x_1y_2 - x_2y_1) + (x_2y_3 - x_3y_2) + (x_3y_1 - x_1y_3)$

= the area of the triangle bounded by ϕ_1, ϕ_2 and ϕ_3.

By substituting (1.13) into (1.12) and then (1.10), the function $\phi(x, y)$ can be evaluated by the three specified quantities ϕ_1, ϕ_2 and ϕ_3, or $\{\phi\}$, to be

$$\phi(x, y) = \{R\}^T[h]\{\phi\} \quad (1.14)$$

The interpolation function, as shown in the above expression, according to the definition, takes the form:

$$N(x, y) = \{R\}^T[h] \quad (1.15)$$

where the matrix $\{R\}^T = \{1 \quad x \quad y\}$ and $[h]$ is given in (1.13).

The interpolation function $N(x, y)$ given in (1.15) is a linear function of x and y, which is the easiest among all such functions known to exist. It has been incorporated in all the subsequent finite element algorithms.

(3) INTERPOLATION FUNCTIONS FOR SERENDIPITY ELEMENTS

It is desirable in some applications to maintain continuity of displacements, not just at the corner nodes of neighboring elements in a discretized structure, but also at the neighboring boundaries i.e. element edges. Serendipity elements were developed for this purpose[24].

The basic shape of the element chosen is quadrilateral, the sides of which can be distorted in a prescribed way.

Let a quadrilateral element be described by a global coordinate system x, y as shown in Figure 1.8.

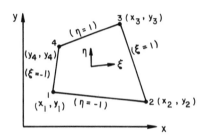

Figure 1.8 Coordinate definition for serendipity elements.

12

The following terminologies are used in describing this type of element:

global coordinates: $\{x, y\}$
nodal coordinates: $\{x_n, y_n\}$
parametric coordinates: $\{\xi, \eta\}$
element quantities: $u(\xi, \eta), v(\xi, \eta)$
nodal quantities: $\{u_n, v_n\}$

The parametric coordinates are chosen in such a way that

$$\eta = -1 \text{ on side } 1\text{-}2; \qquad \eta = +1 \text{ on side } 4\text{-}3$$

$$\xi = -1 \text{ on side } 1\text{-}4; \qquad \xi = +1 \text{ on side } 2\text{-}3$$

Thus by using the nodal parametric coordinates described above, we will find that:

$$\xi = -1, \quad \eta = -1 \qquad \text{at node 1;}$$

$$\xi = +1, \quad \eta = -1 \qquad \text{at node 2;}$$

$$\xi = +1, \quad \eta = +1 \qquad \text{at node 3;}$$

$$\xi = -1, \quad \eta = +1 \qquad \text{at node 4;}$$

The numerical values of ξ and η in between nodes can be determined by interpolating those at the end nodes.

The relationship between $\{x, y\}$ and $\{x_n, y_n\}$ is

$$x = N_1 x_1 + N_2 x_2 + N_3 x_3 + N_4 x_4 = \{N_1 \quad N_2 \quad N_3 \quad N_4\} \begin{Bmatrix} x_1 \\ x_2 \\ x_3 \\ x_4 \end{Bmatrix} = \{N\}^T \{x_n\}$$

$$\text{(1.16a)}$$

$$y = N_1 y_1 + N_2 y_2 + N_3 y_3 + N_4 y_4 = \{N_1 \quad N_2 \quad N_3 \quad N_4\} \begin{Bmatrix} y_1 \\ y_2 \\ y_3 \\ y_4 \end{Bmatrix} = \{N\}^T \{y_n\}$$

$$\text{(1.16b)}$$

in which the isoparametric shape function, $N_i = N_i(\xi, \eta) = 1$ at node i, $= 0$ at other nodes.

The finite element discretization can thus be carried out following the expressions given below:

$$u(\xi, \eta) = N_1 u_1 + N_2 u_2 + N_3 u_3 + N_4 u_4 = \{N\}^T \{u_n\}$$

$$v(\xi, \eta) = N_1 v_1 + N_2 v_2 + N_3 v_3 + N_4 v_4 = \{N\}^T \{v_n\}$$

$$\text{(1.17)}$$

13

Expressions of the shape functions for three common serendipity elements are available:

(a) Linear element:

$$x = a_1 + a_2\xi + a_3\eta$$

$$N_i = \tfrac{1}{4}(1 + \xi_0)(1 + \eta_0) \tag{1.18}$$

where $\xi_0 = \xi\xi_i$ and $\eta_0 = \eta\eta_i$.

The interpolating function for node 3, for example, can be expressed by letting $i = 3$ and $\xi_3 = +1$, $\eta_3 = +1$ to be:

$$N_3 = \tfrac{1}{4}(1 + \xi)(1 + \eta)$$

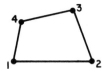

Figure 1.9 A linear serendipity element.

(b) Quadratic element (Figure 1.10): the relationship between the global and parametric coordinates is represented by a quadratic function:

$$x = a_1 + a_2\xi + a_3\eta + a_4\eta\xi + a_5\xi^2 + a_6\eta^2 + a_7\xi^2\eta + a_8\xi\eta^2 \tag{1.18}$$

and for the corner nodes:

$$N_i = \tfrac{1}{4}(\xi_0 + 1)(\eta_0 + 1)(\xi_0 + \eta_0 + 1) \tag{1.19a}$$

For the mid-side nodes:

$$N_i = \tfrac{1}{2}(1 - \xi^2)(1 + \eta_0) \qquad \text{at } \xi_i = 0 \tag{1.19b}$$

$$N_i = \tfrac{1}{2}(1 + \xi_0)(1 - \eta^2) \qquad \text{at } \eta_i = 0 \tag{1.19c}$$

The interpolating function at node number 4, i.e. $i = 4$, $\xi_4 = -1$, $\eta_4 = 1$ in this case can be expressed by:

$$N_4 = \tfrac{1}{4}(\xi\xi_4 + 1)(\eta\eta_4 + 1)(\xi\xi_4 + \eta\eta_4 + 1)$$

$$= \tfrac{1}{4}(1 - \xi)(1 + \eta)(1 + \eta - \xi),$$

and at node number 7, i.e. $i = 7$, $\xi_7 = 0$, $\eta_7 = 1$,

$$N_7 = \tfrac{1}{2}(1 - \xi^2)(1 + \eta)$$

as well as at node number 6, i.e. $i = 6$, $\xi_6 = 1$, $\eta_6 = 0$,

$$N_6 = \tfrac{1}{2}(1 + \xi)(1 - \eta^2)$$

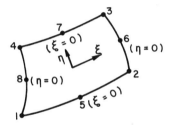

Figure 1.10 A quadratic serendipity element.

(c) Cubic element with nodes at $\frac{1}{3}$, $\frac{2}{3}$ along each side as shown in Figure 1.11: the coordinates expansion is of a cubic function,

$$x = a_1 + a_2\xi + a_3\eta + a_4\eta\xi + a_5\eta^2 + a_6\xi^2$$

$$+ a_7\xi^3 + a_8\xi\eta^2 + a_9\xi\eta^2 + a_{10}\eta^3 + a_{11}\eta^3\xi + a_{12}\eta\xi^3 \quad (1.20)$$

For the corner nodes

$$N_i = \tfrac{1}{32}(1 + \xi_0)(1 + \eta_0)[9(\xi^2 + \eta^2) - 10] \quad (1.21a)$$

For the nodes along the sides $\xi_i = \pm 1$ with $\eta_i = \pm\frac{1}{3}$ (i.e. node number 7, 8, 11, 12):

$$N_i = \tfrac{9}{32}(1 + \xi_0)(1 - \eta^2)(1 + \eta_0) \quad (1.21b)$$

For the nodes along the sides $\eta_i = \pm 1$ with $\xi_i = \pm\frac{1}{3}$ (node number 5, 6, 9, 10):

$$N_i = \tfrac{9}{32}(1 - \xi^2)(1 + \xi_0)(1 + \eta_0) \quad (1.21c)$$

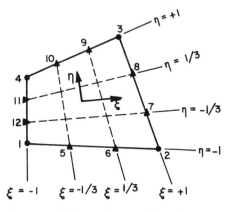

Figure 1.11 A cubic serendipity element.

Step 3: define relationship between actions and reactions

The laws of physics usually provide certain relationships between the *actions* on a substance and the induced *reactions* from the same substance. Table 1.1 indicates some of these physical quantities present in common engineering analysis.

FUNDAMENTALS OF THE FINITE ELEMENT METHOD

Table 1.1 Actions and reactions.

	Actions, $\{P\}$	Reactions, $\{\phi\}$
stress analysis	forces $\{F\}$	displacement $\{u\}$, strains $\{\varepsilon\}$, stresses $\{\sigma\}$
heat conduction	thermal forces $\{Q\}$	temperature $\{T\}$
fluid flow	pressure or head $\{p\}$	velocity $\{V\}$

The physical law which relates $\{F\}$ and $\{u\}$ in the stress analysis case is the *minimization of potential energy* and the *Fourier law* (such as described in Chapter 2) relates $\{Q\}$ and $\{T\}$ in the heat conduction analysis.

Step 4: derive element equations

The element equation relates the reactions in terms of the primary unknown and the action. There are generally two distinct methods used in the FEM for the derivation of these equations, namely the *Rayleigh–Ritz method* and the *weighted residual method* or the *Galerkin method*. The former method is based on the principle of variational calculus and the latter method proves to be more practical for the problems which can be completely described by a set of differential equations. Many heat conduction–convection and fluid flow problems can be handled in this way.

(1) RAYLEIGH-RITZ METHOD

This method is applied on the variation of a suitable functional $\chi(\phi)$ defined in (1.1) with ϕ to be the primary quantity in the solution. This functional can be the potential energy in a solid in equilibrium for a stress analysis problem, or some differential quantities as illustrated in the following example. The variation of the functional in a discretized form, with respect to $\{\phi\}$, as shown in (1.4), can lead to a set of simultaneous algebraic equations describing the status of each element in the system. These equations are called **element equations**.

Example 1.3. A problem which can be described by Poisson's equation:

$$\frac{\partial}{\partial x} k \frac{\partial \phi}{\partial x} + \frac{\partial}{\partial y} k \frac{\partial \phi}{\partial y} + Q = 0 \qquad \text{in domain volume } V$$

and

$$k \frac{\partial \phi}{\partial n} - \bar{q} = 0 \qquad \text{on boundary } S$$

16

In the above expressions, the primary unknown quantity ϕ, coefficient k and other quantities Q are functions of x and y; whereas $\bar{q}$ is a prescribed quantity and n is the outward normal to the surface s.

The functional for this problem can be shown to equal (Chapter 15[3]):

$$\chi(\phi) = \int_v \left[\tfrac{1}{2}k\left(\frac{\partial\phi}{\partial x}\right)^2 + \tfrac{1}{2}k\left(\frac{\partial\phi}{\partial y}\right)^2 - Q\phi \right] dv - \int_s \bar{q}\phi \, ds \qquad (1.22)$$

For a discretized system,

$$\phi = \sum_{i=1}^{n} N_i \phi_i \qquad (1.23)$$

where N_i = interpolation function and ϕ_i are the discrete values of ϕ. Introducing (1.23) into (1.22), one gets

$$\chi(\phi) = \int_v \tfrac{1}{2}k\left(\sum_i \frac{\partial N_i}{\partial x}\phi_i\right)^2 dv + \int_v \tfrac{1}{2}k\left(\sum_i \frac{\partial N_i}{\partial y}\phi_i\right)^2 dv$$

$$- \int_v Q \sum_i N_i \phi_i \, dv - \int_s \bar{q} \sum_i N_i \phi_i \, ds$$

On the variation of χ with respect to ϕ_i, or by differentiating χ with respect to ϕ_i as indicated in (1.2),

$$\frac{\partial\chi}{\partial\phi_j} = \int_v k\left(\frac{\partial N_i}{\partial x}\phi_i\right)\frac{\partial N_j}{\partial x} dv + \int_v k\left(\frac{\partial N_i}{\partial y}\phi_i\right)\frac{\partial N_j}{\partial y} dv$$

$$- \int_v QN_j \, dv - \int_s \bar{q}N_j \, ds \equiv 0$$

Since ϕ_i are discrete values of ϕ and are independent of the coordinates (and for that matter, independent of V and S, and therefore dv and ds), the above expression can be written as

$$\left[\int_v k\left(\frac{\partial N_i}{\partial x}\frac{\partial N_j}{\partial x} + \frac{\partial N_i}{\partial y}\frac{\partial N_j}{\partial y}\right) dv \right]\phi_i = \int_v QN_j \, dv + \int_s \bar{q}N_j \, ds$$

which can be deduced to give

$$[K]\{\phi_i\} = \{F_i\} \qquad (1.24)$$

where $[k] = \int_v k\left(\frac{\partial N_i}{\partial x}\frac{\partial N_j}{\partial x} + \frac{\partial N_i}{\partial y}\frac{\partial N_j}{\partial y}\right) dv$

$$\{F_i\} = \int_v QN_j \, dv + \int_s \bar{q}N_j \, ds$$

Equation (1.24) provides solutions to the discrete values ϕ_i in each element in the system. It is called the **element equation**.

(2) GALERKIN METHOD

This method is sometimes referred to as the "weighted residuals" method, for which the weighting functions are assumed to be identical to the interpolation functions used in the discretization.

Consider a physical problem which can be described mathematically in the form of a differential equation:

$$D(\phi) = 0 \quad \text{in domain } V$$

and the boundary conditions:

$$B(\phi) = 0 \quad \text{on boundary } S$$

This system can be replaced by an integral equation:

$$\int_v WD(\phi) \, dv + \int_s \overline{W}B(\phi) \, ds = 0 \tag{1.25}$$

where W and $\overline{W}$ are arbitrary weighting functions. For a discretized system $\phi \doteq \sum N_i\phi_i$, (1.25) can be approximated to be

$$\int_v W_j D(\sum N_i\phi_i) \, dv + \int_s \overline{W}_j B(\sum N_i\phi_i) \, ds = R \tag{1.26}$$

in which W_j and $\overline{W}_j$ are discretized weighting functions and R is the residual. A good discretization system, of course, should have $R = $ minimum, or $R \rightarrow 0$.

The Galerkin method with weighting function and interpolating function identical (i.e. $W_j = N_j = \overline{W}_j$) is the most popular form used in finite element analysis, although other forms which may provide more stable solutions have been proposed[25].

Thus by replacing the weighting function with the interpolating function in (1.26), one gets

$$\int_v N_i D(\sum N_i\phi_i) \, dv + \int_s N_i B(\sum N_i\phi_i) \, ds \rightarrow 0$$

which can be shown to lead to element equations similar to that shown in (1.24).

For almost all physical problems that can be solved by the finite element method, the element equations can be expressed in a general form:

$$[K_e]\{q\} = \{Q\} \tag{1.27}$$

where $[K_e] = $ element property matrix, $\{q\} = $ vector of primary unknown quantities at the nodes, and $\{Q\} = $ vector of element nodal forcing parameters.

Step 5: derive overall structure stiffness equations

This step of the analysis assembles all individual element equations to provide *stiffness equations* for the entire structure, or mathematically,

$$[K]\{q\} = \{R\} \tag{1.28}$$

in which

$[K]$ = **overall stiffness matrix**

$$= \sum_{1}^{M} [K_e]$$

M = total number of elements in the discretized model,

$\{R\}$ = assemblage of *resultant* vector of nodal forcing parameters.

It should be noted that the elements of the $[K_e]$ matrices common to other elements through nodal connection should be summed up algebraically during the assembly process. The specified boundary conditions on $\{q\}$ or $\{R\}$ should also be imposed on the assembled equations in (1.28).

Step 6: solve for primary unknowns

It is apparent that (1.28) represents a set of simultaneous algebraic equations. The total number of equations is identical to the total number of primary unknowns $q_i (i = 1, 2, \ldots, n)$. Depending on the size and degree of symmetry of the $[K]$ matrix, there are generally two methods which can be used to solve for $\{q\}$, namely

(1) Gaussian elimination (see Chapter 1 in Gallagher *et al.*[12]), or (2) matrix inversion:

$$\{q\} = [K]^{-1}\{R\} \tag{1.29}$$

Step 7: solve for secondary unknowns

Once the primary unknowns $\{q\}$ are found, other unknowns may be calculated from the available physical relations. For example, in the case of elastic stress-deformation analysis, the primary unknowns are the displacement components at the nodes. One may envisage that the corresponding strain components can be calculated from the "displacement–strain" relations following the theory of elasticity. The stress components, of course, can be evaluated by means of the well-known Hooke's law from the computed strain components.

19

Step 8: interpretation of results

Results obtained from the finite element computer code are usually in tabulated form, indicating the calculated primary and all secondary physical quantities at specified nodes and elements under given applied loads. The analyst can then select critical sections of the body and evaluate these results with respect to the established design criteria. In some instances, the computed results may be expressed graphically. Figure 1.12 shows the contours of isoclinic lines in a gear tooth made of steel. These contours were produced by finite element analysis[26] and have shown close correlation to those produced by photoelastic investigation[27] under similar conditions. The critical locations and magnitude of stress concentrations in the gear tooth are readily visible.

Another method of expressing finite element analysis results is the visual display of finite element models under different loading conditions. Figure 1.13 illustrates the shape of a cantilever beam under two different levels of loadings. This comprehensive method of output is particularly suitable for dynamic analyses.

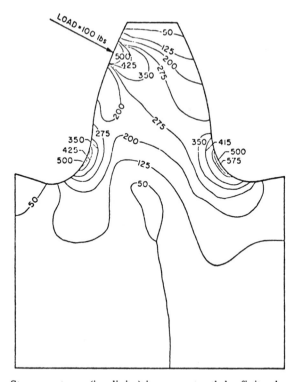

Figure 1.12 Stress contours (isoclinics) in a gear tooth by finite element analysis.

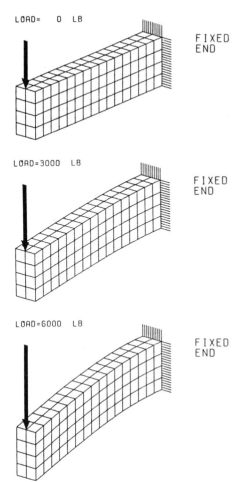

Figure 1.13 Graphic representation of deflection in a cantilever beam.

References

1 Turner, M. J., R. W. Clough, H. C. Martin and L. J. Topp 1956. Stiffness and deflection analysis of complex structure. *J. Aeronaut. Sci.* **23**, 805–23.
2 Zienkiewicz, O. C. and Y. K. Cheung 1967. *The finite element method in structural and continuum mechanics*. New York: McGraw-Hill.
3 Zienkiewicz, O. C. 1971. *The finite element method in engineering science*. New York: McGraw-Hill.
4 Desai, C. S. and J. F. Abel 1972. *Introduction to the finite element method*. New York: Van Nostrand Reinhold.
5 Cook, R. D. 1974. *Concepts and applications of finite element analysis*. New York: John Wiley.
6 Gallagher, R. H. 1975. *Finite element analysis—fundamentals*. Englewood Cliffs, NJ: Prentice-Hall.

7   Segerlind, L. J. 1976. *Applied finite element analysis*. New York: John Wiley.
8   Bathe, K. J. and E. L. Wilson 1976. *Numerical methods in finite element analysis*. Englewood Cliffs, NJ: Prentice-Hall.
9   Desai, C. S. 1979. *Elementary finite element method*. Englewood Cliffs, NJ: Prentice-Hall.
10 Hinton, E. and D. R. J. Owen 1979. *An introduction to finite element computations*. Swansea: Pineridge Press.
11 Akin, J. E. 1982. *Application and implementation of finite element methods*. New York: Academic Press.
12 Gallagher, R. H., J. T. Oden, C. Taylor and O. C. Zienkiewicz (eds.) 1975. *Finite elements in fluids* (3 vols). New York: John Wiley.
13 Owen, D. R. J. and E. Hinton 1980. *Finite elements in plasticity*. Swansea: Pineridge Press.
14 Naylor, D. J. and G. N. Pande 1981. *Finite elements in geotechnical engineering*. Swansea: Pineridge Press.
15 Noor, A. K. 1981. Survey of computer programs for solution of nonlinear structural and solid mechanics problems. *Computers & Structures*. **13**, 425–65.
16 BERSAFE 1982. *Berkeley structural analysis by finite elements*. Berkeley Nuclear Laboratories, Berkeley, Glos., England.
17 Salvadori, M. G. and M. L. Baron 1952. *Numerical methods in engineering*. Englewood Cliffs, NJ: Prentice-Hall.
18 Forsythe, G. E. and W. R. Wasow 1960. *Finite difference methods for partial differential equations*. New York: John Wiley.
19 Fox, L. (ed.) 1962. *Numerical solution of ordinary and partial differential equations*. New York: Macmillan (Pergamon).
20 Richtmyer, R. D. 1957. *Difference methods for initial-value problems*. New York: John Wiley (Interscience).
21 Todd, J. (ed.) 1962. *Survey of numerical analysis*. New York: McGraw-Hill.
22 Hilderbrand, F. B. 1968. *Finite-difference equations and simulations*. Englewood Cliffs, NJ: Prentice-Hall.
23 James, M. L., G. M. Smith and J. C. Wolford 1977. *Applied numerical methods for digital computation*, Chs. 6–8. New York: Harper & Row.
24 Ergatoudis, I., B. M. Irons and O. C. Zienkiewicz 1968. Curved, isoparametric, 'quadrilateral' elements for finite element analysis. *Int. J. Struct.* **4**, 31–42.
25 Heinrich, J. C., P. S. Huyakorn, O. C. Zienkiewicz and A. R. Mitchell 1977. An "upwind" finite element scheme for two-dimensional convective transport equation. *Int. J. Num. Meth. Engng.* **11**, 131–43.
26 Hsu, T.-R. 1976. *Application of finite element technique to the technology transfer design*. ASME Paper 76-DET-74.
27 Dolan, T. J. and E. L. Broghamer 1942. *A photoelastic study of stresses in gear tooth profiles*. Bulletin No. 335, Engineering Experimental Station, University of Illinois.

2

FINITE ELEMENT ANALYSIS IN HEAT CONDUCTION

2.1 Introduction

Heat is a primary quantity in the universe as well as the primary energy source which is an essential element to sustain human civilization. The ever-increasing demand for energy by modern society has made the maximal harnessing of this energy the prime responsibility of engineers. In recent years, the demand for larger and more efficient power generators, including those produced by nuclear fission and engines, has accelerated at an unparalleled rate. Engineers are expected to produce and operate these machines in the best possible economic way, which often requires the use of narrow safety margins in the design process.

Following the basic laws of thermodynamics, it is entirely conceivable that higher operating efficiency of a heat machine will most likely involve high temperature for the machine's structural components. The law of nature on the thermal expansion and contraction of substances immediately indicates that excessive thermal stresses will be the direct consequence if such expansion and contraction of the structural materials cannot be adequately accommodated. In reality, most structures that contain heat are made in certain specific geometries and of finite sizes. Consequently, thermal stresses can be induced in these structural components either with non-uniform temperature distribution (or temperature field) or uniform temperature field but with mechanical constraints such as curved shapes and boundary constraints.

Another potentially serious design consideration related to the thermal effect is the possible creep deformation by the structural material. Creep, although a slow process, can be detrimental to structural integrity. It is one of the primary design concerns for structures to be used in a high temperature environment.

Having assessed the serious, but necessary, design considerations for structures expected to be operated in an elevated temperature environment, it has become evident that the thermal effect which is represented by the temperature field in the structure has to be determined as a necessary first step in the thermal stress and creep analyses. Heat conduction analysis which

23

relates the thermal conditions to the resulting temperature field in the solids has thus become a very important part of thermomechanical analysis.

Classical theory of heat conduction in solids has been well established and well documented, for instance by Özisik[1] and Carslaw and Jaeger[2]. The introduction of the finite element method for this type of analysis[3] has greatly expanded its application to many problems involving complex geometries and thermal conditions. Since the methodology of this technique has been extensively described in[4], it is not the intention of this book to repeat such a description. Instead, only the basic principles and finite element formulations for transient heat conduction on structures of axisymmetric geometry will be presented. The reader will soon discover that other cases of steady-state heat conduction in solids as well as heat conduction in planar geometries can be handled by these formulations as special and degenerated cases.

A general procedure for computer coding will also be outlined. The reader will find it useful when referring to the code listing of the TEPSAC program (an abbreviation of _t_hermo_e_lastic–_p_lastic _s_tress _a_nalysis with _c_reep) given in Appendix 5.

2.2 Review of basic formulations[1]

2.2.1 Fourier law of heat conduction

It is a well-known fact that heat in a solid "flows" from the higher temperature end towards the lower temperature end, or conversely "the flow of heat in a solid is always in the direction of decreasing temperature", which is similar to the common knowledge that water flows in the direction of decreasing elevation. Clearly, then, a definite relation must exist between the heat flow and the temperature variation in the solid. Indeed there is. The basic law which describes this relationship is called the "Fourier law of heat conduction" after a great French mathematician, Jean Baptiste Joseph Fourier (1768–1830).

This law simply states that if heat is to be conducted through a material, there must be a difference in temperature between the two terminal points. The amount of heat to be conducted is proportional to the area of the medium surface, the temperature difference and the duration for which the temperature difference is maintained. It is, however, inversely proportional to the distance between the two terminals. While the total amount of heat flow in a solid has less precise meaning in engineering application, its "intensity" which is represented by a term called "heat flux" defined as the amount of heat flowing through a solid per unit area and time appears to be more useful in the analysis. The mathematical translation of the Fourier law as stated above can be carried out by referring to Figure 2.1 for the heat flux leaving a unit area dA to be expressed as

$$q(\mathbf{r}, t) = -k \nabla T(\mathbf{r}, t) \qquad (2.1)$$

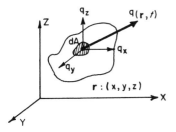

Figure 2.1   Components of a heat flux vector.

where $\mathbf{r}$ = position vector, t = time, and k = thermal conductivity of material.

Since the heat flux q is a vector quantity, it can be decomposed to give:

$$q(x, y, z, t) = q_x(x, y, z, t) \mapsto q_y(x, y, z, t) \mapsto q_z(x, y, z, t)$$

in which the notation "$\mapsto$" denotes summation of vector quantities and

$$q_x(x, y, z, t) = -k_x \frac{\partial T(x, y, z, t)}{\partial x} \tag{2.2a}$$

$$q_y(x, y, z, t) = -k_y \frac{\partial T(x, y, z, t)}{\partial y} \tag{2.2b}$$

$$q_z(x, y, z, t) = -k_z \frac{\partial T(x, y, z, t)}{\partial z} \tag{2.2c}$$

where q_x, q_y, q_z = respective heat fluxes in the x-, y- and z-directions, and k_x, k_y, k_z = respective thermal conductivities in the x-, y- and z-directions.

2.2.2   Newton's cooling law for fluids

In many engineering applications a finite size structure will inevitably have its surfaces in contact with a fluid (gas or liquid). It is important to realize that the mechanism of heat flow in a fluid is somewhat different from that in a solid. The basic law that governs the heat flow in fluids is Newton's law. Mathematically, the heat flux in a fluidic medium can be expressed as

$$q \propto (T_a - T_b)$$

or

$$q = h(T_a - T_b) \tag{2.3}$$

where T_a, T_b = temperature at points a and b respectively, and h = heat transfer coefficient.

2.2.3 Fourier heat conduction equation

One fundamental assumption which has to be imposed on the contents of the entire text is that the structural material is a continuum, which means that it contains no defects or voids. One obvious characteristic of a continuum is that all physical quantities, be they mechanical deformation or temperature, must follow a unique continuous function[5]. In other words, for a solid with or without heat sources (or sinks) and subject to specified thermal boundary conditions, there is a unique (fixed) temperature at any given location and time in the solid. This unique temperature function (sometimes called the temperature field) can be obtained by solving the following differential equation[1]:

$$\rho c \frac{\partial T(\mathbf{r}, t)}{\partial t} = \nabla \cdot [k \nabla T(\mathbf{r}, t)] + Q(\mathbf{r}, t) \tag{2.4}$$

where ρ = density of material, c = specific heat of material, and $Q(\mathbf{r}, t)$ = heat generation by material per unit volume and time.

In Cartesian coordinates, where $\mathbf{r} = (x, y, z)$, (2.4) becomes

$$\rho c \frac{\partial T(x, y, z, t)}{\partial t} = \frac{\partial}{\partial x} k_x \frac{\partial T}{\partial x} + \frac{\partial}{\partial y} k_y \frac{\partial T}{\partial y} + \frac{\partial}{\partial z} k_z \frac{\partial T}{\partial z} + Q(x, y, z, t) \tag{2.5a}$$

where k_x, k_y and k_z are the corresponding thermal conductivities in the x-, y- and z-directions.

The same equation can be shown in the cylindrical polar coordinate system to be

$$\rho c \frac{\partial T(r, \theta, z, t)}{\partial t} = \frac{\partial}{\partial r} k_r \frac{\partial T(r, \theta, z, t)}{\partial r} + \frac{k_r}{r} \frac{\partial T(r, \theta, z, t)}{\partial r} + \frac{1}{r^2} \frac{\partial}{\partial \theta} k_\theta \frac{\partial T(r, \theta, z, t)}{\partial \theta}$$

$$+ \frac{\partial}{\partial z} k_z \frac{\partial T(r, \theta, z, t)}{\partial z} + Q(r, \theta, z, t) \tag{2.5b}$$

with k_r, k_θ, k_z being the thermal conductivities in the r-, θ- and z-directions.

For most engineering problems involving modest temperature variations, thermal conductivities of material can be regarded as independent of temperature and constant[1], (2.4) can thus be simplified to give

$$\nabla^2 T(\mathbf{r}, t) + \frac{Q(\mathbf{r}, t)}{k} = \frac{1}{K} \frac{\partial T}{\partial t} \tag{2.6}$$

with k = constant, and $K = k/\rho c$ = thermal diffusivity.

2.2.4 Initial and boundary conditions

The solution of the Fourier heat conduction equations in (2.4) to (2.6) by classical methods such as described by Özisik[1] and Carslaw and Jaeger[2] involve a number of arbitrary constants to be determined by specific initial and boundary conditions. These conditions are necessary to translate the real physical condition into mathematical expressions. Proper specification and translation of these conditions are important steps in formulating numerical solutions as well, either by the finite difference method[6,7] or by the finite element method as will be described in this chapter.

Initial conditions are required only when dealing with transient heat transfer problems in which the temperature field in a solid changes with elapsing time. These conditions specify the temperature distribution in the solid before, and at the onset of, the changing thermal conditions which cause the change of temperature distribution.

The common initial condition in a solid can be expressed mathematically as:

$$T(\mathbf{r}, t)|_{t=0} = T_0(\mathbf{r}) \tag{2.7}$$

where the temperature field $T_0(\mathbf{r})$ is a function of the spatial coordinates only.

In many practical applications, the initial temperature distribution $T_0(\mathbf{r})$ in (2.7) can be assigned with a constant value for the uniform temperature (isothermal) conditions.

Specific boundary conditions are, however, required in the analysis of all transient or steady-state (time-invariant thermal or temperature conditions) problems involving solids of finite shape. Three types of boundary conditions are commonly used, as described below.

(a) *Prescribed surface temperature, $T_s(t)$.* Quite often, in practice, the temperature at the surface of the solid structure is measured by either attaching thermocouples to the surface or by some non-contact method such as optical pyrometry[8] or infrared scanning. The mathematical expression for this case takes the form

$$T(\mathbf{r}, t)|_{\mathbf{r}_s} = T_s(t) \tag{2.8a}$$

where $\mathbf{r}_s$ = coordinates of the boundary surface where temperatures are specified to be $T_s(t)$.

(b) *Prescribed heat flux, $q_s(t)$.* Many structures have their surface exposed to a heat source or a heat sink. One such example is the heat treatment of a large forged piece (e.g. turbine shaft) in an autoclave in which heat is being supplied to the piece through its outside surface. The mathematical translation of the heat flux into or from a solid surface can be readily carried out by using the Fourier law given in (2.1). The mathematical formulation of the heat flux across a solid boundary surface can be expressed as

$$\frac{\partial T(\mathbf{r}, t)}{\partial n_i}\bigg|_{\mathbf{r} = \mathbf{r}_s} = -q_s(\mathbf{r}_s, t)/k \tag{2.8b}$$

27

where $\partial/\partial n_i$ is the differentiation along the outward-drawn normal at the boundary surface S_i, $q_s(\mathbf{r}_s, t)$, of course, is the known heat flux across this surface specified by the coordinates $\mathbf{r}_s$ in the same direction as n_i.

Example 2.1. The heat flux boundary condition for a long solid cylinder with temperature field $T(r, t)$ illustrated in Figure 2.2 can be expressed, following (2.8b), as

$$\left.\frac{\partial T(r, t)}{\partial r}\right|_{r=a} = -q_s(a, t)/k$$

for the case heat flux $q_s(a, t)$ leaving the surface $r = a$, and

$$\left.\frac{\partial T(r, t)}{\partial r}\right|_{r=a} = q_s'(a, t)/k$$

for the case heat flux $q_s'(a, t)$ entering the surface.

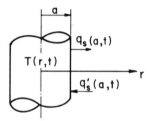

Figure 2.2 Heat flux entering and leaving a solid cylinder.

(c) *Convective boundary conditions.* Many engineering applications involve boundary surfaces of a solid in contact with fluids in gaseous or liquid states. As the mechanism of heat flow in a solid differs from that in a fluid as explained in Sections 2.2.1 and 2.2.2, a special expression has to be derived.

Mathematical formulation of this type of boundary condition can be derived by following the illustration given in Figure 2.3. The solid with an unknown temperature field $T(\mathbf{r}, t)$ has its surface $\mathbf{r}_s$ in contact with a fluid at an average temperature T_0. The heat flux reaching the specified boundary at $\mathbf{r}_s$ from the solid can be expressed by the Fourier law to be

$$q_s = -k\left.\frac{\partial T(\mathbf{r}, t)}{\partial n}\right|_{\mathbf{r}=\mathbf{r}_s}$$

whereas the heat flux leaving $\mathbf{r}_s$ into the fluid at temperature T_0 to be determined by Newton's law as follows:

$$q_f = h[T(\mathbf{r}_s, t) - T_0]$$

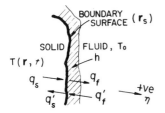

Figure 2.3 Heat flux entering and leaving a solid surface through a boundary layer.

Based on a notion that no heat is to be stored at the interface, i.e. $q_s = q_f$, or

$$-k \left. \frac{\partial T(\mathbf{r}, t)}{\partial n} \right|_{\mathbf{r} = \mathbf{r}_s} = h[T(\mathbf{r}_s, t) - T_0]$$

which leads to

$$\left. \frac{\partial T(\mathbf{r}, t)}{\partial n} \right|_{\mathbf{r} = \mathbf{r}_s} + \frac{h}{k} T(\mathbf{r}_s, t) = \frac{h}{k} T_0 \qquad (2.8c)$$

For the case of reversed heat flow, i.e. heat flows from the fluid to the solid, $q_f = q_s'$, sign changes will be applied to both sides in the expressions for heat flux. The end result is identical to what is shown in (2.8c).

Example 2.2. The convective boundary condition for the solid cylinder described in Figure 2.2 will have the form

$$\left. \frac{\partial T(r, t)}{\partial r} \right|_{r = a} + \frac{h}{k} T(r, t)|_{r = a} = \frac{h}{k} T_0$$

in which h is the heat transfer coefficient between the solid cylinder and the surrounding fluid at temperature T_0 and k is the thermal conductivity of the solid.

There are a number of ways that h can be determined[9]. A physical interpretation of this coefficient is the thermal resistance caused by the boundary layer induced by the fluid circulating over the surface of the solid as shown in the shaded area in Figure 1.3. The higher the velocity of the circulating fluid the thinner the boundary layer and the higher the value for h. On the other hand, in natural convection cases, where the velocity of the fluid is low, small values of h can be expected. Readers are cautioned, however, that significant error may result from assuming $h = 0$ for cases in which h values are expected to be small. The case of $h = 0$ obviously represents an impermeable boundary, as can be seen from (2.8c), which cannot be equivalent to the case of a thermal boundary with small, but nonzero, h values.

2.3   Finite element formulation of transient heat conduction in solids

Consider the problem as illustrated in Figure 2.4, in which the temperature field $T(\mathbf{r}, t)$ in the solid of volume V and surface S is induced by the internal heat generation, $Q(\mathbf{r}, t)$ and heat flux across the boundary, $q(\mathbf{r}, t)$. As the thermal conditions are assumed to be varying with time, the problem is classified as transient with an assumed initial temperature field, $T_0(\mathbf{r}, 0)$.

The finite element formulation of this problem can be achieved either by a variational process via the Rayleigh–Ritz method or by the Galerkin method based on the weighted residual as described in Chapter 1. The use of the latter method will require the discretization of the partial differential equation given in (2.4). The approach adopted here, however, is the Rayleigh–Ritz variational principle with the functional derived by Gurtin[10]. The reader is urged to refer to additional derivations given in Wilson and Nickell[3], in Chapter 5 of Desai[11] and Chapter 11 of Segerlind[4].

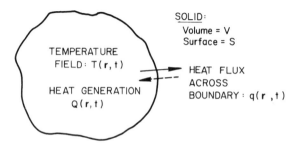

Figure 2.4   Temperature field in a solid induced by heat flow.

Thus, by expressing Gurtin's functional in the form

$$\phi = \tfrac{1}{2} \int_v \{\rho c T * T + T_{,i} * k_{ij} * T_{,j} - 2\rho Q * T - 2\rho c T_0 * T\}(\mathbf{r}, t)\, dv$$

$$- \int_s \{q_i n_i * T\}(\mathbf{r}, t)\, ds \qquad (2.9)$$

where ρ, c = respective density and specific heat of the solid.

   T_0 = initial spatial temperature distribution

   n_i = directional cosine

$$T_{,i} = \frac{\partial T(\mathbf{r}, t)}{\partial \mathbf{r}_i}$$

   k_{ij} = thermal conductivity tensor.

The * sign in the above equation denotes a convolution symbol as defined in Wilson and Nickell[3]. The convective heat flow across the boundary is excluded in q_i in this case. Equation (2.9) may be expressed in a longhand form in the $\mathbf{r} : (x, y, z)$ coordinate system:

$$\phi = \frac{1}{2} \int_v \left(\rho c T^2 + \frac{\partial T}{\partial x} k_x \frac{\partial T}{\partial x} + \frac{\partial T}{\partial y} k_y \frac{\partial T}{\partial y} + \frac{\partial T}{\partial z} k_z \frac{\partial T}{\partial z} - 2QT \right.$$

$$- 2\rho c T_0 T)(x, y, z, t) \, dv - \int_s (q_x n_x + q_y n_y + q_z n_z) T(x, y, z, t) \, ds$$
(2.10)

where k_x, k_y, k_z = thermal conductivities along the respective x-, y-,
z-directions;

n_x, n_y, n_z = directional cosines with respect to the respective x-, y-
and z-coordinates.

For an analysis based on a constant element temperature algorithm, as in this case, the material properties such as thermal conductivities are independent of temperature within an element. Equation (2.10) can thus be written in the form

$$\phi = \int_v \frac{1}{2} \left\{ \rho c T^2 + k_x \left(\frac{\partial T}{\partial x} \right)^2 + k_y \left(\frac{\partial T}{\partial y} \right)^2 + k_z \left(\frac{\partial T}{\partial z} \right)^2 - 2QT \right.$$

$$- 2\rho c T_0 T \Big\} (x, y, z, t) \, dv$$

$$- \int_s (q_x n_x + q_y n_y + q_z n_z) T(x, y, z, t) \, ds$$
(2.11)

or to be expressed in the matrix form

$$\phi = \int_v \frac{1}{2} [\{g\}^T [D] \{g\} - 2QT](x, y, z, t) \, dv$$

$$+ \int_v \frac{1}{2} [\rho c (T - 2T_0) T](x, y, z, t) \, dv - \int_s (\{q\}^T \{n\} T)(x, y, z, t) \, ds$$
(2.12)

in which the matrix notations are

$$\{g\}^T = \left\{ \frac{\partial T}{\partial x} \quad \frac{\partial T}{\partial y} \quad \frac{\partial T}{\partial z} \right\}^T$$
(2.13)

$$[D] = \begin{bmatrix} k_x & 0 & 0 \\ 0 & k_y & 0 \\ 0 & 0 & k_z \end{bmatrix}$$
(2.14)

$$\{q\}^T = \{q_x \quad q_y \quad q_z\}^T$$
(2.15)

$$\{n\}^T = \{n_x \quad n_y \quad n_z\}^T$$
(2.16)

If one follows the discretization procedure described in Section 1.2 to the solid illustrated in Figure 2.4, dividing it into M elements and n nodes as shown in Figure 2.5, the functional in (2.9) and (2.12) for the discretized solid becomes

$$\phi = \phi_1 + \phi_2 + \cdots + \phi_m = \frac{\sum^M}{1}\phi_m \tag{2.17}$$

and further, by assuming the relationship between the temperatures in the element and that in the corresponding nodes to be

$$T_m(\mathbf{r}, t) = [N(\mathbf{r})]\{T(t)\} \tag{2.18}$$

in which the matrix $[N(\mathbf{r})]$ is the interpolation function as described in Step 2, Section 1.3.

Nodal Temp. $\{T\}$ Element Temp. $T(\mathbf{r})$

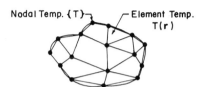

Figure 2.5 Discretization of a solid for thermal analysis.

In Cartesian coordinates, (2.18) takes the form

$$T_m(x, y, z, t) = [N(x, y, z)]\{T(t)\} \tag{2.19}$$

The functional for an individual element m thus becomes

$$\phi_m = \int_v \tfrac{1}{2}[\{g_m\}^T[D_m]\{g_m\} - 2Q_m T_m](x, y, z, t)\, dv$$

$$+ \int_v [\tfrac{1}{2}\rho_m C_m(T_m - 2T_0)T_m](x, y, z, t) - \int_s (\{q_m\}^T\{n\}T)(x, y, z, t)\, ds \tag{2.20}$$

with

$$\{g_m\} = \begin{Bmatrix} \dfrac{\partial T_m}{\partial x} \\[2mm] \dfrac{\partial T_m}{\partial y} \\[2mm] \dfrac{\partial T_m}{\partial z} \end{Bmatrix} = \begin{bmatrix} \dfrac{\partial N(x, y, z)}{\partial x}\{T(t)\} \\[2mm] \dfrac{\partial N(x, y, z)}{\partial y}\{T(t)\} \\[2mm] \dfrac{\partial N(x, y, z)}{\partial z}\{T(t)\} \end{bmatrix} = [B(\mathbf{r})]\{T(t)\} \tag{2.21}$$

in which

$$[B(\mathbf{r})] = \begin{bmatrix} \dfrac{\partial N_1}{\partial x} & \dfrac{\partial N_2}{\partial x} & \cdots & \dfrac{\partial N_n}{\partial x} \\[2ex] \dfrac{\partial N_1}{\partial y} & \dfrac{\partial N_2}{\partial y} & \cdots & \dfrac{\partial N_n}{\partial y} \\[2ex] \dfrac{\partial N_1}{\partial z} & \dfrac{\partial N_2}{\partial z} & \cdots & \dfrac{\partial N_n}{\partial z} \end{bmatrix} \qquad (2.22)$$

By substituting (2.19) and (2.21) into (2.20) and expressing the resulting expression in the general coordinate system

$$\phi_m = \int_v \tfrac{1}{2}([B(\mathbf{r})]\{T\})^T[D_m][B(\mathbf{r})]\{T\}\,dv - \int_v Q_m(\mathbf{r})[N(\mathbf{r})]\{T\}\,dv$$

$$+ \int_v \tfrac{1}{2}\rho_m C_m([N(\mathbf{r})]\{T\})^T[N(\mathbf{r})]\{T\}\,dv$$

$$- \int_v \rho_m C_m([N(\mathbf{r})]\{T_0\})^T[N(\mathbf{r})]\{T\}\,dv$$

$$- \int_s \{q_m(\mathbf{r})\}^T\{n\}[N(\mathbf{r})]\{T\}\,ds \qquad (2.23)$$

The discretized initial temperature in the element, i.e. $T_0(\mathbf{r}, 0) = [N(\mathbf{r})]\{T_0\}$ was used to derive the above expression.

Expanding (2.23) and using the relation that

$$\{T_0\}^T[N(\mathbf{r})]^T[N(\mathbf{r})]\{T\} = \{T\}^T[N(\mathbf{r})]^T[N(\mathbf{r})]\{T_0\}$$

for the symmetrical nature of the product of $[N(\mathbf{r})]^T[N(\mathbf{r})]$, the final expression of ϕ_m is derived to show

$$\phi_m = \int_v \tfrac{1}{2}\{T\}^T[B(\mathbf{r})]^T[D_m][B(\mathbf{r})]\{T\}\,dv$$

$$- \int_v Q_m(\mathbf{r})[N(\mathbf{r})]\{T\}\,dv$$

$$+ \int_v \tfrac{1}{2}\rho_m C_m\{T\}^T[N(\mathbf{r})]^T[N(\mathbf{r})](\{T\} - 2\{T_0\})\,dv$$

$$- \int_s \{q_m(\mathbf{r})\}^T\{n\}[N(\mathbf{r})]\{T\}\,ds \qquad (2.24)$$

The thermal equilibrium condition in the discretized solid requires that the functional ϕ_m in each individual element be kept minimum with respect to the nodal temperatures $\{T\}$, or mathematically by setting-

$$\frac{\partial \phi_m}{\partial \{T\}} \equiv 0$$

which leads to the establishment of the element equations with the argument that

$$\{T\} - \{T_0\} = \{T(t)\} - \{T_0(0)\} = \{\dot{T}(t)\} \tag{2.25}$$

The element equations, after lumping various terms together, takes a similar form to that shown in (1.27):

$$[C_e]\{\dot{T}(t)\} + [K_e]\{T(t)\} = \{F_e\} \tag{2.26}$$

where

$$[K_e] = \text{element thermal conductivity matrix}$$

$$= \int_v [B(\mathbf{r})]^T [D_m][B(\mathbf{r})] \, dv \tag{2.27}$$

$$[C_e] = \text{element heat capacitance matrix}$$

$$= \int_v \rho_m C_m [N(\mathbf{r})]^T [N(\mathbf{r})] \, dv \tag{2.28}$$

$$\{F_e\} = \text{element thermal force matrix}$$

$$= \int_v Q_m(\mathbf{r})[N(\mathbf{r})] \, dv - \int_s \{q_m(\mathbf{r})\}^T \{n\}[N(\mathbf{r})] \, ds \tag{2.29}$$

The reader may be aware that the first term of (2.26) vanishes for the steady-state problems as the temperature everywhere in the solid, including those at the nodes, is time-invariant. Mathematically, it means, of course, that $\{\dot{T}(t)\} = 0$. The remaining terms will be identical to those shown in Segerlind[4], which were derived from a somewhat different functional for the steady-state condition.

2.4 Transient heat conduction in axisymmetric solids

Many structures in industrial applications have a geometry which is symmetrical about one axis. Such a geometry is referred to as axisymmetric. Cylindrical pressure vessels, wheels and stacks, etc., can all be classified as axisymmetric structures. Since, by definition, the geometry is symmetric about the axis of symmetry, a "flat" geometry such as a large slab can be regarded as an axisymmetric solid with an "infinite" radius of curvature. (In reality, of course, this cannot be the case. A sufficiently large value relative to the thickness of the slab is to be added to the r-coordinates of all the nodes.

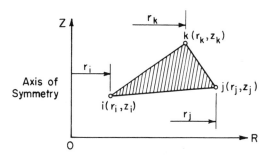

Figure 2.6 Nodal coordinates of a triangular toroidal element.

An axisymmetric structure with such a large artificial radius is a close approximation to a flat slab.) The finite element formulation presented in this chapter for the axisymmetric solids can thus be extended to the solids of planar geometries with the introduction of an artificially large radius of curvature as described above.

Since triangular torus elements are considered to be the basic element geometry for most finite elements analyses,* this element shape has also been adopted in the present analysis, the coordinates (r, z) being shown in Figure 2.6.

Following the procedure outlined in Chapter 1, the analysis begins with the discretization of the continuum into a finite number of triangular torus elements typically as shown in Figure 2.6. The temperature of the element $T_m(r, z, t)$ can be related to its nodal values $\{T(t)\}$ via the interpolation function to be:

$$T_m(r, z, t) = [N(r, z)]\{T(t)\} \tag{2.30}$$

The interpolation function $[N(r, z)]$ can be determined by an assumed temperature distribution function in the element. As illustrated in Step 2, Section 1.3, the simplest form of such function is a linear polynomial of the coordinates. For the case described in Figure 2.6, the following expression is used:

$$T_m(r, z, t) = a_1(t) + a_2(t)r + a_3(t)z \tag{2.31}$$

from which the temperature at the three nodes can be expressed by the substitution of the respective nodal coordinates into (2.31) as follows:

$$T_i(t) = a_1(t) + a_2(t)r_i + a_3(t)z_i$$

$$T_j(t) = a_1(t) + a_2(t)r_j + a_3(t)z_j \tag{2.32}$$

$$T_k(t) = a_1(t) + a_2(t)r_k + a_3(t)z_k$$

where $a_1(t)$, $a_2(t)$ and $a_3(t)$ are arbitrary constants and parameters of time t.

* Elements of quadrilateral cross-sections are treated as a concentration of 4 triangular elements defined by the 2 diagonals.

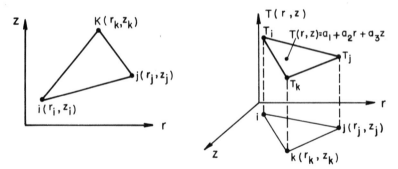

Figure 2.7 Element with linear temperature variation.

Physically a linear polynomial distribution function such as shown in (2.31) results in a uniform (constant) temperature in the element, as illustrated in Figure 2.7.

The constants a_1, a_2 and a_3 in (2.31) can be determined first by expressing (2.32) in matrix form to give

$$\begin{Bmatrix} a_1(t) \\ a_2(t) \\ a_3(t) \end{Bmatrix} = \begin{bmatrix} 1 & r_i & z_i \\ 1 & r_j & z_j \\ 1 & r_k & z_k \end{bmatrix}^{-1} \begin{Bmatrix} T_i(t) \\ T_j(t) \\ T_k(t) \end{Bmatrix}$$

$$\{a(t)\} \quad = \qquad [h] \qquad \{T(t)\} \tag{2.33}$$

where

$$[h] = \begin{bmatrix} 1 & r_i & z_i \\ 1 & r_j & z_j \\ 1 & r_k & z_k \end{bmatrix}^{-1}$$

$$= \frac{1}{2A} \begin{bmatrix} (r_j z_k - r_k z_j) & (r_k z_i - r_i z_k) & (r_i z_j - r_j z_i) \\ z_j - z_k & z_k - z_i & z_i - z_j \\ r_k - r_j & r_i - r_k & r_j - r_i \end{bmatrix}$$

$A =$ the cross-sectional area of the triangular element, or

$$= \tfrac{1}{2}[r_i(z_j - z_k) + r_j(z_k - z_i) + r_k(z_i - z_j)]$$

Another way to express the $[h]$ matrix is as follows:

$$[h] = \begin{bmatrix} h_{1i} & h_{1j} & h_{1k} \\ h_{2i} & h_{2j} & h_{2k} \\ h_{3i} & h_{3j} & h_{3k} \end{bmatrix} \tag{2.34}$$

36

with

$$h_{1i} = \frac{1}{2A}(r_j z_k - r_k z_j) \quad h_{2i} = \frac{1}{2A}(z_j - z_k) \quad h_{3i} = \frac{1}{2A}(r_k - r_j)$$

$$h_{1j} = \frac{1}{2A}(r_k z_i - r_i z_k) \quad h_{2j} = \frac{1}{2A}(z_k - z_i) \quad h_{3j} = \frac{1}{2A}(r_i - r_k)$$

$$h_{1k} = \frac{1}{2A}(r_i z_j - r_j z_i) \quad h_{2k} = \frac{1}{2A}(z_i - z_j) \quad h_{3k} = \frac{1}{2A}(r_j - r_i)$$

$$(2.35)$$

It can be readily seen that the matrix $[h]$ contains elements of the specified nodal coordinates.

Since from (2.33), $\{a(t)\} = [h]\{T(t)\}$ and $[h]$ is given in (2.34),

$$a_1(t) = h_{1i} T_i(t) + h_{1j} T_j(t) + h_{1k} T_k(t)$$

$$a_2(t) = h_{2i} T_i(t) + h_{2j} T_j(t) + h_{2k} T_k(t)$$

$$a_3(t) = h_{3i} T_i(t) + h_{3j} T_j(t) + h_{3k} T_k(t)$$

Substituting $a_1(t)$, $a_2(t)$ and $a_3(t)$ into (2.31) and rearranging the terms gives

$$T_m(r, z, t) = \{a_1(t) \quad a_2(t) \quad a_3(t)\} \begin{Bmatrix} 1 \\ r \\ z \end{Bmatrix} = \begin{Bmatrix} h_{1i} + rh_{2i} + zh_{3i} \\ h_{1j} + rh_{2j} + zh_{3j} \\ h_{1k} + rh_{2k} + zh_{3k} \end{Bmatrix}^{\mathrm{T}} \begin{Bmatrix} T_i(t) \\ T_j(t) \\ T_k(t) \end{Bmatrix}$$

Comparison of the above relation with (2.30) leads to the interpolation (shape) function

$$[N(r, z)]^{\mathrm{T}} = \begin{bmatrix} h_{1i} + rh_{2i} + zh_{3i} \\ h_{1j} + rh_{2j} + zh_{3j} \\ h_{1k} + rh_{2k} + zh_{3k} \end{bmatrix} \quad (2.36)$$

The matrix $[B]$ can be evaluated following (2.22):

$$[B] = \begin{bmatrix} \dfrac{\partial N_1}{\partial r} & \dfrac{\partial N_2}{\partial r} & \dfrac{\partial N_3}{\partial r} \\ \dfrac{\partial N_1}{\partial z} & \dfrac{\partial N_2}{\partial z} & \dfrac{\partial N_3}{\partial z} \end{bmatrix} = \begin{bmatrix} h_{2i} & h_{2j} & h_{2k} \\ h_{3i} & h_{3j} & h_{3k} \end{bmatrix} \quad (2.37)$$

where $N_1 = h_{1i} + rh_{2i} + zh_{3i}$

$N_2 = h_{1j} + rh_{2j} + zh_{3j}$

$N_3 = h_{1k} + rh_{2k} + zh_{3k}$

The $[D_m]$ matrix which relates to the element thermal conductivities for an axisymmetric solid has the form:

$$[D_m] = \begin{bmatrix} k_{rr} & k_{rz} \\ k_{zr} & k_{zz} \end{bmatrix} \tag{2.38}$$

2.5 Computation of the thermal conductivity matrix

The general expression of the element thermal conductivity matrix $[K_e]$ is given in (2.27). For axisymmetric solids, both matrices $[B(r)]$ and $[D_m]$ have been expressed respectively in (2.37) and (2.38). Thus the thermal conductivity matrix can be shown as

$$[K_e] = \int_v [B]^T[D_m][B]\,dv$$

$$= \int_v \begin{bmatrix} h_{2i} & h_{3i} \\ h_{2j} & h_{3j} \\ h_{2k} & h_{3k} \end{bmatrix} \begin{bmatrix} k_{rr} & k_{rz} \\ k_{zr} & k_{zz} \end{bmatrix} \begin{bmatrix} h_{2i} & h_{2j} & h_{2k} \\ h_{3i} & h_{3j} & h_{3k} \end{bmatrix} dv$$

$$= \begin{bmatrix} h_{2i} & h_{3i} \\ h_{2j} & h_{3j} \\ h_{2k} & h_{3k} \end{bmatrix} \begin{bmatrix} k_{rr} & k_{rz} \\ k_{zr} & k_{zz} \end{bmatrix} \begin{bmatrix} h_{2i} & h_{2j} & h_{2k} \\ h_{3i} & h_{3j} & h_{3k} \end{bmatrix} \int_v dv \tag{2.39}$$

For an axisymmetric geometry as illustrated in Figure 2.8,

$$\int_v dv = \int_A 2\pi r\,dA \tag{2.40}$$

The precise integration of the above integral requires very tedious algebraic manipulation. An easier task, however, is to use the area coordinate system first proposed by Felippa[12] with more elaboration available on p. 43 of Segerlind[4]. A brief outline of this system is given in Appendix 1.

By replacing the r variable in (2.40) by the area coordinates described in (A.1.3) of Appendix 1,

$$\int_v dv = 2\pi \int_A (r_i L_1 + r_j L_2 + r_k L_3)\,dA$$

$$= 2\pi \left(r_i \int_A L_1\,dA + r_j \int_A L_2\,dA + r_k \int_A L_3\,dA \right)$$

Using the integrals shown in (A.1.7), the following results are obtained:

$$\int_A L_1\,dA = \int_A L_2\,dA = \int_A L_3\,dA = \frac{A}{3}$$

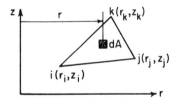

Figure 2.8 Volume integration of a triangular toroidal element.

Thus

$$\int_v dA = 2\pi(r_i + r_j + r_k)\frac{A}{3} = 2\pi r_c A \qquad (2.41)$$

where

$$r_c = (r_i + r_j + r_k)/3$$

Substituting (2.41) into (2.39) results in the element thermal conductivity matrix:

$$[K_e] = 2\pi r_c A([k_r] + [k_z] + [k_{rz}] + [k_{zr}]) \qquad (2.42)$$

where

$$[k_r] = k_r \begin{bmatrix} h_{2i}^2 & h_{2i}h_{2j} & h_{2i}h_{2k} \\ h_{2i}h_{2j} & h_{2j}^2 & h_{2j}h_{2k} \\ h_{2i}h_{2k} & h_{2j}h_{2k} & h_{2k}^2 \end{bmatrix}$$

$$[k_z] = k_z \begin{bmatrix} h_{3i}^2 & h_{3i}h_{3j} & h_{3i}h_{3k} \\ h_{3i}h_{3j} & h_{3j}^2 & h_{3j}h_{3k} \\ h_{3i}h_{3k} & h_{3j}h_{3k} & h_{3k}^2 \end{bmatrix}$$

$$[k_{rz}] = k_{rz} \begin{bmatrix} h_{2i}h_{3i} & h_{2i}h_{3j} & h_{2i}h_{3k} \\ h_{2j}h_{3i} & h_{2j}h_{3j} & h_{2j}h_{3k} \\ h_{2k}h_{3i} & h_{2k}h_{3j} & h_{2k}h_{3k} \end{bmatrix}$$

$$[k_{zr}] = k_{zr} \begin{bmatrix} h_{2i}h_{3i} & h_{2j}h_{3i} & h_{2k}h_{3i} \\ h_{2i}h_{3j} & h_{2j}h_{3j} & h_{2k}h_{3j} \\ h_{2i}h_{3k} & h_{2j}h_{3k} & h_{2k}h_{3k} \end{bmatrix}$$

For most isotropic materials, $K_{rz} = k_{zr} = 0$, $k_r = k_z = k_m$; then

$$[K_e] = 2\pi r_c A k_m \begin{bmatrix} h_{2i}^2 + h_{3i}^2 & h_{2i}h_{2j} + h_{3i}h_{3j} & h_{2i}h_{2k} + h_{3i}h_{3k} \\ & h_{2j}^2 + h_{3j}^2 & h_{2j}h_{2k} + h_{3j}h_{3k} \\ \text{SYM} & & h_{2k}^2 + h_{3k}^2 \end{bmatrix}$$

$$(2.43)$$

with $h_{1i}, h_{2i}, \ldots, h_{3k}$ given in (2.35).

2.6 Computation of the heat capacitance matrix

Following the general expression for the heat capacitance matrix given in (2.28), the special form for an axisymmetric solid can be obtained by the inclusion of the interpolation function in (2.36):

$$[C_e] = \int_v \rho_m C_m [N]^T [N]\, dv = \rho_m C_m \int_v [N]^T [N]\, dv$$

$$= \rho_m C_m \int_v \begin{bmatrix} h_{1i} & h_{2i} & h_{3i} \\ h_{1j} & h_{2j} & h_{3j} \\ h_{1k} & h_{2k} & h_{3k} \end{bmatrix} \begin{Bmatrix} 1 \\ r \\ z \end{Bmatrix} \{1 \quad r \quad z\} \begin{bmatrix} h_{1i} & h_{1j} & h_{1k} \\ h_{2i} & h_{2j} & h_{2k} \\ h_{3i} & h_{3j} & h_{3k} \end{bmatrix} dv$$

$$= \rho_m C_m \int_v [h]^T \begin{Bmatrix} 1 \\ r \\ z \end{Bmatrix} \{1 \quad r \quad z\} [h]\, dv$$

But from (2.33),

$$[h] = \begin{bmatrix} 1 & r_i & z_i \\ 1 & r_j & z_j \\ 1 & r_k & z_k \end{bmatrix}^{-1},$$

and

$$[h]^T = \left(\begin{bmatrix} 1 & r_i & z_i \\ 1 & r_j & z_j \\ 1 & r_k & z_k \end{bmatrix}^{-1} \right)^T = \left(\begin{bmatrix} 1 & r_i & z_i \\ 1 & r_j & z_j \\ 1 & r_k & z_k \end{bmatrix}^{T} \right)^{-1} = \begin{bmatrix} 1 & 1 & 1 \\ r_i & r_j & r_k \\ z_i & z_j & z_k \end{bmatrix}^{-1}$$

Hence, with the aid of the relationships established in (A.1.2), i.e. $1 = L_1 + L_2 + L_3$, we obtain $r = r_i L_1 + r_j L_2 + r_k L_3$ and $z = z_i L_1 + z_j L_2 + z_k L_3$ in (A.1.3) and (A.1.4), where L_1, L_2 and L_3 are the area coordinates defined in Appendix 1.

The substitution of the above expressions into the $[C_e]$ formulation yields

$$[C_e] = \rho_m C_m \int_v \begin{Bmatrix} L_1 \\ L_2 \\ L_3 \end{Bmatrix} \{L_1 \quad L_2 \quad L_3\}\, dv$$

$$= 2\pi \rho_m C_m \int_A \begin{bmatrix} L_1^2 & L_1 L_2 & L_1 L_3 \\ & L_2^2 & L_2 L_3 \\ \text{SYM} & & L_3^2 \end{bmatrix} r\, dA$$

$$= 2\pi \rho_m C_m \int_A \begin{bmatrix} L_1^2 & L_1 L_2 & L_1 L_3 \\ & L_2^2 & L_2 L_3 \\ \text{SYM} & & L_3^2 \end{bmatrix} (r_i L_1 + r_j L_2 + r_k L_3)\, dA$$

By integrating the above matrix and using the relationships given in (A.1.6) and (A.1.7), we obtain the expressions for the $[C_e]$ matrix for an element:

$$[C_e] = \frac{2\pi A \rho_m C_m}{60} \begin{bmatrix} C_{11} & C_{12} & C_{13} \\ & C_{22} & C_{23} \\ SYM & & C_{33} \end{bmatrix} \qquad (2.44)$$

where

$C_{11} = 6r_i + 2r_j + 2r_k$ $C_{12} = 2r_i + 2r_j + r_k$

$C_{13} = 2r_i + r_j + 2r_k$ $C_{22} = 2r_i + 6r_j + 2r_k$

$C_{23} = r_i + 2r_j + 2r_k$ $C_{33} = 2r_i + 2r_j + 6r_k$

2.7 Computation of the thermal force matrix

Again, if we consider only the first term in (2.29), i.e. no heat transfer across the boundaries (e.g. for interior elements), then the thermal force matrix has the form

$$\{F_e\} = \int_v Q[N]^T \, dv = Q \int_V [N]^T \, dv$$

$$\{F_e\} = Q \int_v \begin{bmatrix} h_{1i} + h_{2i}r + h_{3i}z \\ h_{1j} + h_{2j}r + h_{3j}z \\ h_{1k} + h_{2k}r + h_{3k}z \end{bmatrix} dv$$

$$= Q \begin{bmatrix} h_{1i} & h_{2i} & h_{3i} \\ h_{1j} & h_{2j} & h_{3j} \\ h_{1k} & h_{2k} & h_{3k} \end{bmatrix} \int_v \begin{Bmatrix} 1 \\ r \\ z \end{Bmatrix} dv$$

$$= Q[h] \int_z \int_r 2\pi r \begin{Bmatrix} 1 \\ r \\ z \end{Bmatrix} dr\, dz = \frac{2\pi A Q}{12} \begin{Bmatrix} 2r_i + r_j + r_k \\ r_i + 2r_j + r_k \\ r_i + r_j + 2r_k \end{Bmatrix} \qquad (2.45)$$

2.8 Transient heat conduction in the time domain

Thus far, the finite element formulation presented in the foregoing sections was based on the assumption that time variation applies at the nodal temperature in a discretized model. The appearance of the nodal temperature as a function of time, $\{T(t)\}$ and the rate of variation, $\{\dot{T}(t)\}$ in (2.26) requires the discretization of that quantity in the time domain as well.

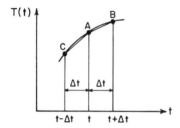

Figure 2.9 Discretization in the time domain.

The discretization in time can be accomplished simply by assuming time increments Δt in the computation of (2.26), as illustrated in Figure 2.9. The expression of the rate of the temperature change $\{\dot{T}(t)\}$, however, requires an optimum finite difference algorithm to achieve both numerical stability and rapid convergence.

There are two time-difference schemes that are frequently used in finite element analysis: (1) the two-level explicit method; and (2) the mid-interval scheme.

2.8.1 Two-level explicit method (p. 113[11])

Following the functional variation illustrated in Figure 2.10, the usual finite difference schemes:

$$\frac{\partial T}{\partial t} \doteq \frac{T(t + \Delta t) - T(t)}{\Delta t}$$

for the forward difference scheme, or

$$\frac{\partial T}{\partial t} \doteq \frac{T(t) - T(t - \Delta t)}{\Delta t}$$

for the backward difference scheme.

Assume that the property matrices, $[K_e]$ and $[C_e]$, are constant within a small time increment Δt, then from the overall structural thermal equilibrium

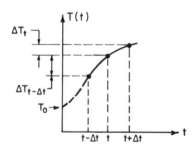

Figure 2.10 Increments in the time domain.

equation similar to the one given in (2.26), the following recurrence relation may be derived:

$$[C_e](\{\dot{T}_{t+\Delta t}\} - \{\dot{T}_t\}) + [K_e](\{T_{t+\Delta t}\} - \{T_t\}) = \{F_{e,t+\Delta t}\} - \{F_{e,t}\} \quad (2.46)$$

Since, by the forward difference scheme,

$$\{\dot{T}_{t+\Delta t}\} = \left\{\frac{\partial T}{\partial t}\right\}_{t+\Delta t} = \frac{\{T_{t+\Delta t}\} - \{T_t\}}{\Delta t}$$

or

$$\{\dot{T}_{t+\Delta t}\} - \{\dot{T}_t\} = \frac{\{T_{t+\Delta t}\} - \{T_t\}}{\Delta t} - \{\dot{T}_t\}$$

and, substituting the above relationship into (2.46),

$$[C_e]\left(\frac{\{T_{t+\Delta t}\} - \{T_t\}}{\Delta t} - \{\dot{T}_t\}\right) + [K_e](\{T_{t+\Delta t}\} - \{T_t\}) = \{F_{e,t+\Delta t}\} - \{F_{e,t}\}$$

which leads the following expression upon re-arranging terms:

$$\left(\frac{[C_e]}{\Delta t} + [K_e]\right)(\{T_{t+\Delta t}\} - \{T_t\}) = [C_e]\{\dot{T}_t\} + (\{F_{e,t+\Delta t}\} - \{F_{e,t}\})$$

Now if we let

$$\{\Delta T_t\} = \{T_{t+\Delta t}\} - \{T_t\}$$

and

$$\{\dot{T}_t\} = \frac{\{T_t\} - \{T_{t+\Delta t}\}}{\Delta t}$$

$$\{\Delta F_{e,t}\} = \{F_{t+\Delta t}\} - \{F_t\}$$

$$\{\Delta T_{t-\Delta t}\} = \{T_t\} - \{T_{t-\Delta t}\}$$

The last equation becomes

$$\left(\frac{[C_e]}{\Delta t} + [K_e]\right)\{\Delta T_t\} = \frac{[C_e]}{\Delta t}\{\Delta T_{t-\Delta t}\} + \{\Delta F_{e,t}\}$$

or, in a more compact form,

$$[K^*]\{\Delta T_t\} = \{F^*\} \quad (2.47)$$

where

$$[K^*] = \text{equivalent conductivity matrix}$$

$$= \frac{[C_e]}{\Delta t} + [K_e] \quad (2.48a)$$

43

$$\{F_e^*\} = \text{equivalent thermal force matrix}$$

$$= \frac{[C_e]}{\Delta t}\{\Delta T_{t-\Delta t}\} + \{\Delta F_e\} \tag{2.48b}$$

Equation (2.47) provides the nodal temperature increase within the time increment Δt.

Numerical solution of (2.47) requires the temperature condition at the end of the previous time step, hence a starting procedure is required.

The equation used for the first time increment is

$$[K^*]\{T_{\Delta t}\} = \{F_{\Delta t}\} \tag{2.49}$$

where $\{T_0\}$ = initial nodal temperature, $\{T_{\Delta t}\}$ = temperature increase for the first time increment, and $\{F_{\Delta t}\}$ = increase of thermal force during the first time increment.

2.8.2 Mid-interval scheme

This finite difference scheme is frequently referred to as the Crank–Nicholson scheme[13]. The essence of this algorithm is to have the function evaluated at the mid-point between the time t and the increment $t + \Delta t$, as illustrated in Figure 2.11. Mathematically, it can be expressed as

$$T(t) = (T_t + T_{t+\Delta t})/2$$

and

$$\dot{T}(t) = \frac{\mathrm{d}T}{\mathrm{d}t} = \frac{T_{t+\Delta t} - T_t}{\Delta t}$$

Substitution of the above relationship into the thermal equilibrium equations in (2.26) results in the following relation:

$$\frac{[C_e]}{\Delta t}(\{T_{t+\Delta t}\} - \{T_t\}) + \frac{[K_e]}{2}(\{T_t\} + \{T_{t+\Delta t}\}) = \tfrac{1}{2}(\{F_{e,t+\Delta t}\} + \{F_{e,t}\})$$

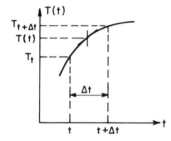

Figure 2.11 Time increments for the mid-interval difference scheme.

Upon combining the appropriate terms,

$$\left(\frac{2[C_e]}{\Delta t} + [K_e]\right)\{T_{t+\Delta t}\} = \left(\frac{2[C_e]}{\Delta t} - [K_e]\right)\{T_t\} + \{F_{e,t+\Delta t}\} + \{F_{e,t}\}$$

$$(2.50a)$$

or, in another form,

$$\left([K_e] + \frac{2[C_e]}{\Delta t}\right)\{T(t)\} = \frac{2}{\Delta t}[C_e]\{T_t\} + \tfrac{1}{2}(\{F_{e,t+\Delta t}\} + \{F_{e,t}\}) \quad (2.50b)$$

Equation (2.50a) can be expressed in yet another form in terms of incremental temperature $\{\Delta T\}$:

$$\left(\frac{2[C_e]}{\Delta t} + [K_e]\right)\{\Delta T\} = -[K_e]\{T_t\} + 2(\{F_{e,t+\Delta t}\} + \{F_{e,t}\}) \quad (2.50c)$$

with $\{\Delta T\} = \{T_{t+\Delta t}\} - \{T_t\}$

Compact forms for (2.50a–c) can be made as follows:

$$[K^*]\{T_{t+\Delta t}\} = \{F^*\} \tag{2.51a}$$

$$[K^*]\{T(t)\} = \{F^{**}\} \tag{2.51b}$$

$$[K^*]\{\Delta T\} = \{F^{***}\} \tag{2.51c}$$

where

$$[K^*] = \frac{2[C_e]}{\Delta t} + [K_e] \tag{2.51d}$$

$$\{F^*\} = \left(\frac{2[C_e]}{\Delta t} - [K_e]\right)\{T_t\} + \{F_{e,t+\Delta t}\} + \{F_{e,t}\} \tag{2.51e}$$

$$\{F^{**}\} = \frac{2}{\Delta t}[C_e]\{T_t\} + \tfrac{1}{2}(\{F_{e,t+\Delta t}\} + \{F_{e,t}\}) \tag{2.51f}$$

$$\{F^{***}\} = -2[K_e]\{T_t\} + (\{F_{e,t+\Delta t}\} + \{F_{e,t}\}) \tag{2.51g}$$

2.9 Boundary conditions

The reader may have become aware at this point that no consideration has been given to incorporating the various boundary conditions described in Section 2.2.4 in the finite element formulation. Matrices such as $[K_e]$, $[C_e]$ and $\{F_e\}$ in (2.42)–(2.45) were derived ignoring the heat flux across the boundary, and also without taking account of the convective heat transfer through the surrounding fluid. These boundary effects are naturally essential in dealing with real problems.

45

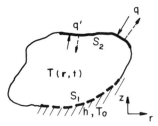

Figure 2.12 Thermal boundaries of a solid.

There are two ways of incorporating these boundary conditions in the computation. The first method is to have these effects included in the element thermal force matrix given in (2.29); the second method is to deal with those boundary conditions for those elements constituting the boundaries of the discretized model. The end results by these two approaches, of course, are identical, as will be demonstrated in a numerical example in Appendix 2.

2.9.1 Boundary conditions by finite element formulation

Figure 2.12 illustrates a typical case with heat flux across the boundary of a solid. Following the mathematical formulation described in Segerlind[4], the boundary conditions can be expressed as

$$k_{rr}\frac{\partial T}{\partial r}l_r + k_{zz}\frac{\partial T}{\partial z}l_z + h(T - T_0) + q = 0$$

where T = temperature at the boundary
  T_0 = bulk fluid temperature
  k_{rr}, k_{zz} = respective thermal conductivities in the r- and z-directions
  l_r, l_z = respective directional cosine along r- and z-coordinates.

By referring to a typical triangular toroidal element with surfaces s_{ij}, s_{jk} and s_{ki} in Figure 2.13, the convective boundary condition can be evaluated by formulating an additional conductivity matrix and thermal force matrix as summarized below. Detailed derivation of these matrices can be found in Segerlind[4].

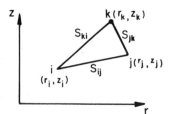

Figure 2.13 Perimeter of a triangular toroidal element.

(A) *The supplementary conductivity matrix* $[K_e']$

$$[K_e'] = \int_s h \begin{pmatrix} N_i \\ N_j \\ N_k \end{pmatrix} \{N_i \quad N_j \quad N_k\}\, ds$$

$$= \int_{s_1} h \begin{bmatrix} N_iN_i & N_iN_j & N_iN_k \\ & N_jN_j & N_jN_k \\ \text{SYM} & & N_kN_k \end{bmatrix} ds$$

where N_i, N_j and N_k are the components of the interpolation function corresponding to $N_i = N_1$, $N_j = N_2$ and $N_k = N_3$ in (2.37).

By using the area coordinate system outlined in Appendix 1, the integral in the above expression can be shown to be

$$[K_e'] = 2\pi \frac{hL_{ij}}{12} \begin{bmatrix} 3r_i + r_j & r_i + r_j & 0 \\ & r_i + 3r_j & 0 \\ \text{SYM} & & 0 \end{bmatrix}$$

$$+ 2\pi \frac{hL_{jk}}{12} \begin{bmatrix} 0 & 0 & 0 \\ & 3r_j + r_k & r_j + r_k \\ \text{SYM} & & r_j + 3r_k \end{bmatrix}$$

$$+ 2\pi \frac{hL_{ki}}{12} \begin{bmatrix} 3r_i + r_k & 0 & r_i + r_k \\ & 0 & 0 \\ \text{SYM} & & r_i + 3r_k \end{bmatrix} \tag{2.52}$$

where L_{ij}, L_{jk} and L_{ki} are the respective length of sides i-j, j-k and k-i.

The resultant conductivity matrix is a simple summation of $[K_e]$ in (2.42) or (2.43) and (2.52).

(B) *The supplementary thermal force matrix for convective boundary condition*

$$\{F_e'\} = \int_{s_1} hT_0[N]^T\, ds = \int_{s_1} hT_0 \begin{pmatrix} N_i \\ N_j \\ N_k \end{pmatrix} ds$$

$$= 2\pi hT_0L_{ij} \begin{Bmatrix} \dfrac{r_i}{3} + \dfrac{r_j}{6} \\[2mm] \dfrac{r_i}{6} + \dfrac{r_j}{3} \\[2mm] 0 \end{Bmatrix} + 2\pi hT_0L_{jk} \begin{Bmatrix} 0 \\[2mm] \dfrac{r_j}{3} + \dfrac{r_k}{6} \\[2mm] \dfrac{r_j}{6} + \dfrac{r_k}{3} \end{Bmatrix} + 2\pi hT_0L_{ki} \begin{Bmatrix} \dfrac{r_k}{6} + \dfrac{r_i}{3} \\[2mm] 0 \\[2mm] \dfrac{r_k}{3} + \dfrac{r_i}{6} \end{Bmatrix} \tag{2.53}$$

(C) *The supplementary thermal force matrix for heat flux boundary*

$$\{F_e''\} = \int_{s_2} q_m [N]^T \, ds = \int_{s_2} q_m \begin{Bmatrix} N_i \\ N_j \\ N_k \end{Bmatrix} ds$$

$$= 2\pi q_m L_{ij} \begin{Bmatrix} \dfrac{r_i}{3} + \dfrac{r_j}{6} \\[2ex] \dfrac{r_i}{6} + \dfrac{r_j}{3} \\[2ex] 0 \end{Bmatrix} + 2\pi q_m L_{jk} \begin{Bmatrix} 0 \\[2ex] \dfrac{r_j}{3} + \dfrac{r_k}{6} \\[2ex] \dfrac{r_j}{6} + \dfrac{r_k}{3} \end{Bmatrix} + 2\pi q_m L_{ki} \begin{Bmatrix} \dfrac{r_k}{6} + \dfrac{r_i}{3} \\[2ex] 0 \\[2ex] \dfrac{r_k}{3} + \dfrac{r_i}{6} \end{Bmatrix}$$

$$\tag{2.54}$$

where q_m is the net heat flux across the boundary s_2 according to Figure 2.12.

2.9.2   Boundary condition by heat flow in boundary elements only

Let nodes i and j be the boundary nodes in a boundary element as illustrated in Figure 2.14. The following common types of boundary conditions, as described in Section 2.2.4, may be formulated.

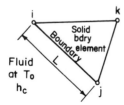

Figure 2.14   Convective boundary of a triangular toroidal element.

(a) *Prescribed surface temperature.* Mathematical expression of this condition is given in (2.8a). This situation can be handled by assuming an imaginary "superconductive" element connected to the nodes at which temperatures are specified.

Suppose the temperature at node i is prescribed to be T_i; then the necessary modifications to (2.47) can be made as shown below:

$$K_{ii}^* = e \tag{2.55a}$$

for the matrix $[K^*]$ and

$$F_i^* = 0 \tag{2.55b}$$

for the matrix $\{F_i\}$

where

$$e = \sum_{m=1}^{N} \beta K^*_{mm}$$

β = a large number, e.g. 10^6
N = total number of nodes with prescribed temperature in the system
K^*_{mm} = diagonal elements in the overall conductivity matrix $[K^*]$ in (2.47)

Large values of K^*_{ii} and zero value of F^*_i so introduced in the respective $[K^*]$ and $\{F^*\}$ matrices for node i in (2.47) will result in $\{\Delta T_i\} = 0$ for that node in the solution, which means no temperature variation at node i during time increment Δt. Consequently, the temperature at this node is maintained at its initially assigned value T_i at all time.

(b) *Free (natural) convection condition.* This condition applies when the fluid surrounding the contacting solid surface is stagnant, or moving at a very low velocity. The heat flow across a boundary layer at the surface of the solid can be shown, according to Newton's law given in (2.3), as follows:

$$F = ah(T_0 - T_s) = ah_c(T_0 - T_s)^\alpha(T_0 - T_s) \qquad (2.56)$$

where

a = area of the contact surface
h = heat transfer coefficient by free convection
h_c = a constant, or time-dependent function of the heat-transfer coefficient
a = index number to be determined by experiments
T_0 = temperature of the bulk fluid
T_s = temperature of the contact surface.

If the contacting surface is between nodes i and j in a finite element analysis as illustrated in Figure 2.14, the total amount of heat flow across this surface may be approximated by assuming that the heat flow through the boundary is concentrated at the nodes i and j. Thus, by assigning the average temperature of the contacting surface to be $(T_i + T_j)/2$, then, according to (2.56),

$$F_i = b_c(T_0 - T_i)$$
$$F_j = b_c(T_0 - T_j) \qquad (2.57)$$

where

$$b_c = \frac{ah_c}{2}\left(T_0 - \frac{T_i + T_j}{2}\right)^\alpha$$

49

The additional heat flow across the boundary elements as given in (2.57) requires necessary modifications of the heat flow equation, e.g. (2.48a), with the following additions:

$$K'_{ii} = K_{ii} + b_c$$
$$K'_{jj} = K_{jj} + b_c$$
$$F'_i = F_i + b_c T_0$$
$$F'_j = F_j + b_c T_0 \tag{2.58}$$

where

K'_{ii}, K'_{jj} = diagonal terms corresponding to nodes i and j, respectively, in the modified overall conductivity matrix

K_{ii}, K_{jj} = diagonal terms corresponding to nodes i and j, respectively, in the conductivity matrix

F'_i, F'_j = rate of heat flow at nodes i and j respectively, in the modified heat flow matrix

F_i, F_j = rate of heat flow at nodes i and j respectively, in the corresponding nodes i and j.

This procedure can be applied repeatedly for all the free convection boundary elements.

(c) *Forced convection condition.* This type of condition applies to a solid surface in contact with a moving fluid. The heat flow across the contacting surface is similar to that shown in (2.56):

$$F = aF(t)^{\beta}(T_0 - T_s)$$

where $F(t)$ = specified time-dependent function, and β = index number determined by experiment.

Using a similar approach to the free-convection case, the rate of heat flow at nodes i and j can be expressed as

$$F_i = b_f(T_0 - T_i)$$
$$F_j = b_f(T_0 - T_j) \tag{2.59}$$

in which

$$b_f = \tfrac{1}{2}aF(t)^{\beta}$$

Again, this type of boundary condition may be accommodated in the analysis by modifying the conductivity and heat flow matrices as described in (2.58), except that b_f is to replace b_c for forced convection problems.

(d) *Radiative condition.* This boundary condition is significant wherever there is a strong thermal radiation present over the surface of the solid. In this case, the Stefan–Boltzmann relation for the thermal radiation between "black" surfaces has been used:

$$F = a\sigma[T_1^4 - T_2^4]$$

where T_1 and T_2 are the thermodynamic (absolute) temperatures of the two surfaces, and σ is the Stefan–Boltzmann constant.

The above equation can also be expressed as:

$$F = ah_r(T_1 - T_2) \qquad (2.60)$$

where

$$h_r = \sigma(T_1^2 + T_2^2)(T_1 + T_2).$$

This thermal radiation condition can be included in the program following similar procedures for the convective condition given in (2.57) and (2.59).

A numerical example which involves all three types of boundary conditions is included in Appendix 2. It will illustrate the application of both approaches of handling these boundary conditions as well as the comparison of the nodal temperature increments computed by both the two-level explicit method and the mid-interval scheme described in Section 2.8.

2.10 Solution procedures for axisymmetric structures

The implementation into a computer code of the finite element formulation presented in the foregoing sections requires certain logical procedures as outlined herein for the axisymmetrical structures. The axisymmetric structural geometry is chosen because the same algorithm can be used for the planar geometry with an artificially large radius of curvature in the analysis.

The following general steps were taken in the thermal analysis part of the TEPSA code (*T*hermo*e*lastic-*p*lastic *S*tress *A*nalysis) for the quasi-coupled[†] thermomechanical stress analysis:

(1) initialize all nodal temperatures;
(2) form $[K_e]$ matrix from (2.42) or (2.43),
   $[C_e]$ matrix from (2.44),
   $\{F_e\}$ matrix from (2.45) for heat generation for all elements;
(3) form the equivalent conductivity matrix $[K^*]$ according to (2.48a);
(4) triangularize $[K^*]$ by the Gaussian elimination technique (see e.g. p. 410–411 of Desai[11]);
(5) modify $[K^*]$ for prescribed surface temperature boundary conditions according to (2.55a);
(6) modify $\{F^*\}$ for applicable thermal boundary conditions, e.g. (2.55b), (2.58), (2.59), or (2.60);
(7) adjust $[K^*]$ corresponding to Step (6);
(8) solve nodal temperature increments $\{\Delta T_t\}$ from (2.47) due to t;
(9) calculate the nodal temperature $\{T_t\} = \{T_{t-\Delta t}\} + \{\Delta T_t\}$;
(10) repeat the same procedure for the next time increment.

† The treatment of the truly coupled thermomechanical analysis will be given in Chapter 9.

References

1 Özisik, M. N. 1968. *Boundary value problems of heat conduction*. Scranton, Penn.: International Textbook Company.
2 Carslaw, H. S. and J. C. Jaeger 1959. *Conduction of heat in solids*. 2nd edn. Oxford: Oxford University Press.
3 Wilson, E. L. and R. E. Nickell 1966. Application of the finite element method to heat conduction analysis. *Nuc. Eng. Des.* **4**, 276-86.
4 Segerlind, L. J. 1976. *Applied finite element analysis*. New York: John Wiley. pp. 71-7 and Chapter 8.
5 Boley, B. A. and J. H. Weiner 1960. *Theory of thermal stresses*. New York: John Wiley. pp. 38-40.
6 Salvadori, M. G. and M. L. Baron 1952. *Numerical methods in engineering*. Englewood Cliffs, NJ: Prentice-Hall. p. 217.
7 Hilderbrand, F. B. 1968. *Finite difference equations and simulations*. Englewood Cliffs, NJ: Prentice-Hall, p. 194.
8 Doebelin, E.O. 1975. *Measurement systems*. New York: McGraw-Hill. Rev. edn., p. 502.
9 Kreith, F. 1965. *Principles of heat transfer*. 2nd edn. New York: International Textbook Company. Chapters 6-9.
10 Gurtin, M. 1964. Variational principles for linear initial-value problems. *Q. Appl. Math.* **22**, 252.
11 Desai, C. S. 1979. *Elementary finite element method*. Englewood Cliffs, NJ: Prentice-Hall. Chapter 4.
12 Felippa, C. A. 1966. *Refined finite element analysis of linear and nonlinear two-dimensional structures*. Rep. No. SESM 66-22, Department of Civil Engineering, University of California, Berkeley.
13 Crank, J. and P. Nicholson 1947. A practical method for numerical evaluation of solutions of partial differential equations of heat conduction type. Proc. Camb. Phil. Soc. **32**, 50-67.

3

THERMOELASTIC–PLASTIC STRESS ANALYSIS

3.1 Introduction

The significant shift in design requirements in recent years for high performance and efficient components in the aerospace and nuclear industries has
made sophisticated design analysis a necessity. In addition to the desired high
performance, the structural integrity of the machine components has become
a prime concern of designers as well as of the general public. For most
structures, its high performance requirement normally requires severe
thermal and mechanical loads to be carried by lighter or nontraditional
materials with high strength-to-weight ratio. The higher performance
requirement can also mean the use of a smaller safety margin in the design
process. The structural integrity is therefore likely to be ensured by a
sophisticated analytical technique such as the finite element method. By
virtue of the inherent flexibility and versatility of this method, coupled with
the ready availability of powerful digital computers, the finite element
method is not only capable of providing credible stress analyses for complex
structural geometries subject to severe loadings, but has also extended the
scope of traditional design practice using elastic stress analysis into the
elastic–plastic regime, resulting in a significant improvement in material
utilization and hence greater economical value.

This chapter will deal with the finite element formulation of elastic–plastic
stress analysis of structures. Thermal effects will be included in the formulation in the form of dilatational stresses (and strains) as well as those induced
by temperature-dependent material properties, including the variation of
plastic yielding criteria due to that effect. The latter behavior is the well-
known material nonlinearity effect in the theory of plastic deformation of
solids.

This chapter will begin with a brief review of terminologies involved in the
subsequent derivations. Constitutive equations based on the two commonly
used work-hardening rules, namely the isotropic and kinematic hardening
rules, will be derived. The latter is considered to be a practical method of
describing the cyclic plastic deformation of solids, a phenomenon which is
common in fatigue analysis.

The theoretical formulation presented in this chapter is intended for three-
dimensional axisymmetric solids and two-dimensional planar structures.

The corresponding computer programs can be found in the listing of the TEPSAC code attached in the Appendix 5.

3.2 Mechanical behavior of materials

It is obvious that the subject of deformable solids cannot be adequately covered in a single section such as presented here. However, it is important for the reader to be aware of some of the fundamental assumptions regarding the behavior of common engineering materials as well as some of the terminologies that will be quoted frequently in the subsequent derivations in order to avoid possible confusion when making references to the literature.

3.2.1 Fundamental assumptions

As in all mathematical derivations, a few assumptions will have to be made here. Unless otherwise specified, these assumptions will apply to all formulations that follow.

(1) The material is treated as a continuous medium, or a continuum. The effects induced by the inherent defects and voids in the material can be neglected.
(2) The material is isotropic, i.e. with its properties uniform in all directions.
(3) The material has no "memory" and hence the effect induced on the material in separate previous events has no importance to the current event.
(4) The material exhibits the same properties in tension and compression.

3.2.2 Definition of stress and strain

There are two sets of definitions for stress and strain used in engineering analyses as will be stated below by an example of a stretching rod as illustrated in Figure 3.1. In this simple example, the rod is stretched by a force acting along its length. The total elongation of the rod is assumed to be Δl.

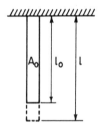

Figure 3.1 Uniaxial elongation of a prismatic bar.

Although the usual definition of strain can be stated as the unit elongation of the rod due to the applied force, the mathematical expressions can be different depending on the reference state used.

(a) *Engineering strain (e)*. This is frequently referred to as the Eulerian strain, as it is defined with reference to the original shape of the solid. Mathematically it is expressed as:

$$e = \Delta l/l_0 = l/l_0 - 1$$

in which l_0 and l are the respective original and deformed lengths of the rod.

(b) *True strain (ε)*. The true strain which is measured on the basis of the immediate preceding length of the rod. It is often called the Lagrangian strain[†]

$$\varepsilon = \Sigma \left(\frac{l_1 - l_0}{l_0} + \frac{l_2 - l_1}{l_1} + \cdots + \frac{l - l_n}{l_n} \right) = \int_{l_0}^{l} \frac{dl}{l} = \ln \left(\frac{l}{l_0} \right)$$

in which $l_0, l_1, \ldots, l_n$ are the instantaneous lengths of the rod.

Another expression for the true strain for a circular rod during the plastic deformation is

$$\varepsilon = \ln \frac{A_0}{A} = 2 \ln \frac{D_0}{D}$$

where A_0, D_0 and A, D are the respective original and instantaneous cross-sectional areas and diameters of the rod.

It is apparent that engineering and true strains are related by the following expression:

$$\varepsilon = \ln (1 + e) \tag{3.1}$$

(c) *Engineering stress (S)*. Like engineering strain, the corresponding engineering stress is defined as

$$S = \frac{\text{instantaneous load}}{\text{original cross-sectional area}} = \frac{P}{A_0}$$

with reference to the original undeformed cross-sectional area of the rod.

(d) *True stress (σ)*. As one may imagine the true stress is determined using the instantaneous cross-sectional area of the rod as in the case of true strain. Thus, by assuming the instantaneous cross-sectional area of the specimen to be A, the true stress can be expressed as:

$$\sigma = P/A \tag{3.2}$$

[†] The reader should not confuse this with the similar term used in Chapter 8 on the finite strain theory.

Again, the reader will find that:

$$\sigma \approx S$$

in elastic regime, and

$$\sigma = (1 + e)S$$

during the plastic deformation.

3.2.3 The stress-strain curve

If one records the strains and computes stresses following every small load increment applied to the rod and plot the corresponding stress-strain curve, two sets of diagrams, namely σ-ε and S-e, can be observed as illustrated in Figure 3.2[†]. Either set of the curves can describe the unique mechanical behavior of the material with the following commonly used terminologies for this purpose:

$$A' = \text{proportional limit}$$
$$A = \text{elastic limit}$$
$$B = \text{yield point}$$
$$m = \text{necking point}$$
$$f = \text{rupture point}$$
$$S_0 = \text{yield strength of material}$$
$$S_u = \text{ultimate tensile strength of material}$$

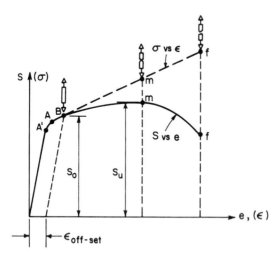

Figure 3.2 Typical stress-strain curves of a material.

[†] The deviation of the σ-ε curve from the S-e curve is somewhat exaggerated in this figure in order to illustrate the difference between these curves. For most parts, the differences between the two sets of stress value is about 2%.

The proportional limit A' represents the end of the straight linear relationship between the stresses and their corresponding strains, whereas the elastic limit is used to determine the end of the elastic behavior of the material, as will be described in the following section. For many materials, the point A coincides with the yield point or yield (strength) designated by B in Figure 3.2. Plastic deformation in the solid begins at this point. Point B is clearly visible for some materials such as carbon steels, but a distinct yield point is not always available for most other materials, especially for nonferrous materials such as aluminum alloys. In such cases, an offset strain of 0.2% is used to determine the point B as illustrated in Figure 3.2. Another point worth noting about point B is that it is the demarcation point for the two sets of stress–strain curves. Note that the σ-ε curve keeps climbing until complete fracture of the specimen occurs, whereas the S-e curve climbs up to the necking point (point m in Figure 3.2) and then drops sharply toward the complete failure of the material. In almost all elastic–plastic stress analyses σ-ε curves are favored, because the slope of these curves, which represents the stiffness of the material, is always positive, and this is a necessary condition to maintain numerical stability in the computation.

3.2.4 Fundamental difference between elastic and plastic deformation of solids

It is obviously next to impossible to outline all the differences in the behavior of materials during elastic and plastic deformation, as each type of deformation behavior warrants a separate book; see References 1–6. The intention here, however, is to present some of the fundamental differences between these two distinct types of material behavior to assist the reader to follow the derivation of the elastic–plastic analysis presented in the remaining part of this chapter.

(a) *Elastic deformation*

(1) Very small deformation with the strain up to about 0.1%.
(2) Straight linear relationship between the stress and strain, resulting in a constant stiffness of the material.
(3) Completely recoverable strain (or deformation) after the applied load is removed (or unloading).

(b) *Plastic deformation*

(1) Larger deformation.
(2) Nonlinear relationship between the stress σ and strain ε. The ratio σ/ε, which represents the stiffness of the material, decreases as further loading takes place.
(3) Results in permanent deformation after the removal of loading.

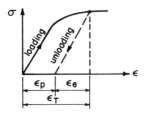

Figure 3.3 Uniaxial elastic–plastic loading and elastic unloading of a material.

(4) No volumetric change in the solid during plastic deformation with a Poisson's ratio at 0.5.

(5) No dilatational or (dimensional) change in the solid. Deformation is due to shear actions on the material. Hence only shape changes can be observed.

(6) The total strain, ε_T, is the sum of the elastic components ε_e and the plastic component ε_p as illustrated in Figure 3.3.

3.2.5 Time-dependent plasticity—creep

Most deformable solids deform only in response to the applied load, as suggested in Section 1.3. Further deformation is possible only if the applied load increases. This one-to-one relationship, however, can be violated under certain conditions, especially when the solid is subject to a high temperature environment. Progressive deformation of solids can be observed even at a constant load or stress under this circumstance. Such behavior is called creep deformation.

Analysis of stresses in solids during creep deformation is a very important part of the design of machine components for high temperature applications. A separate chapter (Ch. 4) will be devoted to this subject.

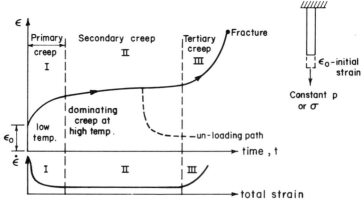

Figure 3.4 A typical creep strain curve.

As illustrated in Figure 3.4, creep deformation of solids is time dependent. It can be generally considered to consist of three stages: (1) the primary stage with relatively high deformation rate; (2) the secondary creep, which has a much slower and constant growth rate and normally occurs at high temperature; (3) the tertiary creep, which is the most detrimental type of creep deformation as the deformation rate becomes extremely high and usually results in catastrophic rupture of the structure in a very short time. More elaborate description of creep deformation of various solids can be found in References 7–11.

3.3   Review of basic formulations in linear elasticity theory

3.3.1   Stresses

When a solid body is subjected to a system of externally applied loads, there exist two notable reactions to the applied load. These are: (1) **deformation**, and (2) **internal resistance**. It is the internal resistance that enables the deformed solid to reach a *new* state of *equilibrium*.

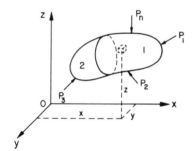

Figure 3.5 Equilibrium condition of a solid under stresses.

Consider a solid situated in a coordinate system x, y, z and loaded as illustrated in Figure 3.5. In order to achieve a new equilibrium state after the application of the forces P_1, P_2, ..., P_n, resistance is induced everywhere in the solid. The *intensity* of this resistance is called **stress**. As one may visualize, both the magnitude and direction of stress varies from point to point in the solid. Thus stress is classified as a **tensor** quantity[†]. The stresses in an infinitesimally small element in a solid can be designated as shown in Figure 3.6. The following rules are commonly used for the designation of stress components[4]:

† A tensor quantity differs from a vector quantity by the fact that the "position" at which this quantity exists has to be specified in addition to the "magnitude" and "direction". The latter are the two necessary conditions to define a vector quantity.

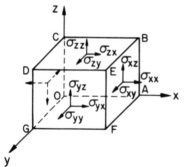

y   **Figure 3.6**   Designation of stress components.

(1)   each component of stress consists of two subscripts;
(2)   the first subscript denotes the axis which is perpendicular to the plane
      of action of these stress components;
(3)   the second subscript represents their directions.

As shown in Figure 3.6, there are altogether nine stress components in a small
element in a stressed solid:

$$\begin{bmatrix} \sigma_{xx} & \sigma_{xy} & \sigma_{xz} \\ \sigma_{yx} & \sigma_{yy} & \sigma_{yz} \\ \sigma_{zx} & \sigma_{zy} & \sigma_{zz} \end{bmatrix} = [\sigma]$$

It is not difficult to prove that for a solid in equilibrium, the following
relations exist:

$$\sigma_{xy} = \sigma_{yx} \qquad \sigma_{yz} = \sigma_{zy} \qquad \sigma_{xz} = \sigma_{zx}$$

Hence, the total number of *independent* stress components is reduced to
six, i.e.

$$\begin{bmatrix} \sigma_{xx} & \sigma_{xy} & \sigma_{xz} \\ & \sigma_{yy} & \sigma_{yz} \\ \text{SYM} & & \sigma_{zz} \end{bmatrix} = [\sigma]$$

Stress components defined in the foregoing description can be conven-
iently expressed in an indicial form as σ_{ij} with the indicies i and j varying
from 1 to 3. In such cases, the coordinates x, y and z are to be replaced by
x_1, x_2 and x_3 respectively. The latter coordinate system, x_i ($i = 1, 2, 3$) is
convenient for the subsequent derivations.

In finite element analysis, however, the stress components in an element
are usually expressed in a *column* matrix:

$$\{\sigma\}^{\mathrm{T}} = \{\sigma_1 \quad \sigma_2 \quad \sigma_3 \quad \sigma_4 \quad \sigma_5 \quad \sigma_6\}^{\mathrm{T}} \qquad (3.3)$$

with

$$\sigma_1 = \sigma_{xx}, \quad \sigma_2 = \sigma_{yy}, \quad \sigma_3 = \sigma_{zz}$$

$$\sigma_4 = \sigma_{xy}, \quad \sigma_5 = \sigma_{yz}, \quad \sigma_6 = \sigma_{xz}$$

3.3.2 Displacements

A solid body is said to be deformed due to external loads when the *relative* positions of points in the body are changed. The displacement of a point is defined as the *vector* distance from the initial to the final location of that point. If we let u, v and w to be the corresponding displacement components along the x, y, and z axis, then

u = displacement component of a point along x-axis
v = displacement component of a point along y-axis
w = displacement component of a point along z-axis

the corresponding displacement components along the new coordinate system can be expressed as u_i ($i = 1, 2, 3$) in the x_i coordinates. The representation of these displacements in matrix form becomes:

$$\{u\}^{\mathrm{T}} = \{u_1 \quad u_2 \quad u_3\}^{\mathrm{T}} \tag{3.4}$$

3.3.3 Strains and strain–displacement relations

Referring to Figure 3.6, the associated unit deformations of the solid corresponding to the six stress components are the six independent strain components. These are:
three normal strain components: ε_{xx}, ε_{yy}, ε_{zz}
and three shearing strain components: $\varepsilon_{xy} = \varepsilon_{yx}$, $\varepsilon_{xz} = \varepsilon_{zx}$ and $\varepsilon_{yz} = \varepsilon_{zy}$
Again, in the coordinate system x_i ($i = 1, 2, 3$), these strain components can be expressed as

$$\varepsilon_{ij} \ (i, j = 1, 2, 3),$$

or, in a matrix form,

$$\{\varepsilon\}^{\mathrm{T}} = \{\varepsilon_1 \quad \varepsilon_2 \quad \varepsilon_3 \quad \varepsilon_4 \quad \varepsilon_5 \quad \varepsilon_6\}^{\mathrm{T}} \tag{3.5}$$

The relationship between strains and displacements can be expressed either in an indicial form[†],

$$\varepsilon_{ij} = \tfrac{1}{2}(u_{i,j} + u_{j,i}),$$

† This expression is both convenient and useful in mathematical derivations. However, the indicial shearing strains so defined account for only half of the real engineering shearing strains.

or in the matrix form

$$
\{\varepsilon\} = \left\{\begin{array}{c} \varepsilon_{xx} \\ \varepsilon_{yy} \\ \varepsilon_{zz} \\ \varepsilon_{xy} \\ \varepsilon_{yz} \\ \varepsilon_{xz} \end{array}\right\} = \left\{\begin{array}{c} \dfrac{\partial u}{\partial x} \\[2mm] \dfrac{\partial v}{\partial y} \\[2mm] \dfrac{\partial w}{\partial z} \\[2mm] \dfrac{\partial u}{\partial y} + \dfrac{\partial v}{\partial x} \\[2mm] \dfrac{\partial v}{\partial z} + \dfrac{\partial w}{\partial y} \\[2mm] \dfrac{\partial w}{\partial x} + \dfrac{\partial u}{\partial z} \end{array}\right\} = \begin{bmatrix} \dfrac{\partial}{\partial x} & 0 & 0 \\[2mm] 0 & \dfrac{\partial}{\partial y} & 0 \\[2mm] 0 & 0 & \dfrac{\partial}{\partial z} \\[2mm] \dfrac{\partial}{\partial y} & \dfrac{\partial}{\partial x} & 0 \\[2mm] 0 & \dfrac{\partial}{\partial z} & \dfrac{\partial}{\partial y} \\[2mm] \dfrac{\partial}{\partial z} & 0 & \dfrac{\partial}{\partial x} \end{bmatrix} \left\{\begin{array}{c} u \\ v \\ w \end{array}\right\} \qquad (3.6)
$$

3.3.4 Generalized Hooke's law—stress-strain relations

It can be readily proved that there exists a unique relationship between stress and strain components in an isotropic solid in an equilibrium state[1,2,4]:

$$
\sigma_{ij} = \lambda \varepsilon_{kk} \delta_{ij} + 2\mu \varepsilon_{ij} \qquad \text{(sum over } k) \qquad (3.7)
$$

where λ = Lamé constant

$$
= \frac{Ev}{(1 + v)(1 - 2v)}
$$

$\mu = G$, shear modulus of elasticity

δ_{ij} = Kronecker delta ($= 0$ for $i \neq j$; $= 1$ for $i = j$)

The corresponding matrix forms suitable for finite element formulations are outlined as follows:

(a) *General three-dimensional cases*

$$
\{\sigma\} = \frac{E}{(1 + v)(1 - 2v)} \begin{bmatrix} (1 - v) & v & v & 0 & 0 & 0 \\ & (1 - v) & v & 0 & 0 & 0 \\ & & (1 - v) & 0 & 0 & 0 \\ & & & \dfrac{1 - 2v}{2} & 0 & 0 \\ & \text{SYM} & & & \dfrac{1 - 2v}{2} & 0 \\ & & & & & \dfrac{1 - 2v}{2} \end{bmatrix} \{\varepsilon\}
$$

$$(3.7a)$$

where E = modulus of elasticity and v = Poisson's ratio.

(b) *Axisymmetric cases*[12]

$$\{\sigma\} = \begin{Bmatrix} \sigma_{rr} \\ \sigma_{zz} \\ \sigma_{\theta\theta} \\ \sigma_{rz} \end{Bmatrix} = \frac{E(1-v)}{(1+v)(1-2v)} \begin{bmatrix} 1 & \dfrac{v}{1-v} & \dfrac{v}{1-v} & 0 \\ & 1 & \dfrac{v}{1-v} & 0 \\ \text{SYM} & & 1 & 0 \\ & & & \dfrac{1-2v}{2(1-v)} \end{bmatrix} \begin{Bmatrix} \varepsilon_{rr} \\ \varepsilon_{zz} \\ \varepsilon_{\theta\theta} \\ \varepsilon_{rz} \end{Bmatrix}$$

(3.7b)

(c) *Plane stress cases*[12]

$$\{\sigma\} = \begin{Bmatrix} \sigma_{rr} \\ \sigma_{zz} \\ \sigma_{rz} \end{Bmatrix} = \frac{E}{1-v^2} \begin{bmatrix} 1 & v & 0 \\ & 1 & 0 \\ \text{SYM} & & \dfrac{1-v}{2} \end{bmatrix} \begin{Bmatrix} \varepsilon_{rr} \\ \varepsilon_{zz} \\ \varepsilon_{rz} \end{Bmatrix}$$

(3.7c)

(d) *Plane strain cases*[12]

$$\{\sigma\} = \begin{Bmatrix} \sigma_{rr} \\ \sigma_{zz} \\ \sigma_{rz} \end{Bmatrix} = \frac{E}{1+v} \begin{bmatrix} \dfrac{1-v}{1-2v} & \dfrac{v}{1-2v} & 0 \\ & \dfrac{1-v}{1-2v} & 0 \\ \text{SYM} & & 1/2 \end{bmatrix} \begin{Bmatrix} \varepsilon_{rr} \\ \varepsilon_{zz} \\ \varepsilon_{rz} \end{Bmatrix}$$

(3.7d)

In general, the stress–strain relations used in elastic stress analysis by the finite element method can be expressed to be:

$$\{\sigma\} = [C_e]\{\varepsilon\} \tag{3.8a}$$

or

$$\{\sigma\} = [C_e]\{\varepsilon\} - \{\sigma_0\} \tag{3.8b}$$

where $\{\sigma_0\}$ = stress components due to thermal strains $\{\varepsilon_0\}$, and $[C_e]$ is the elasticity matrix, which is a function of the material's elastic constants as shown in (3.7a–d).

3.3.5 Strain energy in an elastic solid[13]

The strain energy stored in a deformed element of volume $dV = dx\,dy\,dz$ is

$$\Delta u = \tfrac{1}{2}(\sigma_{xx}\varepsilon_{xx} + \sigma_{yy}\varepsilon_{yy} + \sigma_{zz}\varepsilon_{zz} + \sigma_{xy}\varepsilon_{xy} + \sigma_{xz}\varepsilon_{xz} + \sigma_{yz}\varepsilon_{yz})\,dV$$

Hence the total strain energy stored in the entire solid in an equilibrium condition is:

$$U = \tfrac{1}{2}\int_V (\sigma_{xx}\varepsilon_{xx} + \sigma_{yy}\varepsilon_{yy} + \sigma_{zz}\varepsilon_{zz} + \sigma_{xy}\varepsilon_{xy} + \sigma_{xz}\varepsilon_{xz} + \sigma_{yz}\varepsilon_{yz})\,dV$$

or, in a matrix expression,

$$U = \tfrac{1}{2}\int_V \{\varepsilon\}^T\{\sigma\}\,dV \tag{3.9}$$

3.3.6 Other equations

The following equations are frequently used in linear elastic stress analyses. Readers are referred to appropriate references for detailed derivations.

(1) Equilibrium equations[2]:

$$\sigma_{ij,j} + X_i = 0 \tag{3.10}$$

(2) Navier's displacement equations[2]:

$$Gu_{i,kk} + (\lambda + G)u_{k,ki} + X_i = 0 \qquad \text{(sum over } k) \tag{3.11}$$

where G = shear modulus of elasticity, λ = Lamé constant and X_i = body force components in x_i direction.

(3) Coupled thermoelastic equations[14]:

$$\frac{\partial}{\partial x_i}\left(k_{ij}\frac{\partial T}{\partial x_j}\right) = \rho c\frac{\partial T}{\partial t} + \beta_{ij}T\frac{\partial \varepsilon_{ij}}{\partial t} \tag{3.12}$$

in which

$$\sigma_{ij} = C_{ijkl}\varepsilon_{kl} - \beta_{ij}\Delta T$$

$$= \lambda\varepsilon_{kk}\delta_{ij} + 2\mu\varepsilon_{ij} - \frac{E}{1 - 2v}a\Delta T\delta_{ij} \tag{3.12a}$$

The strain energy per unit volume of the solid can be shown to take the following form[15]:

$$U = \tfrac{1}{2}\sigma_{ij}(\varepsilon_{ij} - a\Delta T\delta_{ij}) \tag{3.12b}$$

in which ε_{ij} = the total thermoelastic strain

$$= \frac{1 + v}{E}\sigma_{ij} - \frac{v}{E}\sigma_{kk}\delta_{ij} + a\Delta T\delta_{ij} \qquad \text{(sum over } k) \tag{3.12c}$$

3.4 Basic formulations in nonlinear elasticity

For many engineering applications, the relationship between the stresses and strains is a straight linear one in the elastic range because the points A', A and B in Figure 3.2 are so close to each other that no distinction can be made between the proportional and elastic limits. However, there are some materials, such as soft rubber and some polymers, that may behave elastically at a deformation which is much beyond that at the proportional limit[16]. The relationship between stresses and strains for these superelastic materials can thus be highly nonlinear. Special mathematical treatment is therefore necessary.

Nonlinear elastic behavior of material is not just confined to superelastic materials. It may be observed in common engineering materials. In such cases, the properties, E, v can vary with the temperature of the material or the rate of deformation by the solid under dynamic loads. In such cases, the matrix $[C_e]$ in (3.8a,b) can no longer be treated as constant and the relationship between $\{\sigma\}$ and $\{\varepsilon\}$ has thus become nonlinear. Detailed treatment of solids undergoing finite elastic deformation is beyond the scope of this book. The reader may refer to published works such as Wyatt and Dew-Hughes[16]. Only a brief outline of the mathematical formulation is given below.

3.4.1 Material nonlinearity

This nonlinear material behavior is primarily due to the fact that the material properties, e.g. E in (3.8a,b), vary with temperature T and strain rates $\dot{\varepsilon}$. Such variations are evident in common engineering materials as described in Chapters 2 and 4 of McGregor Tegart[17]. Figures 3.7 and 3.8 show the respective temperature and strain-rate-dependent mechanical behavior of aluminum alloy 6061-T6.

Although the relationship between $\{\sigma\}$ and $\{\varepsilon\}$ is nonlinear in the global sense, linearity can still be considered to be valid within small increments of strain, i.e. $\{\delta\varepsilon\}$ and the established linear theory of elasticity can thus be used within these increments as a good approximation.

For the situations involving drastic temperature and strain rate changes, the total incremental strain components may be assumed to consist of the following components:

$$\{\delta\varepsilon\} = \{\delta\varepsilon_e\} + \{\delta\varepsilon_{e,T}\} + \{\delta\varepsilon_{e,\dot{\varepsilon}}\} + \{\delta\varepsilon_T\} \qquad (3.13)$$

where $\{\delta\varepsilon_e\}$ = incremental strain components due to elastic deformation;

$\{\delta\varepsilon_{e,T}\}$ = incremental strain components due to temperature dependent material properties (Figure 3.7);

$\{\delta\varepsilon_{e,\dot{\varepsilon}}\}$ = incremental strain components due to strain rate dependent material properties (Figure 3.8);

$\{\delta\varepsilon_T\}$ = incremental thermal strain components.

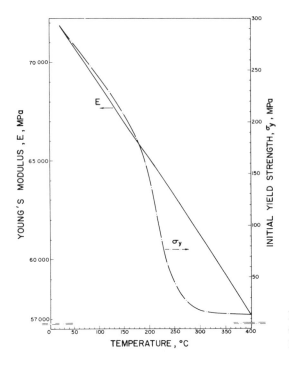

Figure 3.7 Temperature-dependent mechanical properties of steel.

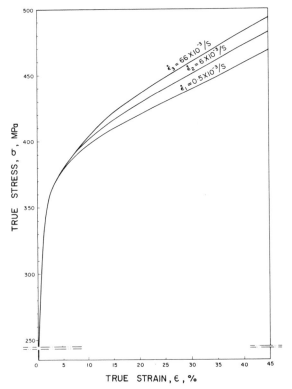

Figure 3.8 Strain-rate dependent stress–strain curves for aluminum alloy.

Functionally (3.13) can be expressed as

$$\{\varepsilon\} = \{\varepsilon\}(\{\sigma\}, T, [C_e]^{-1}(T, \dot{\varepsilon})) \tag{3.14}$$

Applying the chain rule of partial differentiation on (3.14) gives

$$\frac{d\{\varepsilon\}}{dt} = \frac{\partial\{\varepsilon\}}{\partial\{\sigma\}}\frac{\partial\{\sigma\}}{\partial t} + \frac{\partial\{\varepsilon\}}{\partial T}\frac{\partial T}{\partial t} + \frac{\partial\{\varepsilon\}}{\partial [C_e]^{-1}}\left(\frac{\partial [C_e]^{-1}}{\partial T}\frac{\partial T}{\partial t} + \frac{\partial [C_e]^{-1}}{\partial \dot{\varepsilon}}\frac{\partial \dot{\varepsilon}}{\partial t}\right)$$

Since $\{\varepsilon\} = [C_e]^{-1}\{\sigma\}$ from (3.8a) and $\{\varepsilon\} = \{a\}\Delta T$ for thermal strain, where $\{a\}$ is the linear thermal expansion coefficient matrix, the above expression becomes

$$\frac{d\{\varepsilon\}}{dt} = [C_e]^{-1}\frac{\partial\{\sigma\}}{\partial t} + \{a\}\frac{\partial T}{\partial t} + \{\sigma\}^T\left(\frac{\partial [C_e]^{-1}}{\partial T}\frac{\partial T}{\partial t} + \frac{\partial [C_e]^{-1}}{\partial \dot{\varepsilon}}\frac{\partial \dot{\varepsilon}}{\partial t}\right)$$

The constitutive equation for elastic solids with nonlinear material properties can be derived by integrating the above relation over the time interval dt to give:

$$\{\delta\sigma\} = [C_e]\{\delta\varepsilon\} - [C_e]\left[\{a\}\delta T + \left(\frac{\partial [C_e]^{-1}}{\partial T}\delta T + \frac{\partial [C_e]^{-1}}{\partial \dot{\varepsilon}}\delta\dot{\varepsilon}\right)\{\sigma\}\right] \tag{3.15}$$

3.4.2 Geometric nonlinearity

In some cases, a solid is expected to deform by a large amount yet remain elastic, e.g. soft rubbers and polymers. The rotational components in the total displacement will then be too significant to be neglected.

The strain–displacement relation in Section 3.3.3 will have to be modified in order to account for the excessive rotational components:

$$\varepsilon_{xx} = \frac{\partial u}{\partial x} + \frac{1}{2}\left[\left(\frac{\partial u}{\partial x}\right)^2 + \left(\frac{\partial v}{\partial x}\right)^2 + \left(\frac{\partial w}{\partial x}\right)^2\right]$$

$$\varepsilon_{yy} = \frac{\partial v}{\partial y} + \frac{1}{2}\left[\left(\frac{\partial u}{\partial y}\right)^2 + \left(\frac{\partial v}{\partial y}\right)^2 + \left(\frac{\partial w}{\partial y}\right)^2\right]$$

$$\varepsilon_{zz} = \frac{\partial w}{\partial z} + \frac{1}{2}\left[\left(\frac{\partial u}{\partial z}\right)^2 + \left(\frac{\partial v}{\partial z}\right)^2 + \left(\frac{\partial w}{\partial z}\right)^2\right]$$

$$2\varepsilon_{xy} = \frac{\partial u}{\partial y} + \frac{\partial v}{\partial x} + \frac{\partial u}{\partial x}\frac{\partial u}{\partial y} + \frac{\partial v}{\partial x}\frac{\partial v}{\partial y} + \frac{\partial w}{\partial x}\frac{\partial w}{\partial y}$$

$$2\varepsilon_{yz} = \frac{\partial w}{\partial y} + \frac{\partial v}{\partial z} + \frac{\partial u}{\partial y}\frac{\partial u}{\partial z} + \frac{\partial v}{\partial y}\frac{\partial v}{\partial z} + \frac{\partial w}{\partial y}\frac{\partial w}{\partial z}$$

$$2\varepsilon_{zx} = \frac{\partial u}{\partial z} + \frac{\partial w}{\partial x} + \frac{\partial u}{\partial z}\frac{\partial u}{\partial x} + \frac{\partial v}{\partial z}\frac{\partial v}{\partial x} + \frac{\partial w}{\partial z}\frac{\partial w}{\partial x} \tag{3.16}$$

The stress–strain relation becomes

$$\varepsilon_{ij} = \frac{(1 - 2v)}{E} \delta_{ij}\theta + \frac{1}{2G}\sigma'_{ij} \qquad (3.17)$$

where

$$\theta = \tfrac{1}{3}(\sigma_{11} + \sigma_{22} + \sigma_{33})$$

$$\sigma'_{ij} = \text{deviatoric stresses} = \sigma_{ij} - \delta_{ij}\theta \qquad (3.17a)$$

$$\delta_{ij} = \text{Kronecker delta} = 0 \text{ if } i \neq j; \; = 1 \text{ for } i = j$$

The strain energy function per unit volume is

$$U = \int_0^\varepsilon \sigma_{ij}\,\mathrm{d}\varepsilon_{ij}$$

$$= \frac{3E}{2(1 - 2v)}e^2 + G\varepsilon'_{ij}\varepsilon'_{ij} \qquad (3.18)$$

where

$$e = (\varepsilon_{11} + \varepsilon_{22} + \varepsilon_{33})/3$$

$$\varepsilon'_{ij} = \text{deviatoric strains} = \varepsilon_{ij} - \delta_{ij}e. \qquad (3.18a)$$

Further elaboration of this subject will be dealt with in Chapter 8.

3.5 Elements of plasticity theory

The theory of plasticity of solids is a very complex and broad subject. It is, in many ways, a great deal more complex than both linear and nonlinear elasticity. Several excellent books have been published on both the theoretical and practical treatment of solid structures undergoing plastic deformation, e.g. Hill[5], Lin[18] and D'Isa[19]. However, the widespread use of the theory of plasticity in the stress analysis of solid structures was not evident until the development of the finite element method during the 1960s[20–22]. The use of the finite element method for elastic–plastic stress analysis has also been included in several books[6,23,24]. The following subsections will briefly outline only the main ingredients that are relevant to the formulation of combined thermoelastic–plastic stress analysis.

3.5.1 Flow curves

For the reasons mentioned in Section 3.2.3, only true stress (σ) vs. true strain (ε) curves are used in analysis. There are generally three types of such curves as illustrated in Figure 3.9.

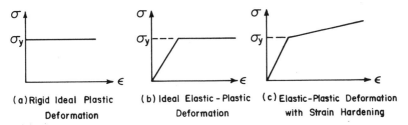

(a) Rigid Ideal Plastic   (b) Ideal Elastic - Plastic   (c) Elastic-Plastic Deformation
Deformation   Deformation   with Strain Hardening

Figure 3.9   Idealized flow curves from uniaxial tension test.

These curves, depicted in Figure 3.9, are idealized for analytical solutions. In reality, however, Type (c) with gradual transition from the elastic to plastic regions should be used.

3.5.2   Stress invariants

The following terminologies are commonly quoted in elastic–plastic stress analysis.

(1)   STRESS INVARIANT

1st stress invariant: $I_1 = \sigma_{xx} + \sigma_{yy} + \sigma_{zz}$ (3.19a)

2nd stress invariant: $I_2 = \begin{vmatrix} \sigma_{xx} & \sigma_{xy} \\ \sigma_{xy} & \sigma_{yy} \end{vmatrix} + \begin{vmatrix} \sigma_{yy} & \sigma_{yz} \\ \sigma_{yz} & \sigma_{zz} \end{vmatrix} + \begin{vmatrix} \sigma_{xx} & \sigma_{xz} \\ \sigma_{xz} & \sigma_{zz} \end{vmatrix}$ (3.19b)

3rd stress invariant: $I_3 = \begin{vmatrix} \sigma_{xx} & \sigma_{xy} & \sigma_{xz} \\ \sigma_{xy} & \sigma_{yy} & \sigma_{yz} \\ \sigma_{xz} & \sigma_{yz} & \sigma_{zz} \end{vmatrix}$ (3.19c)

If the reference axes coincide with the principal stress axes, then:

$$I_1 = \sigma_1 + \sigma_2 + \sigma_3$$

$$I_2 = -(\sigma_1\sigma_2 + \sigma_2\sigma_3 + \sigma_3\sigma_1)$$

$$I_3 = \sigma_1\sigma_2\sigma_3$$

where σ_1, σ_2 and σ_3 are the principal stress components.

By following the definition of deviatoric stresses given in (3.17a), the deviatoric stress invariants are defined as follows.

(2)   DEVIATORIC STRESS INVARIANTS

1st deviatoric stress invariant: $J_1 = \sigma'_{xx} + \sigma'_{yy} + \sigma'_{zz}$ (3.20a)

2nd deviatoric stress invariant:

$$J_2 = - \begin{vmatrix} \sigma'_{xx} & \sigma'_{xy} \\ \sigma'_{xy} & \sigma'_{yy} \end{vmatrix} - \begin{vmatrix} \sigma'_{yy} & \sigma'_{yz} \\ \sigma'_{yz} & \sigma'_{zz} \end{vmatrix} - \begin{vmatrix} \sigma'_{xx} & \sigma'_{xz} \\ \sigma'_{xz} & \sigma'_{zz} \end{vmatrix}$$

$$= \tfrac{1}{2}(\sigma'^2_{xx} + \sigma'^2_{yy} + \sigma'^2_{zz}) + \sigma'^2_{xy} + \sigma'^2_{yz} + \sigma'^2_{xz}$$

$$= \tfrac{1}{6}[(\sigma_{xx} - \sigma_{yy})^2 + (\sigma_{yy} - \sigma_{zz})^2 + (\sigma_{xx} - \sigma_{zz})^2] + \sigma^2_{xy} + \sigma^2_{yz} + \sigma^2_{xz}$$

$$= \tfrac{1}{6}[(\sigma_1 - \sigma_2)^2 + (\sigma_1 - \sigma_3)^2 + (\sigma_2 - \sigma_3)^2] \qquad (3.20\text{b})$$

3rd deviatoric stress invariant:

$$J_3 = \tfrac{1}{3}\sigma'_{ij}\sigma'_{jk}\sigma'_{ki}, \qquad i, j, k = 1, 2, 3. \qquad (3.20\text{c})$$

3.5.3 Effective stress and strain

When a solid is subject to a general state of stress illustrated in Figures 3.5 and 3.6, a total of six independent stress and six strain components may be induced within the material. One way of describing the effect of the general stress or strain state of the material under the combined loading situation is by way of "effective stress" $\bar{\sigma}$, or "effective strain" $\bar{\varepsilon}$ defined as follows:

$$\bar{\sigma} = \frac{1}{\sqrt{2}}[(\sigma_{xx} - \sigma_{yy})^2 + (\sigma_{yy} - \sigma_{zz})^2 + (\sigma_{xx} - \sigma_{zz})^2 + 6(\sigma^2_{xy} + \sigma^2_{yz} + \sigma^2_{xz})]^{1/2}$$

$$= \frac{1}{\sqrt{2}}[(\sigma_1 - \sigma_2)^2 + (\sigma_2 - \sigma_3)^2 + (\sigma_1 - \sigma_3)^2]^{1/2} \qquad (3.21)$$

$$\bar{\varepsilon} = \frac{\sqrt{2}}{3}[(\varepsilon_{xx} - \varepsilon_{yy})^2 + (\varepsilon_{yy} - \varepsilon_{zz})^2 + (\varepsilon_{xx} - \varepsilon_{zz})^2$$

$$+ 6(\varepsilon^2_{xy} + \varepsilon^2_{yz} + \varepsilon^2_{xz})]^{1/2}$$

$$= \frac{\sqrt{2}}{3}[(\varepsilon_1 - \varepsilon_2)^2 + (\varepsilon_2 - \varepsilon_3)^2 + (\varepsilon_1 - \varepsilon_3)^2]^{1/2} \qquad (3.22)$$

The reader is to be reminded that the terms "equivalent stress" and "equivalent strain" are often used in (3.21) and (3.22) in some books, e.g. in Hill[5].

3.5.4 Functional approximation of flow curve

In all stress analysis cases for structures, one important property which has to be included is the material's "stiffness". In elastic stress analysis, this stiffness is a constant represented by the *modulus of elasticity*, E. This property varies, however, when the material is loaded beyond its elastic limit.

A continuous variation of this property can be expected, as shown in the flow curve in Figure 3.2. A function which can describe these curves and thus the continuous change of slopes is called a constitutive law of the material and is a necessary condition for any type of elastic–plastic stress analysis.

By referring to those idealized flow curves shown in Figure 3.2, the following two polynomial functional approaches will enable the analyst to describe a more realistic flow curve with gradual change from the elastic to plastic region with strain- or work-hardening behavior.

(1) *Ramberg–Osgood equation*[25]. This equation was first proposed by Ramberg and Osgood in 1943. It has a relatively simple form as shown below:

$$\varepsilon = \frac{\sigma}{E} + K\left(\frac{\sigma}{E}\right)^n \tag{3.23}$$

where n = shape parameter

$$= 1 + \frac{\log\{m_2(1 - m_1)/[m_1(1 - m_2)]\}}{\log(\sigma_y/\sigma')}$$

and $K = \varepsilon_y\left(\frac{E}{\sigma_y}\right)^n$

with $0 < m_1 < 1$

$$0.7 < m_2 < 1.0 \quad (\sim 0.85)$$

The case of $n \to \infty$ in (3.23) coincides with the flow curve representing perfect plastic materials as illustrated in Figure 3.9a.

(2) *Hsu–Bertels polynomial*[26]. The Ramberg–Osgood equation has been widely used in finite element analysis. However, major shortcomings of this approximation are: (1) it requires knowledge of the σ_y determined at ε_{offset}, and (2) the curve cannot be set straight to describe the approximately constant rate of work-hardening exhibited by most materials beyond the

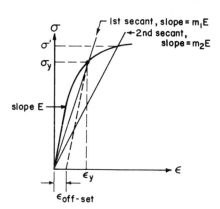

Figure 3.10 Parameters for the Ramberg–Osgood equation.

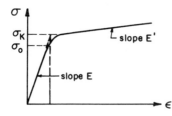

Figure 3.11 Parameters for the Hsu–Bertel's polynomial approximation.

initial yield point, as shown in Figure 3.2. A polynomial function which can describe the gross bilinear nature but with gradual elastic–plastic transition is thus desirable.

The following function which was evolved from a similar work by Richard and Blacklock[27] for elastic-perfect plastic materials has proven to satisfy this need and has been implemented in the TEPSAC code.

The mathematical expression of this polynomial function is given as follows:

$$\sigma = E\varepsilon \bigg/ \left\{ 1 + \left[\frac{E\varepsilon}{(1 - E'/E)\sigma_k + E'\varepsilon} \right]^n \right\}^{1/n} \tag{3.24}$$

where σ_k = stress level at the intersection (kink) of the elastic–plastic curve as illustrated in Figure 3.11;

n = stress power or shape parameter

$\approx \ln 2/\ln (\sigma_k/\sigma_0)$

3.5.5 Functional approximation of the flow curve for the multidimensional case

For almost all practical cases, the only flow curve available to the engineer is one which is derived from a uniaxial tensile test on the material. Yet most analyses involve multiple stress components simultaneously appearing in the structure. Obviously some modifications have to be made to the uniaxial flow curve in order to use it for the real multidimensional stress analysis state. The transformation of the flow curves as illustrated in Figure 3.12 is therefore a necessary first step for this type of analysis.

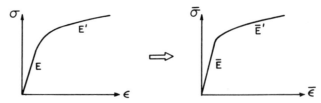

Figure 3.12 Conversion of one-dimensional flow curve to the multidimensional curve.

This transformation can be achieved according to the derivations given in Cheng and Hsu[28] with the following substitutions of $\bar{E}$ and $\bar{E}'$ into (3.24):

$$\bar{E} = \frac{3E}{2(1 + v)} = 3G \qquad (3.25a)$$

$$\bar{E}' = 3E' \Big/ \left[3 - \frac{(1 - 2v)E'}{E} \right] \qquad (3.25b)$$

3.5.6 Yield criteria for the multidimensional state

During a uniaxial tension test of a prism metal bar, the yield point in a σ-ε curve is usually identifiable by an idealization such as that shown in Figure 3.9b & c. However, the situation is quite different when the solid is subject to multiaxial loadings. The yield condition for these cases now can only be defined by the "yield criterion" and there are several such criteria available for this purpose, as follows.

(1) *Tresca yield criterion.* For a solid subjected to the three principal stresses, σ_1, σ_2 and σ_3, the yield condition for the solid satisfies the following equation:

$$[(\sigma_1 - \sigma_2)^2 - \sigma_y^2][(\sigma_1 - \sigma_3)^2 - \sigma_y^2][(\sigma_2 - \sigma_3)^2 - \sigma_y^2] = 0$$

where σ_y = yield strength of the material from a uniaxial tension test.
From the above expression, one may derive the following yield criteria:

$$\sigma_1 - \sigma_2 = \sigma_y \quad \text{or} \quad \sigma_1 - \sigma_3 = \sigma_y \quad \text{or} \quad \sigma_2 - \sigma_3 = \sigma_y$$
$$(3.26)$$

The Tresca yield criterion for a biaxial stress state can be shown graphically as in Figure 3.13.

(2) *Von Mises yield criterion.* This yield criterion was first derived from the distortion energy theory. The mathematical expression of this criterion is simply

$$\bar{\sigma} = \sigma_y = \sqrt{3}\sigma_s \qquad (3.27)$$

or

$$\frac{1}{\sqrt{2}} [(\sigma_{xx} - \sigma_{yy})^2 + (\sigma_{yy} - \sigma_{zz})^2 + (\sigma_{xx} - \sigma_{zz})^2 + 6(\sigma_{xy}^2 + \sigma_{yz}^2 + \sigma_{xz}^2)]^{1/2}$$

$$= \sigma_y = \sqrt{3}\sigma_s$$

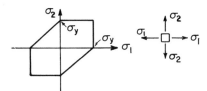

Figure 3.13 A two-dimensional Tresca yield surface.

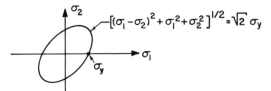

Figure 3.14 A two-dimensional von Mises yield surface.

where σ_s = yield strength of material under pure shear. The two-dimensional graphic expression for the von Mises criterion is an ellipse as shown in Figure 3.14.

If one refers to (3.20b) and (3.27), the von Mises yield criterion can be expressed as

$$J_2 = \sigma_s^2 \tag{3.28}$$

where J_2 = 2nd deviatoric stress invariant.

(3) *Comparison of Tresca, von Mises yield criteria and experimental results.* The Tresca yield criterion is more suitable for the classical solution of elastic–plastic problems due to its obvious simplicity in the mathematical expression in (3.26). However, the discontinuities of the function which appear as corners in Figure 3.13 can induce errors in the solution. The von Mises criterion, on the other hand, is more complex mathematically but behaves as a continuous function, as shown in Figure 3.14. The mathematical complexity can be overcomed by the use of the finite element method. Both criteria have been shown to correlate well with experimental results, as shown in Figure 3.15[29]. The von Mises criterion shows a slightly better correlation in this case.

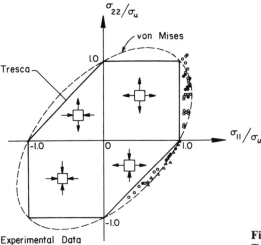

Experimental Data

o Steel
o Copper
▵ Aluminum

Figure 3.15 Comparison of Tresca and von Mises yield criteria.

3.6 Strain hardening

When a material is loaded beyond its elastic limit, initial plastic yielding takes place as described in Section 3.2.3. Theoretically, the material starts to "flow" without any additional load. This behavior is depicted in Figure 3.9b. In reality, however, the fact is that most material retains some of its original stiffness after the initial yielding. Further plastic deformation by these materials requires additional loads, a behavior illustrated in Figure 3.9c. Another interesting phenomenon to observe is that the material becomes "harder" after some plastic deformation as a higher applied load is required to cause the same material to deform plastically again after the completion of one previous loading cycle.

Two types of strain-hardening schemes are commonly used in finite element analysis. These are: (1) isotropic hardening; (2) kinematic hardening.

3.6.1 Isotropic hardening scheme

Figure 3.16 illustrates the principle of isotropic hardening of a material. The material is first uniaxially loaded beyond its initial yield strength σ_y to an instantaneous strain ε_1 and followed by an unloading upon reaching point A. A permanent strain ε_2 is expected to be introduced in the material after the completion of the unloading. If the same bar is loaded again, the material is found to yield at a higher yield strength σ_y' which coincides with the stress at the last unloading level. The isotropic strain-hardening behavior for biaxial stress states can be graphically represented by the uniform expansion of the initial yield surface of Figure 3.14, or as shown in Figure 3.17.

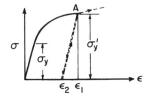

Figure 3.16 Paths for loading, unloading and reloading for an isotropic hardening solid.

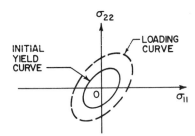

Figure 3.17 Biaxial loading surfaces for isotropic hardening solids.

75

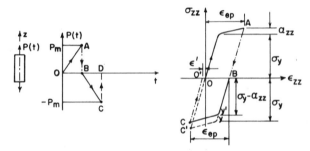

Figure 3.18 Paths for a uniaxially loaded bar under kinematic hardening plastic deformation.

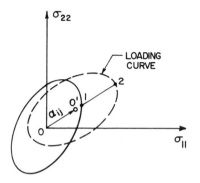

Figure 3.19 Shift of loading surfaces with kinematic hardening rule under biaxial loading condition.

3.6.2 Kinematic hardening scheme

When some materials are plastically deformed in tension, their compressive yield strength in the subsequent compression reduces by the same amount of the increment in the previous tensile loading. This behavior was first discovered by J. Bauschinger in 1881 and is called the *Bauschinger effect*. The result of this effect is a strain hysterisis observed after a complete tension-compression load cycle, as shown in Figure 3.18.

Again, for the multiaxial loading situation, the loading curves not only expand as in the case of isotropic hardening, but also shift as shown in Figure 3.19 for a biaxial stress situation.

3.7 Plastic potential (yield) function

The strain-hardening behavior of most engineering materials during plastic deformation indicates the fact that the yield strength of a material is not a fixed value but a function of stresses and strains. Mathematically it can be shown to be

$$F = F(\sigma_{ij}, \varepsilon_{ij}^p) \tag{3.29}$$

76

Strictly speaking the loading path or history is also important, i.e. $F = F(\sigma_{ij}, \varepsilon_{ij}^p, \text{history})$. But for the sake of simplicity and following the assumption made in Section 3.2.1, the form in (3.29) is adopted here as the yield function.

For a von Mises type of material, the plastic potential (yield) function for an isotropic strain-hardening material may take the following forms:

$$F(\sigma_{ij}) = J_2 - \sigma_s^2$$
$$= J_2 - \tfrac{1}{3}\sigma_y^2$$
$$= \tfrac{1}{3}\bar{\sigma}^2 - \tfrac{1}{3}\sigma_y^2 \qquad (3.30)$$

where σ_s, σ_y = respective yield strength of material by pure shear and uniaxial tension;

J_2 = 2nd deviatoric stress invariant, (3.20b);

$\bar{\sigma}$ = effective stress, (3.21).

It is obvious that the plastic deformation starts when

$$dF(\sigma_{ij}) = 0$$

but for material with strain-hardening capability, any further plastic deformation requires

$$dF(\sigma_{ij}) > 0$$

or, by straight differentiation,

$$\frac{\partial F(\sigma_{ij})}{\partial \sigma_{ij}} d\sigma_{ij} > 0 \qquad (3.31)$$

3.8 Prandtl–Reuss relation

As described in Section 3.6, additional stress has to be applied to materials with strain-hardening behavior in order to produce further plastic deformation. Variation of stresses, of course, means variation of yield function (plastic potential function) in a σ space. This variation means that the incremental plastic strain is directly proportional to the rate of change of the plastic potential function: mathematically,

$$d\varepsilon_{ij}^p \propto \frac{\partial F(\sigma_{ij})}{\partial \sigma_{ij}}$$

or

$$d\varepsilon_{ij}^p = d\lambda \frac{\partial F(\sigma_{ij})}{\partial \sigma_{ij}} = \sigma_{ij}' \, d\lambda \qquad (3.32)$$

where ε_{ij}^{p} = plastic strain

$d\lambda$ = proportionality constant, a scalar quantity which can be determined experimentally but may vary throughout the loading process.

A very useful relationship which can be expressed as

$$\frac{\partial F(\sigma_{ij})}{\partial \sigma_{ij}} = \sigma_{ij}' \qquad (3.33)$$

is valid only for von Mises type of materials, as used in the above derivation.

The above expression in (3.32) may be expanded to give the Prandtl–Reuss relation as follows:

$$\frac{d\varepsilon_{xx}^{p}}{\sigma_{xx}'} = \frac{d\varepsilon_{yy}^{p}}{\sigma_{yy}'} = \frac{d\varepsilon_{zz}^{p}}{\sigma_{zz}'} = \frac{d\varepsilon_{xy}^{p}}{\sigma_{xy}'} = \frac{d\varepsilon_{xz}^{p}}{\sigma_{xz}'} = \frac{d\varepsilon_{yz}^{p}}{\sigma_{yz}'} = d\lambda \qquad (3.34)$$

3.9   Derivation of plastic stress–strain relations

The following assumptions are necessary in deriving the generalized plastic stress and strain relations:

(1)   The infinitesimal increments of plastic strains are linearly proportional to the infinitesimal increment of stress, i.e.

$$d\varepsilon_{ij}^{p} = f_{ijkl}\, d\sigma_{kl}$$

where f_{ijkl} are the material parameters.

This assumption serves as the basis of the theory of incremental plasticity.

(2)   The loading surface in stress space must be convex with respect to the origin, as illustrated in Figure 3.20.

From the observations made in Section 3.2.4, the total strain increment can be considered to be the sum of the incremental elastic and plastic components:

$$d\varepsilon_{ij} = d\varepsilon_{ij}^{e} + d\varepsilon_{ij}^{p} \qquad (3.35)$$

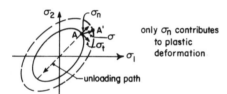

only σ_n contributes to plastic deformation

Figure 3.20   Incremental plastic loading path for a solid under biaxial stress states.

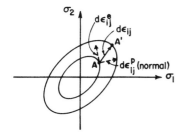

Figure 3.21 Vectorial representation of incremental strain components.

in which the incremental elastic strain components are

$$d\varepsilon_{ij}^{e} = C_{ijkl}\, d\sigma_{kl} = \frac{1+\nu}{E}\, d\sigma_{ij} - \frac{\nu}{E}\, \delta_{ij}\, d\sigma_{kk} \qquad (3.36a)$$

and the corresponding incremental plastic strain components are

$$d\varepsilon_{ij}^{p} = \sigma_{ij}'\, d\lambda \qquad (3.36b)$$

where C_{ijkl} = material compliance.

Graphically, the total strain increment can be expressed as shown in Figure 3.21.

Equation (3.32) can be expanded to give

$$
\begin{aligned}
d\varepsilon_{xx}^{p} &= \tfrac{2}{3}\, d\lambda\, [\sigma_{xx} - \tfrac{1}{2}(\sigma_{yy} + \sigma_{zz})] \\
d\varepsilon_{yy}^{p} &= \tfrac{2}{3}\, d\lambda\, [\sigma_{yy} - \tfrac{1}{2}(\sigma_{zz} + \sigma_{xx})] \\
d\varepsilon_{zz}^{p} &= \tfrac{2}{3}\, d\lambda\, [\sigma_{zz} - \tfrac{1}{2}(\sigma_{xx} + \sigma_{yy})] \\
d\varepsilon_{xy}^{p} &= d\lambda\, \sigma_{xy}\,;\ d\varepsilon_{yz}^{p} = d\lambda\, \sigma_{yz}\,;\ d\varepsilon_{xz}^{p} = d\lambda\, \sigma_{xz}
\end{aligned}
\qquad (3.37)
$$

Comparison of (3.37) with (3.7a) leads to the fact that the Poisson's ratio ν for all materials during plastic deformation is 0.5.

3.10 Constitutive equations for thermoelastic–plastic stress analysis

The behavior of materials during plastic deformation has been reviewed in Sections 3.5 to 3.9. As the relationship between the stresses and strains in the material is not linear, the stiffness of the material cannot be regarded as constant. Appropriate constitutive equations which describe such relationships can be derived by either of the following methods:

(1) Iteration from its initial slope (or stiffness), e.g. the Young's modulus E of the σ–ε curve by various numerical iteration techniques e.g. the Newton–Raphson method. This process is called "initial stress" or "initial strain" [30]

(2) By assuming piecewise linear relation within small stress and strain increments.

The second approach, i.e. the "incremental plasticity" approach, has been adopted here for its relative simplicity in computation. The assumption of a linear relationship between the incremental stresses and strains allows extended use of the well-established theory of linear elasticity.

Following (3.13), the total thermoelastic–plastic strain increment in a solid can be expressed as

$$\{\delta\varepsilon\} = \{\delta\varepsilon_e\} + \{\delta\varepsilon_{e,T}\} + \{\delta\varepsilon_{e,\dot{\varepsilon}}\} + \{\delta\varepsilon_T\} + \{\delta\varepsilon_p\} \tag{3.38}$$

where $\{\delta\varepsilon_p\}$ denotes the incremental plastic strain components.

Considering now a solid undergoing elastic–plastic deformation and subject to significant variation of temperature, T and strain rate, $\dot{\varepsilon}$, the plastic potential function F can no longer be treated as a function of stresses alone but has to incorporate the effects of strain (or work) hardening rates, the temperature and the strain rates. Mathematically it should now be expressed as:

$$F = F(\{\sigma\}, K, T, \dot{\varepsilon}) \tag{3.39a}$$

where K is the work-hardening parameter of the material.

A similar postulation, but without the effect of $\dot{\varepsilon}$, was proposed by Ueda and Yamakawa[31]. For materials obeying the von Mises yield criterion under multiaxial stress state, the function F which given in (3.30) has the form

$$F = J_2 - \sigma_s^2 = \tfrac{1}{3}\bar{\sigma}^2 - \tfrac{1}{3}\sigma_y^2 \tag{3.39b}$$

where

$$J_2 = \tfrac{1}{2}\sigma'_{ij}\sigma'_{ij} = \tfrac{1}{3}\bar{\sigma}^2 \tag{3.39c}$$

as shown in (3.20b).

It can be proved that the following relation exists for von Mises type materials:

$$\{\sigma'\} = \{\partial F/\partial\sigma'\} \tag{3.40}$$

The plastic strain increments in (3.38) can be expressed in terms of the Prandtl–Reuss stress–strain relation given in (3.32) to be

$$\{d\varepsilon_p\} = d\lambda\{\sigma'\} = d\lambda\{\partial F/\partial\sigma'\} \tag{3.41}$$

where $d\lambda$ = proportionality factor.

If we differentiate F in (3.39a) by the chain rule of partial differentiation, we obtain

$$dF = \left\{\frac{\partial F}{\partial\sigma}\right\}^T \{d\sigma\} + \frac{\partial F}{\partial K}dK + \frac{\partial F}{\partial T}dT + \frac{\partial F}{\partial\dot{\varepsilon}}d\dot{\varepsilon} \tag{3.42}$$

and, since the hardening parameter K is considered to be represented by the amount of plastic work done to the solid during the plastic deformation, the second term in the right-hand side of (3.42) can be expressed in terms of plastic strains $\{d\varepsilon_p\}$ as

$$\frac{\partial F}{\partial K} dK = \frac{\partial F}{\partial K} \left\{\frac{\partial K}{\partial \varepsilon_p}\right\}^T \{d\varepsilon_p\}$$

Substituting this relation into (3.42) gives

$$dF = \left\{\frac{dF}{d\sigma}\right\}^T \{d\sigma\} + \frac{\partial F}{\partial K}\left\{\frac{\partial K}{\partial \varepsilon_p}\right\}^T \{d\varepsilon_p\} + \frac{\partial F}{\partial T} dT + \frac{\partial F}{\partial \dot{\varepsilon}} d\dot{\varepsilon}$$

Equilibrium conditions for the solid during a small incremental plastic deformation requires that the plastic energy variation be stationary, which implies that $dF = 0^{22,30}$, or

$$\left\{\frac{dF}{d\sigma}\right\}^T \{d\sigma\} + \frac{\partial F}{\partial K}\left\{\frac{\partial K}{\partial \varepsilon_p}\right\}^T \{d\varepsilon_p\} + \frac{\partial F}{\partial T} dT + \frac{\partial F}{\partial \dot{\varepsilon}} d\dot{\varepsilon} \equiv 0 \qquad (3.43)$$

Now, if we rearrange the terms in (3.38) in the following form for the elastic strain components:

$$\{d\varepsilon_e\} = \{d\varepsilon\} - \{d\varepsilon_T\} - \{d\varepsilon_{e,T}\} - \{d\varepsilon_{e,\dot{\varepsilon}}\} - \{d\varepsilon_p\}$$

and follow Hooke's law for elastic stress and strain, and express the total stress increments as

$$\{d\sigma\} = [C_e](\{d\varepsilon\} - \{d\varepsilon_T\} - \{d\varepsilon_{e,T}\} - \{d\varepsilon_{e,\dot{\varepsilon}}\} - \{d\varepsilon_p\}) \qquad (3.44)$$

Substituting (3.44) into (3.43) and (3.32) for $\{d\varepsilon_p\}$ leads to the following equation:

$$\left\{\frac{\partial F}{\partial \sigma}\right\}^T [C_e](\{d\varepsilon\} - \{d\varepsilon_T\} - \{d\varepsilon_{e,T}\} - \{d\varepsilon_{e,\dot{\varepsilon}}\} - \{d\varepsilon_p\})$$

$$+ \frac{\partial F}{\partial K}\left\{\frac{\partial K}{\partial \varepsilon_p}\right\}^T \left\{\frac{\partial F}{\partial \sigma'}\right\} d\lambda + \frac{\partial F}{\partial T} dT + \frac{\partial F}{\partial \dot{\varepsilon}} d\dot{\varepsilon} \equiv 0$$

From which the expression for the proportionality factor, $d\lambda$ is obtained:

$$d\lambda = \left(\{dF/d\sigma\}^T [C_e](\{d\varepsilon\} - \{d\varepsilon_T\} - \{d\varepsilon_{e,T}\} - \{d\varepsilon_{e,\dot{\varepsilon}}\}) \right.$$

$$\left. + \frac{\partial F}{\partial T} dT + \frac{\partial F}{\partial \dot{\varepsilon}} d\dot{\varepsilon} \right) \bigg/ \left(\left\{\frac{\partial F}{\partial \sigma}\right\}^T [C_e] \left\{\frac{\partial F}{\partial \sigma}\right\} - \frac{\partial F}{\partial K}\left\{\frac{\partial F}{\partial \varepsilon_p}\right\}^T \left\{\frac{\partial F}{\partial \sigma}\right\} \right)$$

Now, if we let

$$S = \left\{\frac{\partial F}{\partial \sigma}\right\}^{T} [C_e] \left\{\frac{\partial F}{\partial \sigma}\right\} - \frac{\partial F}{\partial K}\left\{\frac{\partial K}{\partial \varepsilon_p}\right\}^{T}\left\{\frac{\partial F}{\partial \sigma}\right\} \tag{3.45}$$

and recognize the following relationship previously described in Section 3.4 and (3.15),

$$\{d\varepsilon_T\} = \{a\}\,dT \tag{3.45a}$$

$$\{d\varepsilon_{e,T}\} = \frac{\partial [C_e]^{-1}}{\partial T}\{\sigma\}\,dT \tag{3.45b}$$

$$\{d\varepsilon_{e,\dot{\varepsilon}}\} = \frac{\partial [C_e]^{-1}}{\partial \dot{\varepsilon}}\{\sigma\}\,d\dot{\varepsilon} \tag{3.45c}$$

then on substituting (3.45) and (3.45a–c) into the above expression for $d\lambda$, a more condensed form results:

$$d\lambda = \frac{1}{S}\left([\partial F/\partial \sigma][C_e]\left(\{d\varepsilon\} - \{a\}\,dT - \frac{\partial [C_e]^{-1}}{\partial T}\{\sigma\}\,dT - \frac{\partial [C_e]^{-1}}{\partial \dot{\varepsilon}}\{\sigma\}\,d\dot{\varepsilon}\right)\right.$$
$$\left. + \frac{\partial F}{\partial T}dT + \frac{\partial F}{\partial \dot{\varepsilon}}d\dot{\varepsilon}\right) \tag{3.46}$$

Let us now take a look at (3.43) again and substitute all incremental strain components given in (3.45) and $d\lambda$ in (3.46) to (3.44), the total incremental stress components then become

$$\{d\sigma\} = [C_e]\{d\varepsilon\} - [C_e]\left(\{a\}\,dT + \frac{\partial [C_e]^{-1}}{\partial T}\{\sigma\}\,dT + \frac{\partial [C_e]^{-1}}{\partial \dot{\varepsilon}}\{\sigma\}\,d\dot{\varepsilon}\right)$$
$$- \frac{[C_e]}{S}\left\{\frac{\partial F}{\partial \sigma}\right\}\left(\left\{\frac{\partial F}{\partial \sigma}\right\}[C_e]\left(\{d\varepsilon\} - \{a\}\,dT - \frac{\partial [C_e]^{-1}}{\partial T}\{\sigma\}\,dT\right.\right.$$
$$\left.\left. - \frac{\partial [C_e]^{-1}}{\partial \dot{\varepsilon}}\{\sigma\}\,d\dot{\varepsilon}\right) + \frac{\partial F}{\partial T}dT + \frac{\partial F}{\partial \dot{\varepsilon}}d\dot{\varepsilon}\right)$$

Upon rearranging the terms,

$$\{d\sigma\} = [C_e]\{d\varepsilon\} - [C_e]\left(\{a\}\,dT + \frac{\partial [C_e]^{-1}}{\partial T}\{\sigma\}\,dT + \frac{\partial [C_e]^{-1}}{\partial \dot{\varepsilon}}\{\sigma\}\,d\dot{\varepsilon}\right)$$
$$- \frac{1}{S}[C_e]\left\{\frac{\partial F}{\partial \sigma}\right\}\left\{\frac{\partial F}{\partial \sigma}\right\}^{T}[C_e]\{d\varepsilon\}$$
$$+ \frac{1}{S}[C_e]\left\{\frac{\partial F}{\partial \sigma}\right\}\left\{\frac{\partial F}{\partial \sigma}\right\}^{T}[C_e]\left(\{a\}\,dT + \frac{\partial [C_e]^{-1}}{\partial T}\{\sigma\}\,dT + \frac{\partial [C_e]^{-1}}{\partial \dot{\varepsilon}}\{\sigma\}\,d\dot{\varepsilon}\right)$$
$$- \frac{1}{S}[C_e]\left\{\frac{\partial F}{\partial \sigma}\right\}\left(\frac{\partial F}{\partial T}dT + \frac{\partial F}{\partial \dot{\varepsilon}}d\dot{\varepsilon}\right) \tag{3.47}$$

Now, by defining the **plasticity matrix** [C_p] to be

$$[C_p] = \frac{1}{S}[C_e]\left\{\frac{\partial F}{\partial \sigma}\right\}\left\{\frac{\partial F}{\partial \sigma}\right\}^T[C_e] = \frac{1}{S}[C_e]\{\sigma'\}\{\sigma'\}^T[C_e] \quad (3.48)$$

with the second form resulting from use of (3.33).

Further, by substituting the above newly defined [C_p] into (3.47),

$$\{d\sigma\} = [C_e]\{d\varepsilon\} - [C_e]\left(\{a\}\,dT + \frac{\partial[C_e]^{-1}}{\partial T}\{\sigma\}\,dT + \frac{\partial[C_e]^{-1}}{\partial\dot{\varepsilon}}\{\sigma\}\,d\dot{\varepsilon}\right)$$

$$= [C_p]\{d\varepsilon\} + [C_p]\left(\{a\}\,dT + \frac{\partial[C_e]^{-1}}{\partial T}\{\sigma\}\,dT + \frac{\partial[C_e]^{-1}}{\partial\dot{\varepsilon}}\{\sigma\}\,d\dot{\varepsilon}\right)$$

$$- \frac{1}{S}[C_e]\left\{\frac{\partial F}{\partial\sigma}\right\}\left(\frac{\partial F}{\partial T}\,dT + \frac{\partial F}{\partial\dot{\varepsilon}}\,d\dot{\varepsilon}\right)$$

Also, for the sake of convenience, one may define the **elasto-plasticity matrix** to be:

$$[C_{ep}] = [C_e] - [C_p] \quad (3.49)$$

The thermoelastic–plastic constitutive equation then reads

$$\{d\sigma\} = [C_{ep}]\{d\varepsilon\} - [C_{ep}]\left(\{a\}\,dT + \frac{\partial[C_e]^{-1}}{\partial T}\{\sigma\}\,dT + \frac{\partial[C_e]^{-1}}{\partial\dot{\varepsilon}}\{\sigma\}\,d\dot{\varepsilon}\right)$$

$$- \frac{[C_e]\{\sigma'\}}{S}\left(\frac{\partial F}{\partial T}\,dT + \frac{\partial F}{\partial\dot{\varepsilon}}\,d\dot{\varepsilon}\right) \quad (3.50)$$

The reader will find that (3.50) is identical to what was given in Yamada *et al.*[22] without the effects of the thermal- and strain-rate-dependent material properties.

3.11 Derivation of the [C_{ep}] matrix

The derivation of the constitutive equation as given in (3.50) is a very important first step towards the finite element formulation for thermoelastic–plastic stress analysis. The question remains, however, of which exact form of the [C_p] matrix in (3.48) is suitable for this type of analysis. Since, from (3.49), the matrix [C_{ep}] = [C_e] − [C_p] with the expressions of [C_e] included in (3.7), only [C_p] needs to be derived. From (3.48),

$$[C_p] = \frac{[C_e]\{\sigma'\}\{\sigma'\}^T[C_e]}{S}$$

and one may find that [C_e] = $2G$ during the plastic deformation of a solid, due to the fact that the first deviatoric strain invariant vanishes[18]. Hence the following relationship holds:

$$\{\sigma'\}^T[C_e] = 2G\{\sigma'_{11}\ \ \sigma'_{22}\ \ \sigma'_{33}\ \ \sigma'_{12}\ \ \sigma'_{13}\ \ \sigma'_{23}\}^T$$

THERMOELASTIC-PLASTIC STRESS ANALYSIS

which leads to the exact expression for the numerator of the $[C_p]$ matrix:

$$[C_e]\{\sigma'\}\{\sigma'\}^{\mathrm{T}}[C_e] = 2G \left\{ \begin{array}{c} \sigma'_{11} \\ \sigma'_{22} \\ \sigma'_{33} \\ \sigma'_{12} \\ \sigma'_{13} \\ \sigma'_{23} \end{array} \right\} 2G\{\sigma'_{11} \quad \sigma'_{22} \quad \sigma'_{33} \quad \sigma'_{12} \quad \sigma'_{13} \quad \sigma'_{23}\}$$

$$= 4G^2[\mathrm{SYM}] \tag{3.51}$$

where the symmetric matrix [SYM] is:

$$\begin{bmatrix} \sigma'^2_{11} & \sigma'_{11}\sigma'_{22} & \sigma'_{11}\sigma'_{33} & \sigma'_{11}\sigma'_{12} & \sigma'_{11}\sigma'_{13} & \sigma'_{11}\sigma'_{23} \\ & \sigma'^2_{22} & \sigma'_{22}\sigma'_{33} & \sigma'_{22}\sigma'_{12} & \sigma'_{22}\sigma'_{13} & \sigma'_{22}\sigma'_{23} \\ & & \sigma'^2_{33} & \sigma'_{33}\sigma'_{12} & \sigma'_{33}\sigma'_{13} & \sigma'_{33}\sigma'_{23} \\ & \mathrm{SYM} & & \sigma'^2_{12} & \sigma'_{12}\sigma'_{13} & \sigma'_{12}\sigma'_{23} \\ & & & & \sigma'^2_{13} & \sigma'_{13}\sigma'_{23} \\ & & & & & \sigma'^2_{23} \end{bmatrix} \tag{3.51a}$$

Now to evaluate S in (3.45), using the following form,

$$S = \left\{\frac{\partial F}{\partial \sigma}\right\}^{\mathrm{T}} [C_e] \left\{\frac{\partial F}{\partial \sigma}\right\} - \frac{\partial F}{\partial K}\left(\frac{\partial K}{\partial \varepsilon_p}\right)^{\mathrm{T}}\left(\frac{\partial F}{\partial \sigma}\right)$$

With the aid of (3.33), and $[C_e] = 2G$, the first term becomes

$$\left\{\frac{\partial F}{\partial \sigma}\right\}^{\mathrm{T}} [C_e] \left\{\frac{\partial F}{\partial \sigma}\right\} = \{\sigma'\}^{\mathrm{T}}[C_e]\{\sigma'\}$$

$$= 2G(\sigma'^2_{11} + \sigma'^2_{22} + \sigma'^2_{33} + \sigma'^2_{12} + \sigma'^2_{13} + \sigma'^2_{23})$$

$$= 2G\frac{2\bar{\sigma}^2}{3} = \frac{4G}{3}\bar{\sigma}^2 \tag{3.52}$$

with the use of (3.39c).

The work-hardening parameter K of the material is taken to be represented by the amount of plastic work done during the plastic deformation. Mathematically, we may express its variation as the fluctuation of plastic strain energy, or

$$dK = \sigma_{11} d\varepsilon^P_{11} + \sigma_{22} d\varepsilon^P_{22} + \cdots = \{\sigma\}^{\mathrm{T}}\{d\varepsilon_p\} \tag{3.53}$$

Substituting (3.41) into the above gives

$$dK = d\lambda \{\sigma\}^{\mathrm{T}}\left\{\frac{\partial F}{\partial \sigma}\right\}$$

84

But from (3.53),

$$\left\{\frac{\partial K}{\partial \varepsilon_p}\right\}^T = \{\sigma\}^T$$

and from (3.33),

$$\left\{\frac{\partial F}{\partial \sigma}\right\} = \{\sigma'\}$$

Also during the plastic deformation of materials with work hardening, we have learned that the yield strength increases with the load, as $\sigma_y = \bar{\sigma}$ as expressed in (3.30). The definition of the plastic potential function in (3.30) can also be used to derive the following useful relation:

$$\frac{\partial F}{\partial K} = \frac{\partial}{\partial K}\left(J_2 - \tfrac{1}{3}\bar{\sigma}^2\right) = -\tfrac{2}{3}\bar{\sigma}\frac{\partial \bar{\sigma}}{\partial K} \tag{3.54}$$

Note that $\partial J_2/\partial K = 0$ was used in the above derivation.

Now, if we refer to a typical flow curve of a material as shown in Figure 3.22, the work done on a multiaxially loaded solid during the plastic deformation can be determined by the plastic strain energy, or shown mathematically to be

$$\text{work done} = dK = \bar{\sigma}\,d\bar{\varepsilon}_p$$

or

$$\frac{d\bar{\varepsilon}_p}{dK} = \frac{1}{\bar{\sigma}} \tag{3.55}$$

But since

$$\frac{\partial \bar{\sigma}}{\partial K} = \frac{\partial \bar{\sigma}}{\partial \bar{\varepsilon}_p}\frac{d\bar{\varepsilon}_p}{dK} = H'\frac{1}{\bar{\sigma}} = \frac{H'}{\bar{\sigma}}$$

in which H' is the plastic modulus of the material in a multiaxial stress state as shown in Figure 3.22, (3.55) has therefore become

$$\frac{\partial F}{\partial K} = -\tfrac{2}{3}\bar{\sigma}\left(\frac{H'}{\bar{\sigma}}\right) = -\frac{2H'}{3}$$

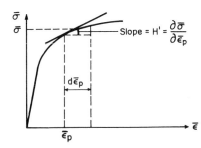

Figure 3.22 Stiffness of a multiaxially loaded solid during plastic deformation.

which leads to the following relationship:

$$\frac{\partial F}{\partial K}\left(\frac{\partial K}{\partial \varepsilon_p}\right)^{\mathrm{T}}\left\{\frac{\partial F}{\partial \sigma}\right\} = \left(-\frac{2H'}{3}\right)(\{\sigma\}^{\mathrm{T}})(\{\sigma'\}) = -\frac{2H'}{3}\{\sigma\}^{\mathrm{T}}\{\sigma'\}$$

One may further simplify the last expression by using the relationship

$$\{\sigma\}^{\mathrm{T}}\{\sigma'\} = \tfrac{2}{3}\bar{\sigma}^2$$

to show that

$$\frac{\partial F}{\partial K}\left(\frac{\partial K}{\partial \varepsilon_p}\right)^{\mathrm{T}}\left\{\frac{\partial F}{\partial \sigma}\right\} = -\frac{2H'}{3}\tfrac{2}{3}\bar{\sigma}^2 = -\tfrac{4}{9}\bar{\sigma}^2 H'$$

Substituting the above expression and (3.52) into (3.45) results in the following simple expression for S in terms of material properties and current stress state $\bar{\sigma}$:

$$S = \tfrac{4}{3}G\bar{\sigma}^2 - (-\tfrac{4}{9}\bar{\sigma}^2 H') = \tfrac{4}{9}\bar{\sigma}^2(3G + H') \qquad (3.56)$$

Accordingly, the plasticity matrix can be determined by combining (3.51), (3.52) and (3.56) as follows:

$$[C_{\mathrm{p}}] = \frac{[C_{\mathrm{e}}]\{\sigma'\}\{\sigma'\}^{\mathrm{T}}[C_{\mathrm{e}}]}{S} = \frac{4G^2[\mathrm{SYM}]}{\tfrac{4}{9}\bar{\sigma}^2(3G + H')} = \frac{2G}{S_0}[\mathrm{SYM}] \qquad (3.57)$$

where the matrix [SYM] is given in (3.51a) with

$$S_0 = \tfrac{2}{3}\bar{\sigma}^2\left(1 + \frac{H'}{3G}\right) \qquad (3.58)$$

The complete form of $[C_{\mathrm{p}}]$ is therefore

$$[C_{\mathrm{p}}] = 2G
\begin{bmatrix}
\dfrac{\sigma_{11}'^2}{S_0} & \dfrac{\sigma_{11}'\sigma_{22}'}{S_0} & \dfrac{\sigma_{11}'\sigma_{33}'}{S_0} & \dfrac{\sigma_{11}'\sigma_{12}'}{S_0} & \dfrac{\sigma_{11}'\sigma_{13}'}{S_0} & \dfrac{\sigma_{11}'\sigma_{23}'}{S_0} \\[2ex]
 & \dfrac{\sigma_{22}'^2}{S_0} & \dfrac{\sigma_{22}'\sigma_{33}'}{S_0} & \dfrac{\sigma_{22}'\sigma_{12}'}{S_0} & \dfrac{\sigma_{22}'\sigma_{13}'}{S_0} & \dfrac{\sigma_{22}'\sigma_{23}'}{S_0} \\[2ex]
 & & \dfrac{\sigma_{33}'^2}{S_0} & \dfrac{\sigma_{33}'\sigma_{12}'}{S_0} & \dfrac{\sigma_{33}'\sigma_{13}'}{S_0} & \dfrac{\sigma_{33}'\sigma_{23}'}{S_0} \\[2ex]
 & \mathrm{SYM} & & \dfrac{\sigma_{12}'^2}{S_0} & \dfrac{\sigma_{12}'\sigma_{13}'}{S_0} & \dfrac{\sigma_{12}'\sigma_{23}'}{S_0} \\[2ex]
 & & & & \dfrac{\sigma_{13}'^2}{S_0} & \dfrac{\sigma_{13}'\sigma_{23}'}{S_0} \\[2ex]
 & & & & & \dfrac{\sigma_{23}'^2}{S_0}
\end{bmatrix} \qquad (3.59)$$

It can be seen from the above derivation that the plasticity matrix $[C_p]$ is a function of $\{\sigma'\}$ and the material stiffness, H'. It is obvious from (3.57) that the value of $[C_p]$ increases as the level of stress increases, together with decreasing H', as illustrated in Figure 3.22. Since $[C_e]$ is constant during the elastic–plastic loading process, an increasing $[C_p]$ will result in the reduction of the elasto–plasticity matrix $[C_{ep}]$ as given in (3.49). This behavior explains the fact that the overall stiffness of the material decreases as it is further loaded into the plastic deformation.

3.12 Determination of material stiffness (H')

It is clear from (3.57) that determination of the elasto–plasticity matrix $[C_{ep}]$ requires a value of the material stiffness which is sometimes called **equivalent plastic modulus** for the multidimensional stress state. This term, in the practical sense, is similar to E' for the uniaxial case.

Since H' is defined to be the slope of the flow curve of a material in the plastic range as shown in Figure 3.22, the following expression appears to be appropriate:

$$H' = d\bar{\sigma}/d\bar{\varepsilon}_p \qquad (3.60)$$

It is obvious that functions which describe the relationships between $\bar{\sigma}$ and $\bar{\varepsilon}_p$ must be available in order to determine H' in (3.60). As already indicated in Section 3.5.4, approximate functions are available for the description of the material's flow curve, i.e. the Ramberg–Osgood equation in (3.23), or the Hsu–Bertels polynomial in (3.24).

Since in a general case, as illustrated in Figure 3.23,

$$\bar{\varepsilon} = \bar{\varepsilon}_e + \bar{\varepsilon}_p$$

and following what is given in (3.35),

$$d\bar{\varepsilon} = d\bar{\varepsilon}_e + d\bar{\varepsilon}_p \qquad (3.61)$$

The reader should, however, be reminded that although the above relation, which was derived from differentiation of the total strain, appears to be correct, in reality it is not always true. The validity of the last equation can only be regarded as a good approximation if the slope E' of the σ–ε curve or H' of the $\bar{\sigma}$–$\bar{\varepsilon}$ curve does not deviate significantly from a straight line. Thus by following the illustration given in Figure 3.23, one may derive the following relations:

$$d\bar{\varepsilon}_e = d\bar{\sigma}/\bar{E} \qquad \text{and} \qquad d\bar{\varepsilon}_p = d\bar{\sigma}/H'$$

with

$$d\bar{\sigma}/d\bar{\varepsilon} = E_t$$

or

$$d\bar{\varepsilon} = d\bar{\sigma}/E_t$$

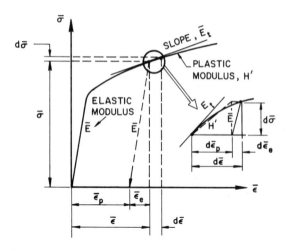

Figure 3.23 Graphic representation of material stiffness during elastic–plastic deformation.

Substituting the above relations into (3.61), we obtain

$$H' = \frac{1}{(1/E_t) - (1/\bar{E})} \tag{3.62}$$

where $E_t = d\bar{\sigma}/d\bar{\varepsilon}$ = slope of $\bar{\sigma}$–$\bar{\varepsilon}$ curve as illustrated in Figure 3.23. If one uses (3.24) for the flow curve, then

$$E_t = \frac{\bar{E}\left\{1 + \left[\dfrac{\bar{E}\bar{\varepsilon}}{(1 - \bar{E}'/\bar{E})\bar{\sigma}_K + \bar{E}'\bar{\varepsilon}}\right]^{n+1} \dfrac{\bar{E}'}{\bar{E}}\right\}}{\left\{1 + \left[\dfrac{\bar{E}\bar{\varepsilon}}{(1 - \bar{E}'/\bar{E})\bar{\sigma}_K + \bar{E}'\bar{\varepsilon}}\right]^n\right\}^{(n+1)/n}} \tag{3.63}$$

with $\bar{E}$ and $\bar{E}'$ given in (3.25).

Strictly speaking, the effective plastic stiffness H' and the tangent modulus E_t in the above equations should be determined by the current value of $\bar{\varepsilon}$ determined by the elastic–plastic strain components only. In other words, all other strain components such as those shown in (3.45a,b,c) should be excluded in this process.

Since the current value of $\bar{\varepsilon}_i$, where the subscript i denote the loading step, is not available in the incremental approach, one may evaluate H' and E_t on the basis of $\bar{\varepsilon}_{i-1}$. However, a scheme which leads to a more accurate description of the material stiffness has been adopted in the present analysis:

$$\bar{\varepsilon} = \bar{\varepsilon}_{i-1} + (\bar{\varepsilon}_{i-1} - \bar{\varepsilon}_{i-2})/2$$

where $\bar{\varepsilon}$ is to be used to determine H' and E_t in (3.62) and (3.63).

3.12.1 Numerical example

The following numerical example is presented to illustrate the relative magnitudes of various components of the stress increment given in the thermoelastic–plastic constitutive equation (3.50). For the sake of simplicity, the case involved here concerns the uniaxial post-yielding stretching of a cylindrical rod made of A514 steel. The rod was first stretched to a 0.015 strain at room temperature. A moderate temperature rise of 130°F was applied to facilitate a further stretching to a final strain at 0.02. The following mechanical properties of A514 steel at room temperature were found from a handbook[32]:

Young's modulus, $E = 29\,000$ ksi;
shear modulus of elasticity, $G = 11\,000$ ksi;
initial yield strength at 0.2% offset strain, $\sigma_y = 100.05$ ksi from Figure 3.24a[32];
rate of change of E with temperature at 70°F, $dE/dT = -6042$ psi/°F Figure 3.24b[32];
rate of change of yield strength with temperature at 70°F = $dF/dT = -55.6$ psi/°F from Figure 3.24b[32];
linear thermal expansion coefficient, $a = 7.5 \times 10^{-6}$/°F[33].

The incremental thermomechanical loading conditions for this case are

$$\{d\varepsilon\} = \varepsilon_2 - \varepsilon_1 = 0.02 - 0.015 = 0.005$$

$$\{dT\} = T_2 - T_1 = 200 - 70 = 130°\text{F}.$$

Since the loadings are assumed to be uniform and one-dimensional, the following quantities, required to calculate the stress increment in (3.50), can be simplified to give

from (3.21), $\bar{\sigma} = \sigma$, initial stress at 70°F = 102 ksi at $\varepsilon = 0.015$ from Figure 3.24a[32]

from (3.24), $H' = E' = 0.4 \times 10^6$ psi from Figure 3.24a[32]

from (3.8), $[C_e] = E = 29 \times 10^6$ psi

from which $[C_e]^{-1} = \dfrac{1}{E}$

and

$$\frac{d[C_e]^{-1}}{dT} = \frac{d}{dT}\left(\frac{1}{E}\right) = -\frac{1}{E^2}\frac{dE}{dT} = 7.18 \times 10^{-12}/\text{psi-°F}$$

The total stress increment induced by the above incremental thermomechanical load increment can be calculated by using (3.50), which is rearranged, in the absence of strain-rate effect, in the form

$$\{d\sigma\} = \{d\sigma_M\} + \{d\sigma_T\} + \{d\sigma_{Te}\} + \{d\sigma_F\} \tag{3.64}$$

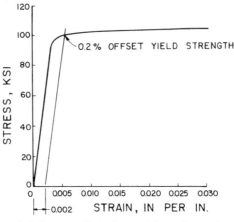

(a) Stress vs. Strain Curve at Room Temperature

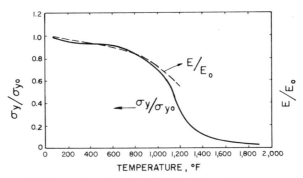

(b) Temperature Dependent Mechanical Properties

Figure 3.24 Mechanical properties of A514 steel:

where

$\{d\sigma_M\}$ = mechanical stress

$\qquad = [C_{ep}]\{d\varepsilon\}$ (3.65a)

$\{d\sigma_T\}$ = thermal stress = $-[C_{ep}]\{a\}\,dT$ (3.65b)

$\{d\sigma_{Te}\}$ = auxiliary stress due to temperature-dependent mechanical

$\qquad$ properties = $-[C_{ep}]\dfrac{\partial[C_e]^{-1}}{\partial T}\{\sigma\}\,dT$ (3.65c)

$\{d\sigma_F\}$ = auxiliary stress due to temperature-dependent yield function =

$\qquad -\dfrac{[C_e]\{\sigma\}}{S}\dfrac{\partial F}{\partial T}\,dT$ (3.65d)

Numerical evaluations of the last three incremental stress components in (3.64) require the determination of the following quantities:
from (3.56),

$$S = \tfrac{4}{9}\bar{\sigma}^2(3G + H') = 154.44 \times 10^{15}$$

from (3.58),

$$S_0 = \tfrac{2}{3}\bar{\sigma}^2\left(1 + \frac{H'}{3G}\right) = 7019 \times 10^6$$

from (3.59),

$$[C_p] = 2G\frac{\sigma_{11}'^2}{S_0} = \frac{2G}{S_0}\left(\sigma_{11} - \frac{\sigma_{11} + 0 + 0}{3}\right) = 14.49 \times 10^6 \text{ psi}$$

from (3.49),

$$[C_{ep}] = [C_e] - [C_p] = 29 \times 10^6 - 14.49 \times 10^6 = 14.51 \times 10^6 \text{ psi}$$

All components of the stress increment in (3.64) and (3.65) can be computed to be

$$\{d\sigma_M\} = (14.51 \times 10^6)(0.005) = 72\,550 \text{ psi}$$

$$\{d\sigma_T\} = -(14.51 \times 10^6)(7.5 \times 10^{-6})(130) = -14\,147 \text{ psi}$$

$$\{d\sigma_{Te}\} = -(14.51 \times 10^6)(7.18 \times 10^{-12})(102 \times 10^3)(130) = -1381 \text{ psi}$$

$$\{d\sigma_F\} = -\frac{(29 \times 10^6)(68 \times 10^3)}{154.44 \times 10^{15}}(-55.6)(130) = 0.0922 \text{ psi}$$

The total stress increment, of course, can be calculated by summing up all above components to be $57\,022$ psi. It is seen that the total contribution of the thermal effects is 27.2%. A 2.42% error is introduced if the effect of $\{d\sigma_{Te}\}$ is neglected. The contribution of $\{d\sigma_F\}$ is insignificant in this particular case. However, this effect can be highly significant at higher temperatures where a drastic drop of yield strength of the material is expected, as depicted in Figure 3.7.

3.13 Thermoelastic–plastic stress analysis with kinematic hardening rule

As described in Section 3.6, the Bauschinger effect in metallic solids became known in 1881. This effect explains the fact that many metallic materials may exhibit strain hysteresis loops after cyclic loadings which are common in engineering applications. The kinematic hardening rule proposed by Prager[34] with Ziegler's modification[35] has been widely used for this purpose due to its relative simplicity compared with other methods (e.g. that of

91

Mroz[36]) when applied to finite element analysis. The essence of the Prager–Ziegler kinematic hardening model is a continuous shift of yield surface during plastic loadings. An illustration of kinematic hardening in conjunction with the von Mises yield criterion in (3.28) for a biaxial stress state is shown in Figure 3.25a. As can be seen from this figure, the loading curve shifts by an amount a_{ij} upon a change of stress state from 1 to 2.

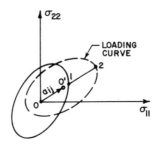

(a) Loading Path for Kinematic Hardening

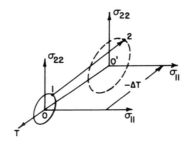

(b) Yield Surface as Function of Temperature

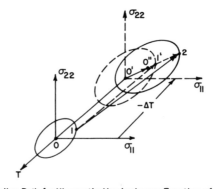

(c) Loading Path for Kinematic Hardening as Function of Temperature

Figure 3.25 Thermoelastic–plastic loading path with kinematic hardening rule.

For cases involving significant temperature variation, the yield function and flow curves become temperature dependent. The change of the yield surface of a von Mises type material due to a temperature increment ΔT alone is illustrated in Figure 3.25b. For a structure subject to combined cyclic thermal and mechanical loadings, both types of loading path variations have to be included. While no general acceptance is presently available for the description of this kind of general plastic behavior of materials, the loading path from stress state 1 to state 2 is assumed to be from 1 to 1' for a finite temperature increment ΔT, and from 1' to 2 due to kinematic hardening. This loading pattern is illustrated in Figure 3.25c.

No rigid rule on the proper sequence of the shift and the expansion or contraction of the yield surfaces is necessary, as the effects by both these variations in the mathematical model are additive, as will be shown in the following derivations.

The original kinematic hardening theory proposed by Prager was based on an important postulation that the increments of translation of the loading surface in the nine-dimensional stress space take place in the direction of the exterior normal to the surface at an instantaneous stress state as illustrated in Figure 3.26a. However, inconsistencies were observed when it was expanded into the stress space, as has been demonstrated by Ziegler[35]. Consequently, modifications had to be made for the elastic–plastic analysis in the following form:

$$da_{ij} = d\mu(\sigma_{ij} - a_{ij}) \qquad d\mu > 0 \tag{3.66}$$

and

$$da_{ij} = C(T, \dot{\varepsilon})\, d\varepsilon_{ij}^{p} \tag{3.67}$$

where $d\mu$ = multiplier, $C(T, \dot{\varepsilon})$ = material parameter, and $d\varepsilon_{ij}^{p}$ = incremental plastic strain components.

Geometrically, the incremental shift in (3.66) is represented by $da_{ij}^{(Z)}$ in Figure 3.26a and the equation represents the loading path in Figure 3.26b.

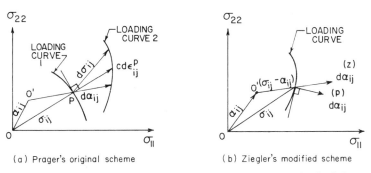

(a) Prager's original scheme   (b) Ziegler's modified scheme

Figure 3.26 Loading path for modified kinematic hardened plastic deformation.

The flow rule which includes both kinematic hardening and the temperature and strain rate dependence is similar to that for isotropic hardening behavior, i.e. similar to (3.32):

$$d\varepsilon_{ij}^p = d\lambda \frac{\partial F}{\partial \sigma_{ij}}$$

The plastic potential function F in the above equation, however, has to be modified in order to accommodate the shift of the yield surface for the kinematic hardening material behavior:

$$F = F(\sigma_{ij}, a_{ij}, T, \dot{\varepsilon})$$

Again, the total differential of F is

$$dF = \frac{\partial F}{\partial \sigma_{ij}} d\sigma_{ij} + \frac{\partial F}{\partial a_{ij}} da_{ij} + \frac{\partial F}{\partial T} dT + \frac{\partial F}{\partial \dot{\varepsilon}} d\dot{\varepsilon} = 0 \qquad (3.68)$$

for the plastic yielding condition.

The following relationship can be derived from (3.66):

$$\frac{\partial F}{\partial a_{ij}} = -\frac{\partial F}{\partial \sigma_{ij}} \qquad (3.69a)$$

$$\frac{\partial \sigma_{ij}}{\partial a_{ij}} = -1 \qquad (3.69b)$$

Now, if one substitutes the above expressions into (3.68), the following expression is obtained:

$$\frac{\partial F}{\partial \sigma_{ij}} (d\sigma_{ij} - da_{ij}) + \frac{\partial F}{\partial T} dT + \frac{\partial F}{\partial \dot{\varepsilon}} d\dot{\varepsilon} = 0$$

or

$$\frac{\partial F}{\partial \sigma_{ij}} da_{ij} = \frac{\partial F}{\partial \sigma_{ij}} d\sigma_{ij} + \frac{\partial F}{\partial T} dT + \frac{\partial F}{\partial \dot{\varepsilon}} d\dot{\varepsilon}$$

Further substitution of (3.66) into the above expression yields

$$d\mu = \left\{ \frac{\partial F}{\partial \sigma_{ij}} d\sigma_{ij} + \frac{\partial F}{\partial T} dT + \frac{\partial F}{\partial \dot{\varepsilon}} d\dot{\varepsilon} \right\} \bigg/ \left\{ (\sigma_{kl} - a_{kl}) \frac{\partial F}{\partial \sigma_{kl}} \right\} \qquad (3.70)$$

The proportionality factor $d\lambda$ can be derived by combining (3.67) and the flow rule given in (3.32) as follows:

$$d\lambda = \frac{1}{C(T, \dot{\varepsilon})} \left\{ \frac{\partial F}{\partial \sigma_{ij}} d\sigma_{ij} + \frac{\partial F}{\partial T} dT + \frac{\partial F}{\partial \dot{\varepsilon}} d\dot{\varepsilon} \right\} \bigg/ \left\{ \left(\frac{\partial F}{\partial \sigma_{kl}} \right) \left(\frac{\partial F}{\partial \sigma_{kl}} \right) \right\} \qquad (3.71)$$

Since the principle of the kinematic hardening rule is a continuous shift of the yielding and loading surfaces in the stress space, an approach which is based on the shifted stress tensors has proven to be expedient in deriving the constitutive equation. As a result of such shifts, the analysis becomes physically identical to that for the isotropic hardening materials. The procedure established in the foregoing sections can thus be followed for the present purpose.

Now if we define the following new terms for the subsequent derivation[37]:

translated stress tensor: $\quad \sigma_{ij}^* = \sigma_{ij} - a_{ij}$ $\qquad$ (3.72)

translated stress deviators: $\quad \bar{\sigma}_{ij}^* = \sigma_{ij}^* - \frac{1}{3}\sigma_{kk}^* \delta_{ij}$ $\qquad$ (3.73)

$$(\text{sum over } k = 1, 2, 3)$$

The yield function becomes

$$F = \frac{3}{2}\bar{\sigma}_{ij}^* \bar{\sigma}_{ij}^* \qquad (3.74)$$

which leads to

$$\frac{\partial F}{\partial \sigma_{ij}} = 3\bar{\sigma}_{ij}^* \frac{\partial \bar{\sigma}_{ij}^*}{\partial \sigma_{ij}} = 6\bar{\sigma}_{ij}^* \qquad (3.75)$$

By substituting (3.75) into (3.70), (3.71) and (3.32), the following expressions have been derived for the evaluation of the parameters given below:

$$d\mu = \frac{6\bar{\sigma}_{ij}^* d\sigma_{ij} + (\partial F/\partial T) dT + (\partial F/\partial \dot{\varepsilon}) d\dot{\varepsilon}}{6\bar{\sigma}_{ij}^* \bar{\sigma}_{ij}^*} \qquad (3.76)$$

$$d\lambda = \frac{1}{C(T, \dot{\varepsilon})} \frac{6\bar{\sigma}_{ij}^* d\sigma_{ij} + (\partial F/\partial T) dT + (\partial F/\partial \dot{\varepsilon}) d\dot{\varepsilon}}{36\bar{\sigma}_{ij}^* \bar{\sigma}_{ij}^*} \qquad (3.77)$$

$$d\varepsilon_{ij}^P = 6\bar{\sigma}_{ij}^* d\lambda \qquad (3.78)$$

Now, by using Hooke's law for the elastic portion of the total strain components and by repeating (3.38) but in indicial notation:

$$d\sigma_{ij} = C_{ijkl}(d\varepsilon_{ij} - d\varepsilon_{ij}^{e,T} - d\varepsilon_{ij}^{e,T} - d\varepsilon_{ij}^{e,\dot{\varepsilon}} - d\varepsilon_{ij}^P)$$

Substituting (3.78) for $d\varepsilon_{ij}^P$ and (3.45) for other incremental strain components into the above expression leads to the following expression for the incremental stress component:

$$d\sigma_{ij} = C_{ijkl}^e d\varepsilon_{kl} - C_{ijkl}^e(a_{kl} dT + D_{kl}^T dT + D_{kl}^{\dot{\varepsilon}} d\dot{\varepsilon} + 6\bar{\sigma}_{kl}^* d\lambda) \qquad (3.79)$$

with

$$D_{kl}^T = \frac{\partial [C_e]^{-1}}{\partial T} \sigma_{kl} \qquad (3.80a)$$

and

$$D_{kl}^{\dot{\varepsilon}} = \frac{\partial [C_e]^{-1}}{\partial \dot{\varepsilon}} \sigma_{kl} \qquad (3.80b)$$

The incremental shift of the yield surface, on substituting (3.78) into (3.67), is

$$d a_{ij} = 6C(T, \dot{\varepsilon})\bar{\sigma}_{ij}^* d\lambda$$

which, when substituted into (3.68) with the aid of (3.69a), leads to

$$dF = [d\sigma_{ij} - 6C(T, \dot{\varepsilon})\bar{\sigma}_{ij}^* d\lambda] \frac{\partial F}{\partial \sigma_{ij}} + \frac{\partial F}{\partial T} dT + \frac{\partial F}{\partial \dot{\varepsilon}} d\dot{\varepsilon} = 0$$

The substitution of $d\sigma_{ij}$ in (3.79) into the above expression results in the following expression for $d\lambda$:

$$d\lambda = \frac{\dfrac{\partial F}{\partial \sigma_{ij}}[C_{ijkl}^e d\varepsilon_{kl} - C_{ijkl}^e(a_{kl} dT + D_{kl}^T dT + D_{kl}^{\dot{\varepsilon}} d\dot{\varepsilon})] + \dfrac{\partial F}{\partial T} dT + \dfrac{\partial F}{\partial \dot{\varepsilon}} d\dot{\varepsilon}}{6[C(T, \dot{\varepsilon})\bar{\sigma}_{ij}^* + C_{ijkl}^e\bar{\sigma}_{kl}^*] \dfrac{\partial F}{\partial \sigma_{ij}}} \tag{3.81}$$

Again, the substitution of (3.81) into (3.79) gives

$$d\sigma_{ij} = C_{ijkl}^e d\varepsilon_{kl} - C_{ijkl}^e(a_{kl} dT + D_{kl}^T dT + D_{kl}^{\dot{\varepsilon}} d\dot{\varepsilon})$$

$$- C_{ijkl}\bar{\sigma}_{kl}^* \frac{\dfrac{\partial F}{\partial \sigma_{ij}}[C_{ijkl} d\varepsilon_{kl} - C_{ijkl}(a_{kl} dT + D_{kl}^T dT + D_{kl}^{\dot{\varepsilon}} d\dot{\varepsilon})] + \dfrac{\partial F}{\partial T} dT + \dfrac{\partial F}{\partial \dot{\varepsilon}} d\dot{\varepsilon}}{[C(T, \dot{\varepsilon})\bar{\sigma}_{ij}^* + C_{ijkl}^e\bar{\sigma}_{kl}^*] \dfrac{\partial F}{\partial \sigma_{ij}}}$$

$$\tag{3.82}$$

Now let

$$S = [C(T, \dot{\varepsilon})\bar{\sigma}_{ij}^* + C_{ijkl}^e \bar{\sigma}_{kl}] \frac{\partial F}{\partial \sigma_{ij}}$$

$$= 6\bar{\sigma}_{ij}^* \bar{\sigma}_{ij}^* C(T, \dot{\varepsilon}) + 6\bar{\sigma}_{ij}^* C_{ijkl}^e \bar{\sigma}_{kl}^* \tag{3.83}$$

With the relationship given in (3.75), (3.82) becomes

$$d\sigma_{ij} = C_{ijkl}^e d\varepsilon_{kl} - C_{ijkl}^e(a_{kl} dT + D_{kl}^T dT + D_{kl}^{\dot{\varepsilon}} d\dot{\varepsilon})$$

$$- C_{ijkl}^e\bar{\sigma}_{kl}^* \frac{\partial F}{\partial \sigma_{ij}} \frac{1}{S} [C_{ijkl}^e d\varepsilon_{kl} - C_{ijkl}^e(a_{kl} dT + D_{kl}^T dT + D_{kl}^{\dot{\varepsilon}} d\dot{\varepsilon})]$$

$$- \frac{1}{S} \left\{ C_{ijkl}^e\bar{\sigma}_{kl}^* \left(\frac{\partial F}{\partial T} dT + \frac{\partial F}{\partial \dot{\varepsilon}} d\dot{\varepsilon} \right) \right\}$$

$$= \left[C_{ijkl}^e - \frac{1}{S} \left(C_{ijkl}^e\bar{\sigma}_{kl}^* \frac{\partial F}{\partial \sigma_{ij}} C_{ijkl}^e \right) \right] d\varepsilon_{ij}$$

$$- \left[C_{ijkl}^e - \frac{1}{S} \left(C_{ijkl}^e\bar{\sigma}_{kl}^* \frac{\partial F}{\partial \sigma_{ij}} C_{ijkl}^e \right) \right] (a_{kl} dT + D_{kl}^T dT + D_{kl}^{\dot{\varepsilon}} d\dot{\varepsilon})$$

$$- \frac{1}{S} C_{ijkl}^e\bar{\sigma}_{kl}^* \left(\frac{\partial F}{\partial T} dT + \frac{\partial F}{\partial \dot{\varepsilon}} d\dot{\varepsilon} \right) \tag{3.83a}$$

Further, by letting

$$\frac{1}{S}\left(C_{ijkl}^{e}\bar{\sigma}_{kl}^{*}\frac{\partial F}{\partial \sigma_{ij}}C_{ijkl}^{e}\right) = \frac{1}{S}(6C_{ijkl}^{e}\bar{\sigma}_{kl}^{*}\bar{\sigma}_{ij}^{*}C_{ijkl}^{e}) = C_{ijkl}^{p} \tag{3.84}$$

and

$$C_{ijkl}^{ep} = C_{ijkl}^{e} - C_{ijkl}^{p} \tag{3.85}$$

the constitutive equation for the solid subject to thermoelastic–plastic deformation and kinematic hardening rule can be derived by the substitution of (3.83)–(3.85) into (3.83a), resulting in the form

$$d\sigma_{ij} = C_{ijkl}^{ep}\,d\varepsilon_{kl} - C_{ijlk}^{ep}(a_{kl}\,dT + D_{kl}^{T}\,dT + D_{kl}^{\dot{\varepsilon}}\,d\dot{\varepsilon})$$

$$- \frac{1}{S}C_{ijkl}^{e}\bar{\sigma}_{kl}^{*}\left(\frac{\partial F}{\partial T}dT + \frac{\partial F}{\partial \dot{\varepsilon}}d\dot{\varepsilon}\right) \tag{3.86}$$

In order to facilitate the subsequent finite element formulation, translations of the above key equations in indicial notation into appropriate matrix forms is necessary. Such translation can be readily done and the result is summarized as follows:

(1) Constitutive equation (3.86):

$$\{d\sigma\} = [C_{ep}]\{d\varepsilon\} - [C_{ep}](\{a\}\,dT + \{D^{T}\}\,dT + \{D^{\dot{\varepsilon}}\}\,d\dot{\varepsilon})$$

$$- \frac{1}{S}[C_{e}]\{\bar{\sigma}^{*}\}\left(\frac{\partial F}{\partial T}dT + \frac{\partial F}{\partial \dot{\varepsilon}}d\dot{\varepsilon}\right) \tag{3.87}$$

(2) Elasto–plastic matrix (3.84, 3.83):

$$[C_{ep}] = [C_{e}] - 6[C_{e}]\{\bar{\sigma}^{*}\}\{\bar{\sigma}^{*}\}^{T}[C_{e}]/S \tag{3.88}$$

where $S = 6\{\bar{\sigma}^{*}\}\{\bar{\sigma}^{*}\}C(T, \dot{\varepsilon}) + 6\{\bar{\sigma}^{*}\}[C_{e}]\{\bar{\sigma}^{*}\}$.

(3) Proportionality factor (3.77):

$$d\lambda = \frac{1}{C(T, \dot{\varepsilon})}\ \frac{6\{\bar{\sigma}^{*}\}^{T}\{d\sigma\} + \dfrac{\partial F}{\partial T}dT + \dfrac{\partial F}{\partial \dot{\varepsilon}}d\dot{\varepsilon}}{36\{\bar{\sigma}^{*}\}^{T}\{\bar{\sigma}^{*}\}} \tag{3.89}$$

(4) The multiplier (3.76):

$$d\mu = \frac{6\{\bar{\sigma}^{*}\}^{T}\{d\sigma\} + \dfrac{\partial F}{\partial T}dT + \dfrac{\partial F}{\partial \dot{\varepsilon}}d\dot{\varepsilon}}{6\{\sigma^{*}\}^{T}\{\bar{\sigma}^{*}\}} \tag{3.90}$$

(5) The plastic strain increments (3.78):

$$\{d\varepsilon_{p}\} = 6\{\bar{\sigma}^{*}\}\,d\lambda \tag{3.91}$$

(6)   The shift of yield surface (3.66), (3.72)–(3.74):

$$\{da\} = \{\sigma_{ij}^*\}\, d\mu \qquad\qquad (3.92a)$$

where

$$F = \tfrac{3}{2}\{\bar{\sigma}*\}^{\mathrm{T}}\{\bar{\sigma}*\} \qquad\qquad (3.92b)$$

$$\{\sigma*\} = \{\sigma\} - \{a\} \qquad\qquad (3.92c)$$

and

$$\{\bar{\sigma}*\} = \{\sigma*\} - \{\sigma_{\mathrm{m}}^*\} \qquad\qquad (3.92d)$$

$$\{D^{\mathrm{T}}\} = \frac{\partial [C_e]^{-1}}{\partial T}\{\sigma\}; \qquad\qquad (3.92e)$$

$$\{D^{\dot{\varepsilon}}\} = \frac{\partial [C_e]^{-1}}{\partial \dot{\varepsilon}}\{\sigma\} \qquad\qquad (3.92f)$$

The coefficient $C(T, \dot{\varepsilon})$ in the Prager–Ziegler model in (3.67), designated as a material parameter, physically means the stiffness of the material during the plastic deformation. It has a similar meaning to the H' parameter in the case of isotropic hardening. A difficulty arises, however, in the case of kinematic hardening, as the stress $(\bar{\sigma})$ vs. strain $(\bar{\varepsilon})$ curve becomes anisotropic due to the postulated shift of the yield surfaces. The numerical value of H' which is derived from this curve varies from one direction to another as a result of these shifts. A tensor quantity, H'_{ij}, has to be used instead of a single value of H'. Once again, for the expedience of the finite element formulation, a weighted average value of H' was used as shown below[37]:

$$\frac{1}{C(T, \dot{\varepsilon})} = \frac{1}{H_{ij}(T, \dot{\varepsilon})}\left(\frac{\sigma_{ij}}{\sigma_{\mathrm{w}}}\right)^2 \qquad\qquad (3.93)$$

where

$$2\sigma_{\mathrm{w}}^2 = B_{ij}(\sigma_{ij})^2 + \delta_{ij}(\sigma_{ij})^2 \quad \text{(sum over } i, j = 1, 2, 3)$$

$$B_{ij} = 1$$

$$H_{ij} = \text{weighted } H \text{ defined in (3.62)}.$$

3.14   Finite element formulation of thermoelastic–plastic stress analysis

The essence of the theory of incremental plasticity is the assumption of a piecewise linear relation between the stress and strain in the solid during a small load increment. The basic formulation presented in Section 3.3 for the linear elastic theory can thus be used for the nonlinear thermoelastic–plastic deformation of solids induced by small load increments. All physical quantities, including the displacements, strains, stresses and strain energy, etc., will be dealt with in the incremental sense[†].

Thus, by following the usual discretization procedure as described in Chapter 1 for finite element analysis, the incremental displacement components in an element, $\{\Delta U\}$, by the expression

$$\{\Delta U(\mathbf{r})\} = [N(\mathbf{r})]\{\Delta u\} \tag{3.94}$$

where $[N(\mathbf{r})]$ = interpolation function, and $\mathbf{r}$ = position vector or spatial coordinates.

The incremental displacement vector $\{\Delta u\}$ at the nodes in (3.94) can be related to the incremental strains in an element by the same matrix shown in (3.6):

$$\{\Delta\varepsilon\} = [B(\mathbf{r})]\{\Delta u\} \tag{3.95}$$

The incremental stress–strain relationship is given by the constitutive equations, i.e. (3.50) for isotropic hardening solids and (3.87) for kinematic hardening.

The constitutive equations will be expressed in a somewhat different form as follows for the subsequent derivation of the element equations in the case of isotropic hardening materials:

$$\{\Delta\sigma\} = [C_{\text{ep}}](\{\Delta\varepsilon\} - \{\Delta\varepsilon^{\text{T}}\} - \{\Delta\varepsilon^{\dot{\varepsilon}}\}) = [C_{\text{ep}}]\{\Delta\varepsilon'\} \tag{3.96}$$

in which

$$\{\Delta\varepsilon'\} = \{\Delta\varepsilon\} - \{\Delta\varepsilon^{\text{T}}\} - \{\varepsilon^{\dot{\varepsilon}}\}$$

$$= \text{total incremental thermomechanical strains} \tag{3.97a}$$

$\{\Delta\varepsilon^{\text{T}}\}$ = overall incremental thermal strain components

$$= \{a\}\,\Delta T + \frac{\partial[C_e]^{-1}}{\partial T}\{\sigma\}\,\Delta T + \frac{[C_{\text{ep}}]^{-1}[C_e]\{\sigma\}}{S}\frac{\partial F}{\partial T}\Delta T \tag{3.97b}$$

$\{\Delta\varepsilon^{\dot{\varepsilon}}\}$ = overall incremental strain components due to strain rate effect

$$= \frac{\partial[C_e]^{-1}}{\partial\dot{\varepsilon}}\{\sigma\}\,\Delta\dot{\varepsilon} + \frac{[C_{\text{ep}}]^{-1}[C_e]\{\sigma\}}{S}\frac{\partial F}{\partial\dot{\varepsilon}}\Delta\dot{\varepsilon} \tag{3.97c}$$

The incremental strain energy in the element can be formulated in a similar way as described in Section 3.3.5:

$$\Delta U = \tfrac{1}{2}\int_V \{\Delta\varepsilon'\}^{\text{T}}\{\Delta\sigma\}\,dv$$

[†] In theory, of course, the term "incremental" should really mean "infinitesimal", i.e. $d\sigma$, $d\varepsilon$ for incremental stress and strain. In reality, however, it can only mean "small", not infinitesimal. The sign Δ is therefore used instead of d.

Replacing $\{\Delta\varepsilon'\}$ and $\{\Delta\sigma\}$ respectively by (3.97a) and (3.96), and $\{\Delta\varepsilon\}$ by (3.95), the above expression for ΔU can be expressed in terms of $\{\Delta u\}$ as

$$\Delta U = \tfrac{1}{2}\{\Delta u\}^{\mathrm{T}}\left(\int_V [B]^{\mathrm{T}}[C_{\mathrm{ep}}][B]\, dv\right)\{\Delta u\}$$

$$- \{\Delta u\}^{\mathrm{T}}\int_V [B]^{\mathrm{T}}[C_{\mathrm{ep}}](\{\Delta\varepsilon^{\mathrm{T}}\} + \{\Delta\varepsilon^{\dot{e}}\})\, dv$$

$$+ \tfrac{1}{2}\int_V (\{\Delta\varepsilon^{\mathrm{T}}\} + \{\Delta\varepsilon^{\dot{e}}\})^{\mathrm{T}}[C_{\mathrm{ep}}](\{\Delta\varepsilon^{\mathrm{T}}\} + \{\Delta\varepsilon^{\dot{e}}\})\, dv \qquad (3.98)$$

The fundamental condition for the equilibrium in a deformed solid such as illustrated in Figure 3.5 is that its stored potential energy be kept to a minimum. This condition, when applied to a discretized solid in Figure 1.2, means that the potential energy in individual elements can be used as the functional which is to be minimized for the derivation of the element equations as demonstrated in (1.4).

The above principle can be applied to the nonlinear thermoelastic–plastic analysis during a small load increment. All physical quantities are to be interpreted, of course, in an incremental sense.

Thus, for a deformed element with volume v and surface s in equilibrium, the incremental potential energy stored as a result of the application of incremental body forces $\{\Delta f\}$ and surface tractions $\{\Delta t\}$ can be expressed as:

$$\Delta\pi = \Delta U + \Delta W$$

in which ΔW = work done on the element

$$= -\int_v \{\Delta U\}^{\mathrm{T}}\{\Delta f\}\, dv - \int_s \{\Delta U\}^{\mathrm{T}}\{\Delta t\}\, ds$$

$$= -\{\Delta u\}^{\mathrm{T}}\int_v [N]^{\mathrm{T}}\{\Delta f\}\, dv - \{\Delta u\}^{\mathrm{T}}\int_s [N]^{\mathrm{T}}\{\Delta t\}\, ds \qquad (3.99)$$

Upon substituting ΔU in (3.98) and ΔW in (3.99) into the expression for $\Delta\pi$ and applying the variational process to this quantity, with respect to $\{\Delta u\}$, leads to the following set of equations:

$$\frac{\partial(\Delta\pi)}{\partial\{\Delta u\}} = \left(\int_v [B]^{\mathrm{T}}[C_{\mathrm{ep}}][B]\, dv\right)\{\Delta u\} - \int_v [B]^{\mathrm{T}}[C_{\mathrm{ep}}](\{\Delta\varepsilon^{\mathrm{T}}\} + \{\Delta\varepsilon^{\dot{e}}\})\, dv$$

$$- \int_v [N]^{\mathrm{T}}\{\Delta f\}\, dv - \int_s [N]^{\mathrm{T}}\{\Delta t\}\, ds = 0$$

The second portion of the above equality can be rearranged to give the following element equations:

$$[K_{\mathrm{e}}]\{\Delta u\} = \{\Delta F\} \qquad (3.100)$$

where

$$[K_e] = \text{elastic–plastic stiffness matrix}$$

$$= \int_v [B(\mathbf{r})]^T [C_{ep}][B(\mathbf{r})] \, dv \qquad (3.101\text{a})$$

$$\{\Delta F\} = \text{incremental thermomechanical load matrix}$$

$$= \{\Delta F_T\} + \{\Delta F_M\}$$

$$\{\Delta F_M\} = \text{incremental mechanical load matrix}$$

$$= \int_v [N(\mathbf{r})]^T \{\Delta f\} \, dv + \int_s [N(\mathbf{r})]^T \{\Delta t\} \, ds \qquad (3.101\text{b})$$

The incremental thermal load matrix $\{\Delta F_T\}$ includes all the effects of thermal and strain-rate-dependent material properties. Thus, by substituting into the above derivation the definitions of $\{\Delta \varepsilon^T\}$ and $\{\Delta \varepsilon^t\}$ in (3.97b) and (3.97c), the exact expression of $\{\Delta F_T\}$ takes the form

$$\{\Delta F_T\} = \int_v [B(\mathbf{r})]^T [C_{ep}] \left(\{a\} \Delta T + \frac{\partial [C_e]^{-1}}{\partial T} \{\sigma\} \Delta T + \frac{[C_{ep}]^{-1}[C_e]\{\sigma\}}{S} \frac{\partial F}{\partial T} \Delta T \right) dv$$

$$+ \int_v [B(\mathbf{r})]^T [C_{ep}] \left(\frac{\partial [C_e]^{-1}}{\partial T} \{\sigma\} \Delta \dot{\varepsilon} + \frac{[C_{ep}]^{-1}[C_e]\{\sigma\}}{S} \frac{\partial F}{\partial \dot{\varepsilon}} \Delta \dot{\varepsilon} \right) dv$$
$$(3.101\text{c})$$

The overall structure equation can then be readily assembled from the element equations as outlined in Step (5) of Section 1.3:

$$[K]\{\Delta u\} = \{\Delta R\} \qquad (3.102)$$

where

$$[K] = \text{overall stiffness matrix}$$

$$= \sum_1^M [Ke]$$

$$[\Delta R] = \text{overall incremental nodal forces}$$

$$= \sum_1^M \{\Delta F\}$$

in which M = total number of elements in the structure.

Although the element equation in (3.100) and (3.101) was derived on the basis of the constitutive equation for the isotropic hardening case, the reader will find that a similar element equation can be readily made available for kinematic hardening cases, as the constitutive equations for both cases are virtually identical, except for the introduction of the translated stresses defined in (3.72) and (3.73) for the latter hardening scheme.

3.15 Finite element formulation for the base TEPSAC code

A computer code under the name of TEPSAC[†] has been developed based on the finite element formulations derived in Chapter 2 for the thermal analysis and those presented in the foregoing sections for the integrated thermoelastic–plastic stress analysis. The effect of creep deformation of the material will be described in detail in Chapter 4. The reader will find that only minor modifications are necessary to the base thermoelastic–plastic stress analysis code in order to accommodate this effect.

The basic element used in the TEPSAC code is the simplex element with triangular toroidal geometry as illustrated in Figure 3.27a.* Only key formulae that are used in the program will be presented here. Detailed description of the code structure will be given in Appendix 4.

3.15.1 Nodal displacements

By referring to Figure 3.27, we may assume that the components of the displacement at the three representative nodes i, j, k of a particular triangular toroidal element are linear functions of the coordinates, i.e.:

$$U_r(r, z) = b_1 + b_2 r + b_3 z$$

$$U_z(r, z) = b_4 + b_5 r + b_6 z$$

where U_r and U_z are the respective radial and axial displacement components of the element, and $b_1, b_2, \ldots, b_6$ are arbitrary constants.

The above function may be expressed in matrix form as

$$\{U\} = \begin{Bmatrix} U_r \\ U_z \end{Bmatrix} = \begin{bmatrix} 1 & r & z & 0 & 0 & 0 \\ 0 & 0 & 0 & 1 & r & z \end{bmatrix} \begin{Bmatrix} b_1 \\ b_2 \\ b_3 \\ b_4 \\ b_5 \\ b_6 \end{Bmatrix} = [R]\{b\} \quad (3.103)$$

The nodal displacement components,

$$\{u\}^T = \{u_r^i \ \ u_r^j \ \ u_r^k \ \ u_z^i \ \ u_z^j \ \ u_z^k\}^T$$

can be expressed by substituting the nodal coordinates, e.g. (r_i, z_i), into (3.103). These components can then be shown in matrix form as follows:

$$\begin{Bmatrix} u_r^i & u_z^i \\ u_r^j & u_z^j \\ u_r^k & u_z^k \end{Bmatrix} = \begin{bmatrix} 1 & r_i & z_i \\ 1 & r_j & z_j \\ 1 & r_k & z_k \end{bmatrix} \begin{Bmatrix} b_1 & b_4 \\ b_2 & b_5 \\ b_3 & b_6 \end{Bmatrix} \quad (3.104)$$

[†] An acronym for *T*hermo *E*lastic-*P*lastic *S*tress *A*nalysis with *C*reep.
* Refer to the footnote on p. 35.

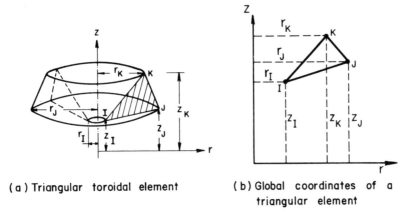

(a) Triangular toroidal element

(b) Global coordinates of a triangular element

Figure 3.27 Geometry and coordinate system of a toroidal element for axisymmetric structures.

or

$$\{u\} = [A]\{b\}$$

$$\{b\} = [A]^{-1}\{u\} = [h]\{u\} \tag{3.105}$$

where

$$[h] =$$

$$\frac{1}{\beta}\begin{bmatrix} r_i z_k - r_k z_j & 0 & r_k z_i - r_i z_k & 0 & r_i z_j - r_j z_i & 0 \\ z_j - z_k & 0 & z_k - z_i & 0 & z_i - z_j & 0 \\ r_k - r_j & 0 & r_i - r_k & 0 & r_j - r_i & 0 \\ 0 & r_j z_k - r_k z_j & 0 & r_k z_i - r_i z_k & 0 & r_i z_j - r_j z_i \\ 0 & z_j - z_k & 0 & z_k - z_i & 0 & z_i - z_j \\ 0 & r_k - r_j & 0 & r_i - r_k & 0 & r_j - r_i \end{bmatrix}$$

$$\beta = r_j(z_k - z_i) + r_i(z_j - z_k) + r_k(z_i - z_j) \tag{3.106}$$

Combining (3.103) and (3.104), the following relation between element displacements and nodal displacements is obtained:

$$\{U\} = [R][h]\{u\} \tag{3.107}$$

Since the product of matrices $[R]$ and $[h]$ in (3.107) relates the element and nodal displacements, it fits the definition of the interpolation function, or

$$[N] = [R][h] \tag{3.108}$$

103

The $[R]$ matrix, as shown in (3.103), involves the global coordinates as follows:

$$[R] = \begin{bmatrix} 1 & r & z & 0 & 0 & 0 \\ 0 & 0 & 0 & 1 & r & z \end{bmatrix} \tag{3.109}$$

and the matrix $[h]$ is related to the nodal coordinates only as shown in (3.106).

3.15.2 Strain-displacement relation

The average strain components of an element as illustrated in Figure 3.27b may be obtained from (3.6):

$$\{\varepsilon\} = \left\{ \begin{array}{c} \varepsilon_{rr} \\ \varepsilon_{zz} \\ \varepsilon_{\theta\theta} \\ \varepsilon_{rz} \end{array} \right\} = \left\{ \begin{array}{c} \dfrac{\partial U_r}{\partial r} \\ \dfrac{\partial U_z}{\partial z} \\ U_r/r \\ \dfrac{\partial U_r}{\partial z} + \dfrac{\partial U_z}{\partial r} \end{array} \right\}$$

By substituting the displacement function given in (3.103) into the above expression, we get

$$\left\{ \begin{array}{c} \varepsilon_{rr} \\ \varepsilon_{zz} \\ \varepsilon_{\theta\theta} \\ \varepsilon_{rz} \end{array} \right\} = \begin{bmatrix} 0 & 1 & 0 & 0 & 0 & 0 \\ 0 & 0 & 0 & 0 & 0 & 1 \\ \dfrac{1}{r} & 1 & \dfrac{z}{r} & 0 & 0 & 0 \\ 0 & 0 & 1 & 0 & 1 & 0 \end{bmatrix} \left\{ \begin{array}{c} b_1 \\ b_2 \\ b_3 \\ b_4 \\ b_5 \\ b_6 \end{array} \right\}$$

or

$$\{\varepsilon\} = [G]\{b\} \tag{3.110}$$

Again, the elements of the $[G]$ matrix involve global coordinates as can be seen from (3.110).

The relationship between the element strains and the nodal displacements can be obtained by substituting (3.105) into (3.110), resulting in

$$\{\varepsilon\} = [B]\{u\} \tag{3.111}$$

where the matrix $[B]$ is the product of $[G]$ and $[h]$.

104

The elements in the $[B]$ matrix are given below:

1st row: $\quad B_{11} = z_j - z_k, \qquad B_{12} = 0, \qquad\qquad B_{13} = z_k - z_i$

$\qquad\qquad B_{14} = 0, \qquad\qquad B_{15} = z_i - z_j, \qquad B_{16} = 0$

2nd row: $\quad B_{21} = 0, \qquad\qquad B_{22} = r_k - r_j, \qquad B_{23} = 0$

$\qquad\qquad B_{24} = r_i - r_k \qquad B_{25} = 0, \qquad\qquad B_{26} = r_j - r_i$

3rd row: $\quad B_{31} = \dfrac{1}{r}(r_j z_k - r_k z_j) + (z_j - z_k) + \dfrac{z}{r}(r_k - r_j)$

$\qquad\qquad B_{32} = 0$

$\qquad\qquad B_{33} = \dfrac{1}{r}(r_k z_i - r_i z_k) + (z_k - z_i) + \dfrac{3}{r}(r_i - r_k)$

$\qquad\qquad B_{34} = 0$

$\qquad\qquad B_{35} = \dfrac{1}{r}(r_i z_j - r_j z_i) + (z_i - z_j) + \dfrac{z}{r}(r_j - r_i)$

$\qquad\qquad B_{36} = 0$

4th row: $\quad B_{41} = r_k - r_j, \qquad B_{42} = z_j - z_k, \qquad B_{43} = r_i - r_k$

$\qquad\qquad B_{44} = z_k - z_i, \qquad B_{45} = r_j - r_i, \qquad B_{46} = z_i - z_j \qquad (3.112)$

For structures subject to thermal load,

$$\{\varepsilon\} = [G][h]\{u\} + \{\varepsilon_0\} \qquad (3.113)$$

in which $\{\varepsilon_0\}$ are the components of the strain due to temperature change, e.g. thermal expansion, crystal growth, etc.

3.15.3 The $[C_{ep}]$ matrix

The elastic–plasticity matrix $[C_{ep}]$ is evaluated following its definition given in (3.49) with the generalized $[C_p]$ shown in (3.59). The elements of the $[C_e]$ matrix can be determined from the generalized Hooke's law described in Section 3.3.4. The exact form of the entries of the $[C_{ep}]$ matrix can thus be derived, e.g. as shown below[12,38] for three common types of structures.

(a) *Plane stress case*

$$[C_{ep}] = \frac{\bar{E}}{1 - v^2}
\begin{bmatrix}
1 & v & 0 \\
v & 1 & 0 \\
0 & 0 & \dfrac{1-v}{2}
\end{bmatrix}
-
\begin{bmatrix}
\dfrac{S_1^2}{S_0} & \dfrac{S_1 S_2}{S_0} & \dfrac{S_1 S_3}{S_0} \\
& \dfrac{S_2^2}{S_0} & \dfrac{S_2 S_3}{S_0} \\
\text{SYM} & & \dfrac{S_3^2}{S_0}
\end{bmatrix}
\qquad (3.114)$$

$$S_0 = \tfrac{4}{9}\bar{\sigma}^2 H' + (S_1\sigma'_{rr} + S_2\sigma'_{zz} + 2S_3\sigma'_{rz})$$

$$S_1 = \frac{\bar{E}}{1-v^2}(\sigma'_{rr} + v\sigma'_{zz}), \qquad S_2 = \frac{\bar{E}}{1-v^2}(v\sigma'_{rr} + \sigma'_{zz}), \qquad S_3 = \frac{\bar{E}}{1+v}\sigma'_{rz}$$

The out-of-plane strain, $\varepsilon_{\theta\theta}$ from the r-z plane does not vanish. It can be readily computed by the following expression:

$$\Delta\varepsilon_{\theta\theta} = -\frac{v}{E}[\Delta\sigma_{rr} + \Delta\sigma_{zz}] - (\Delta\varepsilon^P_{rr} + \Delta\varepsilon^P_{zz})$$

where the incremental plastic strain components, $\Delta\varepsilon^P_{rr}$ and $\Delta\varepsilon^P_{zz}$ may be determined by using (3.32).

(b) *Plane strain case*

$$[C_{ep}] = \frac{\bar{E}}{1+v}\begin{bmatrix} \dfrac{1-v}{1-2v} & \dfrac{v}{1-2v} & 0 \\[2mm] & \dfrac{1-v}{1-2v} & 0 \\[2mm] \text{SYM} & & \tfrac{1}{2} \end{bmatrix} - \frac{2G}{S_0}\begin{bmatrix} \sigma'^2_{rr} & \sigma'_{rr}\sigma'_{zz} & \sigma'_{rr}\sigma'_{rz} \\[2mm] & \sigma'^2_{zz} & \sigma'_{zz}\sigma'_{rz} \\[2mm] \text{SYM} & & \sigma'^2_{rz} \end{bmatrix}$$

(3.115)

where

$$S_0 = \tfrac{2}{3}\left(1 + \frac{H'}{3G} - \frac{3\bar{E}}{4G\bar{\sigma}^2}\sigma'^2_{\theta\theta}\right)\bar{\sigma}^2$$

$$\Delta\sigma_{\theta\theta} = v(\Delta\sigma_{rr} + \Delta\sigma_{zz}) - [(\sigma'_{rr} + v\sigma'_{\theta\theta})\Delta\sigma_{rr} + (\sigma'_{zz} + v\sigma'_{\theta\theta})\Delta\sigma_{zz} + 2\sigma'_{rz}\Delta\sigma_{rz}]\sigma'_{\theta\theta}/R$$

$$R = \sigma'^2_{\theta\theta} - \tfrac{4}{9}\bar{\sigma}^2\frac{H'}{\bar{E}}$$

(c) *Axisymmetric case*

$$[C_{ep}] = 2G\begin{bmatrix} \dfrac{1-v}{1-2v} & \dfrac{v}{1-2v} & \dfrac{v}{1-2v} & 0 \\[2mm] & \dfrac{1-v}{1-2v} & \dfrac{v}{1-2v} & 0 \\[2mm] & & \dfrac{1-v}{1-2v} & \tfrac{1}{2} \\[2mm] \text{SYM} \end{bmatrix}$$

$$- \frac{2G}{S_0}\begin{bmatrix} \sigma'^2_{rr} & \sigma'_{rr}\sigma'_{zz} & \sigma'_{rr}\sigma'_{\theta\theta} & \sigma'_{rr}\sigma'_{rz} \\[2mm] & \sigma'^2_{zz} & \sigma'_{zz}\sigma'_{\theta\theta} & \sigma'_{zz}\sigma'_{rz} \\[2mm] \text{SYM} & & \sigma'^2_{\theta\theta} & \sigma'_{\theta\theta}\sigma'_{rz} \\[2mm] & & & \sigma'^2_{rz} \end{bmatrix}$$

(3.116)

where $S_0 = \tfrac{2}{3}\bar{\sigma}^2\left(1 + \dfrac{H'}{3G}\right)$.

106

3.15.4 Element stiffness matrix

The element stiffness matrix can be evaluated by (3.101a) to be:

$$[K_e] = \int_v [B]^T [C_{ep}][B]\,dv$$

with the $[B]$ matrix given in (3.112) and $[C_{ep}]$ in (3.114) for the plane stress case, (3.115) for plane strain case or (3.116) for axisymmetric geometry.

Another way to evaluate $[K_e]$ in the above expression is to break up the $[B]$ matrix into $[G][h]$ as described in Section 3.15.2. The integral for the element stiffness matrix can thus be expressed as

$$[K_e] = [h]^T \left(\int_v [G]^T [C_{ep}][G]\,dv \right)[h] \tag{3.117}$$

where the integrand is

$[G]^T[C_{ep}][G] =$

$$
\begin{bmatrix}
\frac{1}{r^2}C_{33} & \frac{1}{r}(C_{31}+C_{33}) & \frac{z}{r^2}C_{33}+\frac{1}{r}C_{34} & 0 & \frac{1}{r}C_{34} & \frac{1}{r}C_{32} \\[2mm]
 & C_{11}+C_{13}+C_{33}+C_{31} & \frac{z}{r}(C_{13}+C_{33})+C_{14}+C_{34} & 0 & C_{14}+C_{34} & C_{12}+C_{32} \\[2mm]
 & & \frac{z^2}{r^2}C_{33}+\frac{z}{r}(C_{34}+C_{43})+C_{44} & 0 & \frac{z}{r}C_{34}+C_{44} & \frac{z}{r}C_{32}+C_{42} \\[2mm]
(SYM) & & & 0 & 0 & 0 \\[2mm]
 & & & & C_{44} & C_{42} \\[2mm]
 & & & & & C_{22}
\end{bmatrix}
$$

$$\tag{3.118}$$

in which the coefficients $C_{11}, C_{12}, \ldots, C_{44}$ are the elements of the $[C_{ep}]$ matrix shown in (3.116) for axisymmetric solids. For planar structures, the coefficients in the third column and row in (3.118) are set to zero.

3.15.5 Element load matrices

The load matrices consist of the mechanical components, $\{\Delta F_M\}$, in (3.101b) and the nonmechanical components, $\{\Delta F_T\}$, in (3.101c).

In the TEPSA code[†], the $[N]$ matrix was replaced by $[R][h]$ in (3.108) and the $[B]$ matrix by $[G][h]$. The load matrices thus become

$$\{\Delta F\} = \{\Delta F_M\} + \{\Delta F_T\}$$

† TEPSAC code without creep module.

with

$$\{\Delta F_M\} = [h]^T \int_v [R]^T \{\Delta f\} \, dv + [h]^T \int_s [R]^T \{\Delta t\} \, ds \qquad (3.119a)$$

$$\{\Delta F_T\} = [h]^T \int_v [G]^T [C_{ep}]\{a\} \, \Delta T \, dv$$

$$+ [h]^T \int_v [G]^T [C_{ep}] \frac{\partial [C_e]^{-1}}{\partial T} \{\sigma\} \, \Delta T \, dv$$

$$+ [h]^T \int_v \frac{[G]^T [C_e]\{\sigma\}}{S} \frac{\partial F}{\partial T} \Delta T \, dv$$

$$+ [h]^T \int_v [G]^T [C_{ep}] \frac{\partial [C_e]^{-1}}{\partial \dot{\varepsilon}} \{\sigma\} \, \Delta \dot{\varepsilon} \, dv$$

$$+ [h]^T \int_v \frac{[G]^T [C_e]\{\sigma\}}{S} \frac{\partial F}{\partial \dot{\varepsilon}} \Delta \dot{\varepsilon} \, dv \qquad (3.119b)$$

3.15.6 Numerical evaluation of matrices $[K_e]$ and $\{\Delta F\}$

It is quite clear from (3.117) to (3.119) that the required integration for the numerical values of the matrices $[K_e]$ and $\{\Delta F\}$ is extremely tedious. An approximate method has been used in the TEPSAC code, by introducing the centroidal coordinates of an element:

$$\bar{r} = (r_i + r_j + r_k)/3,$$

$$\bar{z} = (z_i + z_j + z_k)/3,$$

and assuming these to adequately represent the coordinates of all points in the element. Integrations are then reduced to a volume determination only, and the element stiffness and load matrices become

$$[K_e] = [h]^T [\bar{G}]^T [C_{ep}][\bar{G}][h](\Delta \text{vol}), \qquad (3.120)$$

$$\{\Delta F\} = [h]^T [\bar{G}]^T \{\sigma_0\}(\Delta \text{vol}), \qquad (3.121)$$

in which $\{\sigma_0\}$ represent the various components in (3.119a) and (3.119b).

The "barred" matrices in the above equations are based on centroidal coordinates. This approximation will be accurate if small elements are used.

3.15.7 Elastic–plastic stress analysis by incremental deformation theory

The main advantage of adopting the incremental deformation theory is that no iterative procedure is necessary as in the case of the initial stress approach[30]. The adoption of the Hsu–Bertels polynomial constitutive relation for the material given in (3.24) and (3.25) makes the present incremental approach unique by the fact that the thermoelastic–plastic constitutive equations in (3.50) and (3.87) are valid for the entire range of the flow curve. Efforts to identify the elastic or plastic status in each element such as described in Marcal and King[21] can thus be avoided. A disadvantage of the incremental approach is the necessity of using very small load increments in order to insure the convergence of the solution. Another major shortcoming is that the stiffness matrix $[K_e]$ in (3.117) has to be computed at every load increment.

Following the theoretical formulation presented in Sections 3.14 and 3.15, the nodal displacements are calculated by the accumulation of the incremental values computed from each loading step:

$$\{u\} = \{u\} + \{\Delta u\} \qquad (3.122)$$

where the incremental displacement components can be calculated from

$$[K]\{\Delta u\} = \{\Delta R\} \qquad (3.102)$$

in which $\quad [K]$ = overall current elastic–plastic stiffness matrix

$$= \sum_{1}^{M} [K_e^{ep}]$$

$[K_e^{ep}]$ = current element elastic–plastic stiffness matrix;

$\{\Delta F\}$ = incremental load matrix.

The incremental strain components can be obtained from the corresponding expression in (3.111), i.e.

$$\{\Delta \varepsilon\} = [B]\{\Delta u\} \qquad (3.123)$$

and the incremental stresses by the constitutive equations given in (3.50) for the isotropic hardening and (3.87) for kinematic hardening materials.

The total stresses and strains can thus be calculated from the corresponding incremental values, i.e.

$$\{\sigma\} = \{\sigma\} + \{\Delta \sigma\} \qquad (3.124)$$

$$\{\varepsilon\} = \{\varepsilon\} + \{\Delta \varepsilon\} \qquad (3.125)$$

3.15.8 Unloading criteria

Many structural components may be subjected to not only radically different thermal and stress fields but the signs of the stresses can be opposite. Unloading can take place in some parts of the structure if certain unloading criteria are met while the remaining parts are still in the loading process. Since the local unloading process in an elastic–plastically loaded solid follows an elastic path (thus H' in (3.50) is replaced by $\bar{E}$), the computational algorithm for unloading is obviously different than for loading part. It is essential then that some unloading criteria be established in order to check and identify the locality of such local unloading.

There are two sets of criteria which one may use for such a purpose in the finite element analysis:

(1) Distinguish between every two consecutive steps the numerical variation of the proportionality factor $d\lambda$ in (3.46) with S given in (3.56) for an isotropic hardening material and (3.83) for the case of kinematic hardening. The unloading criterion becomes (see Hill[5], p. 38)

$$d\lambda > 0 \quad \text{loading}$$

$$d\lambda < 0 \quad \text{unloading}$$

(2) Sometimes it is more convenient to use the following criterion in the finite element analysis to check if

$$dF > 0 \quad \text{for loading}$$

and $\qquad\qquad dF < 0 \quad \text{for unloading}$

where the plastic potential function $F = \frac{1}{3}\bar{\sigma}^2 - \frac{1}{3}\sigma_y^2$.

3.16 Solution procedure for the base TEPSA code

(1) Select thermal and mechanical load increments.
(2) Perform thermal analysis following Section 2.4, and calculate nodal temperatures, $\{T_t\}$ at time t.
(3) Select material properties based on the average element temperature $=$ $(T_i + T_j + T_k)/3$ and the strain rate, $\dot{\varepsilon}$ determined from the previous loading step.
(4) Determine H' from (3.62) for the isotropic hardening case, and $C(T, \dot{\varepsilon})$ from (3.93).
(5) Compute $[G]$ from (3.110) and $[h]$ from (3.106).
(6) Form the elastic–plastic matrix $[C_{ep}]$ from (3.114) to (3.116) for the isotropic hardening case, and from (3.88) for the kinematic hardening case.
(7) Form the element stiffness matrix $[K_e^{ep}]$ similar to (3.120):

$$[K_e^{ep}] = [h]^T [\bar{G}]^T [C_{ep}][\bar{G}] \, [h](\Delta \text{vol})$$

(8) Form the element force matrix according to (3.121).
(9) Assemble the overall stiffness matrix $[K]$ and construct the overall structural equilibrium equations (3.102):

$$[K]\{\Delta u\} = \{\Delta R\}$$

(10)   Modify $\{\Delta F\}$ for applicable boundary conditions.
(11)   Adjust $[K]$ corresponding to Step (9).
(12)   Solve for $\{\Delta u\}$ by the Gaussian elimination technique and therefore the total displacement components by $\{u\} = \{u\} + \{\Delta u\}$.
(13)   Compute $\{\Delta \varepsilon\}$ from (3.123).
(14)   Compute $\{\Delta \sigma\}$ from (3.50) for the isotropic hardening case, and from (3.87) for the kinematic hardening case.
(15)   Compute the total element stresses and strains by

$$\{\sigma\} = \{\sigma\} + \{\Delta \sigma\} \qquad \text{and} \qquad \{\varepsilon\} = \{\varepsilon\} + \{\Delta \varepsilon\}$$

(16)   Update nodal coordinates by

$$r = r + \Delta u_r \qquad \text{and} \qquad z = z + \Delta u_z$$

and proceed to the next load increment.
(17)   Repeat Steps (1)–(15) until the end criteria are met.

References

1   Love, A. E. H. 1944. *A treatise on the mathematical theory of elasticity.* 4th edn. New York: Dover.
2   Sokolnikoff, I. S. 1956. *Mathematical theory of elasticity.* 2nd edn. New York: McGraw-Hill.
3   Muskhelishvili, N. I. 1963. *Some basic problems of the mathematical theory of elasticity.* 4th edn. Translated by J. R. M. Radok. Amsterdam: Noordhoff.
4   Timoshenko, S. and J. N. Goodier 1951. *Theory of elasticity.* 2nd edn. New York: McGraw-Hill.
5   Hill, R. 1950. *The mathematical theory of plasticity.* Oxford: Oxford University Press.
6   Owen, D. R. J. and E. Hinton 1980. *Finite elements in plasticity.* New York: Pineridge Press.
7   Penny, R. K. and D. L. Marriott 1971. *Design for creep.* New York: McGraw-Hill.
8   Argon, A. S. (ed.) 1975. *Constitutive equations in plasticity.* Cambridge, Mass.: MIT Press.
9   Bernasconi, G. and G. Piatti (eds.) 1978. *Creep of engineering materials and structures.* London: Applied Science Publishers.
10   Gittus, J. 1975. *Creep, viscoelasticity and creep fracture in solids.* London: Applied Science Publishers.
11   Mukherjee, S. 1982. *Boundary element methods in creep and fracture.* London: Applied Science Publishers.
12   Wu, R. and T.-R. Hsu 1978. *Finite element formulations on thermoelastoplastic analysis of planar structures.* Univ. Manitoba, Thermomechanics Lab. Rep. 78-6-56.
13   Volterra, E. and J. H. Gaines 1971. *Advanced strength of materials.* Englewood Cliffs, NJ: Prentice-Hall.
14   Fung, Y. C. 1967. *Foundations of solid mechanics.* Englewood Cliffs, NJ: Prentice-Hall.
15   Boley, B. A. and J. H. Weiner 1960. *Theory of thermal stresses.* New York: John Wiley. p. 263.

16 Wyatt, O. H. and D. Dew-Hughes 1974. *Metals, ceramics and polymers.* Cambridge: Cambridge University Press. Chapter 8.
17 McGregor Tegart, W. J. 1966. *Elements of mechanical metallurgy.* New York: Macmillan.
18 Lin, T. H. 1968. *Theory of inelastic structures.* New York: John Wiley.
19 D'Isa, F. A. 1968. *Mechanics of metals.* Reading, Mass: Addison-Wesley.
20 Marcal, P. V. 1965. A stiffness method for elastic–plastic problems. *Int. J. Mech. Sci.* **7**, 229–38.
21 Marcal, P. V. and I. P. King 1967. Elastic–plastic analysis of two dimensional stress systems by the finite element method. *Int. J. Mech. Sci.* **9**, 143–55.
22 Yamada, Y., N. Yoshimura and T. Sakurai 1968. Plastic stress–strain matrix and its application for the solution of elastic–plastic problems by the finite element method. *Int. J. Mech. Sci.* **10**, 343–54.
23 Zienkiewicz, O. C. 1967. *The finite element method in structural and continuum mechanics.* New York: McGraw-Hill.
24 Desai, C. S. and J. F. Abel 1972. *Introduction to the finite element method.* New York: Van Nostrand Reinhold.
25 Ramberg, W. and W. R. Osgood 1943. *Description of stress–strain curves by three parameters.* Technical Note No. 902, National Advisory Committee for Aeronautics.
26 Hsu, T.-R. and A. W. M. Bertels 1974. An improved approximation of constitutive elasto-plastic stress–strain relationship for finite element analysis. *AIAA J.* **12**(10), 1450–2.
27 Richard, R. M. and J. R. Blacklock 1969. Finite element analysis of inelastic structures. *AIAA J.* **7**(3), 432–8.
28 Cheng, S. Y. and T.-R. Hsu 1978. On elasto-plastic stress–strain relationship for multi-axial stress states. *Int. J. Num. Meth. Engng.* **12**, 1617–22.
29 Popov, E. P. 1976. *Mechanics of materials.* 2nd edn. Englewood Cliffs, NJ: Prentice-Hall. p. 298.
30 Zienkiewicz, O. C., S. Valliappan and I.P. King 1969. Elasto-plastic solutions of engineering problems "initial stress", finite element approach, *Int. J. Num. Meth. Engng.* **1**, 75–100.
31 Ueda, Y. and T. Yamakawa 1972. Thermal nonlinear behaviour of structures. In *Advances in computational methods in structural mechanics and design*, J. T. Oden, R. W. Clough and Y. Yamamoto (eds.). Huntsville: University of Alabama Press. pp. 375–92.
32 Merritt, F. S. 1972. *Structural steel designer's handbook.* New York: McGraw-Hill. pp. 1-10 to 1-13.
33 Cubberly, W. H. *et al.* (ed.) 1978. *Metals handbook*, Vol. 1, 9th edn. Metals Park, Ohio: American Society for Metals.
34 Prager, W. 1956. A new method of analyzing stress and strains in work-hardening plastic solids. *J. Appl. Mech.* **23**, 493.
35 Ziegler, H. 1959. A modification of Prager's hardening rule. *Q. Appl. Math.* **17**(1), 55.
36 Mroz, Z. 1969. An attempt to describe the behaviour of metals under cyclic loads using a more general workhardening model. *Acta Mech.* **7**, 199–212.
37 Hsu, T.-R. and A. W. M. Bertels 1976. Propagation and opening of a through crack in a pipe subject to combined cyclic thermomechanical loadings. *J. Press. Vess. Technol., ASME Trans.*, February, 17–25.
38 Hsu, T.-R. 1974. *Transient thermoelastic-plastic stress analysis by finite element method.* Thermomechanics Lab. report 74-2-15, University of Manitoba.

4

CREEP DEFORMATION OF SOLIDS BY FINITE ELEMENT ANALYSIS

4.1 Introduction

Creep deformation of solid structures has become of increasing concern to the design and application of engineering structures in a high temperature environment. Most nuclear reactor core and high temperature, high performance engine components are vulnerable to creep failure. Creep can also cause catastrophic failure of pressure vessels.

Metallurgically, creep deformation is primarily induced by a thermally activated deformation process[1] which causes, in the microscopic sense: (1) slip or sliding of blocks of crystals; (2) sub-grain formation; and (3) grain boundary sliding. Macroscopically, of course, creep deformation appears to be a deformation of the solid under constant load. Associated with this deformation is the relaxing of stresses in the material.

Creep damage for most materials becomes significant at a temperature above $0.4T_m$, where T_m is the melting temperature of the material, e.g. about $260°C$ for aluminum alloys; $520°C$ for steels and between $-50°C$ to $0°C$ for ice[2].

As described in Section 3.2.5, creep deformation is regarded as time-dependent inelastic deformation of solids. The three stages of creep deformation have varying degrees of significance in engineering applications. For instance, secondary creep, or steady-state creep, is a very important aspect in low cycle fatigue analysis[3], whereas the third stage, or tertiary creep, is mainly responsible for high temperature structural instability, such as fuel sheath ballooning during a loss of coolant accident in a nuclear reactor.

4.2 Theoretical background

Although many physical phenomena related to creep deformation of solids have been known for centuries, common occurrences such as the "natural" elongation of a fine wire by a dead weight, sinking of heavy structures into ground, etc., are common knowledge to persons who are not necessarily technically oriented. Despite the recognition of this outstanding problem and grave concern about its consequences, very little was known about the design methodology to cope with it until about mid-1930s[4] and henceforth it has received much needed attention by scientists and engineers. An extensive description of the development of creep mechanics is given in[5].

As creep in solids is generally regarded as one form of inelastic deformation, intuitively one would conclude that much of the theory developed in Chapter 3 for plastic deformation can be equally well adapted for this type of problem. Indeed, the reader will find that use of that theory has been made in the following derivations. Reality, however, shows that many of such assertions have been made without concrete experimental evidence, as will be discussed in the conclusion of this chapter.

4.2.1 Material creep law under uniaxial stress

As demonstrated in Chapter 3, the most fundamental input required to establish the constitutive equations given in (3.50) is the material's equivalent stiffness H' which can only be derived from the relevant stress–strain curve via functional representations such as given in (3.23) or (3.24). These functional representations are called constitutive laws for the material. The stress analysis of solids involving creep requires similar constitutive law to describe the material behavior under loads. Such a constitutive law is often called the "creep law".

A typical creep strain curve for a solid subject to constant uniaxial load was shown in Figure 3.4.

A general form of function describing this type of curve can be expressed as follows[6]:

$$\varepsilon^c = f(\sigma, t, T) = f_1(\sigma) f_2(t) f_3(T) \tag{4.1}$$

in which σ, t, T are, respectively, applied stress, time and temperature.

There are, of course, various forms of the functions $f_1(\sigma)$, $f_2(t)$ and $f_3(T)$ proposed by researchers. Some of the commonly used forms are presented below:

(1) Common forms of $f_1(\sigma)$ are

Norton[7]:	$f_1(\sigma) = K\sigma^n$	(4.2a)
McVetty:	$= A \sinh(\sigma/\sigma_0)$	(4.2b)
Soderberg:	$= B[\exp(\sigma/\sigma_0) - 1]$	(4.2c)
Dorn:	$= C \exp(\sigma/\sigma_0)$	(4.2d)
Johnson:	$= D_1\sigma^{m_1} + D_2\sigma^{m_2}$	(4.2e)
Garofalo:	$= A[\sinh(\sigma/\sigma_0)]^m$	(4.2f)

where K, A, B, C, D, m, n, m_1 and m_2 are material constants.

(2) Common forms of $f_2(t)$ are

Andrade: $f_2(t) = (1 + bt^{1/3}) \exp(kt) - 1$ (4.3a)

Bailey: $= Ft^n (\frac{1}{3} \leqslant n \leqslant \frac{1}{2})$ (4.3b)

McVetty: $= G(1 - e^{-qt}) + Ht$ (4.3c)

Graham and Walles: $= \sum a_i t^{n_i}$ (4.3d)

where F, G, H, a_i, b, k, n, n_i, q are material constants.

(3) A common form of $f_3(T)$ appears to be[8]

$$f_3(T) = A \exp(-Q/RT)$$ (4.4)

where Q = activation energy, R = Boltzmann's constant, and T = absolute temperature.

For most engineering applications, Norton's law is used. Referring to (4.1), (4.2) and (4.4), the typical form of creep law becomes

$$\varepsilon^c = K_c \sigma^n t \exp(-Q/RT)$$ (4.5)

or

$$\dot{\varepsilon}^c = K_c \sigma^n \exp(-Q/RT)$$ (4.6)

where K_c and n are material constants.

4.2.2 Creep law under multi-axial stresses

Creep test data for most engineering materials have been derived from various types of uniaxial creep tests. Many of these tests have been described in Penny and Marriott[6]. Constitutive laws which correlate the stresses and the corresponding strains in the material as functions of time and temperature can be derived from the results of these tests. Several forms of constitutive laws for some common materials have been presented in (4.1) to (4.6). In reality, however, engineers are likely to have to deal with structures subject to multiaxial stress states. It is logical to expect that relevant constitutive laws be derived from multidimensional loadings. Unfortunately, creep tests involving multidimensional loads are costly and hard to operate. Common practice is therefore simply to express these constitutive laws in terms of effective stress and strain for multidimensional stress cases. The general form of Norton's law in (4.5) and (4.6) for multi-dimensional analysis can thus be expressed in the form

$$\bar{\varepsilon}^c = K_c \bar{\sigma}^n t \exp(-Q/RT)$$ (4.7)

or

$$\dot{\bar{\varepsilon}}^c = K_c \bar{\sigma}^n \exp(-Q/RT)$$ (4.8)

in which $\bar{\varepsilon}^c$ = effective creep strain (following (3.22))

$$= \frac{\sqrt{2}}{3}[(\varepsilon_{xx}^c - \varepsilon_{yy}^c)^2 + (\varepsilon_{yy}^c - \varepsilon_{zz}^c)^2 + (\varepsilon_{xx}^c - \varepsilon_{zz}^c)^2$$

$$+ 6(\varepsilon_{xy}^{c\,2} + \varepsilon_{yz}^{c\,2} + \varepsilon_{xz}^{c\,2})]^{1/2} \tag{4.9}$$

and $\bar{\sigma}$ = effective stress defined in (3.21).

Other forms of rate equations similar to the one shown in (4.8) can be found in Pizzo[9], e.g. Norton's law for 304 stainless steel can be expressed as follows:

$$\dot{\varepsilon}^c = 4.63 \times 10^{34} \frac{D}{GT}\left(\frac{\bar{\sigma}}{G}\right)^3 \quad /\text{hr} \tag{4.10}$$

where $D = 0.37 \times 10^{-4} \exp\left(\frac{3.37 \times 10^4}{T}\right) \quad \text{m}^2/\text{s}$

$G = 8.1 \times 10^4[1 - 4.7 \times 10^{-4}(T - 300)] \quad \text{MPa}$

T = material temperature $\quad °\text{K}$

4.2.3 Flow rule

Similar to the Prandtl–Reuss relation described in Section 3.8 for plastic deformation, incremental creep strain can be expressed in terms of a creep potential function[6]:

$$\{\dot{\varepsilon}^c\} = \beta \frac{\partial\psi(\{\sigma\})}{\partial\{\sigma\}} \tag{4.11}$$

where β is a positive parameter depending on the loading history and $\psi(\{\sigma\})$ is the creep potential function similar to the plastic potential function in (3.29).

If the material is assumed to be homogeneous, initially isotropic, incompressible and obeys von Mises yield criterion (see Section 3.5.6), then

$$\{\sigma'\} = \frac{\partial\psi(\{\sigma\})}{\partial\{\sigma\}} = \frac{\partial J_2}{\partial\{\sigma\}} \tag{4.12}$$

in which $\{\sigma'\}$ = deviatoric stress components defined in (3.17a), and J_2 = second deviatoric stress invariant given in (3.20b). Since the effective creep strain can be expressed in accordance with (4.9) to be

$$\dot{\bar{\varepsilon}}^c = (\tfrac{2}{3}\{\dot{\varepsilon}^c\}^T\{\dot{\varepsilon}^c\})^{1/2} \tag{4.13}$$

116

and by combining (4.11) and (4.13), the incremental creep strain rate and strain can be computed from the expressions

$$\dot{\beta} = \frac{3}{2}\frac{\dot{\bar{\varepsilon}}^c}{\bar{\sigma}}$$

and

$$\{\dot{\varepsilon}^c\} = \frac{3}{2}\frac{\dot{\bar{\varepsilon}}^c}{\bar{\sigma}}\{\sigma'\} \qquad (4.14)$$

or

$$\{d\varepsilon_c\} = \tfrac{3}{2}(d\bar{\varepsilon}^c/\bar{\sigma})\{\sigma'\} \qquad (4.15)$$

in which the numerical values of $\dot{\bar{\varepsilon}}$ are determined from the experimentally derived creep law as shown in (4.6) or (4.10).

4.3 Constitutive equations for thermoelastic–plastic creep stress analysis

Following the derivation in Section 3.10, the total incremental strain components can be expressed as follows as suggested in p. 47 of Penny and Marriott[6]:

$$\{d\varepsilon\} = \{d\varepsilon_{ep}\} + \{d\varepsilon_c\} \qquad (4.16)$$

where $\{d\varepsilon_{ep}\}$ represent thermoelastic–plastic strain components given in (3.38) and $\{d\varepsilon_c\}$ are the incremental creep strains.

The constitutive relation for the thermoelastic–plastic deformation of solids with isotropic hardening is shown in (3.50):

$$\{d\sigma\} = [C_{ep}]\{d\varepsilon_{ep}\} - [C_{ep}]\left(\{a\}\,dT + \frac{\partial[C_e]^{-1}}{\partial T}\{\sigma\}\,dT\right.$$

$$\left. + \frac{\partial[C_e]^{-1}}{\partial\bar{\varepsilon}}\{\sigma\}\,d\bar{\varepsilon}\right) - \frac{[C_e]\{\sigma'\}}{S}\left(\frac{\partial F}{\partial T}\,dT + \frac{\partial F}{\partial\bar{\varepsilon}}\,d\bar{\varepsilon}\right)$$

Rearrangement of the above equation to express $\{d\varepsilon_{ep}\}$ in terms of $\{d\sigma\}$ gives

$$\{d\varepsilon_{ep}\} = [C_{ep}]^{-1}\{d\sigma\} + [I]\left(\{a\}\,dT + \frac{\partial[C_e]^{-1}}{\partial T}\{\sigma\}\,dT + \frac{\partial[C_e]^{-1}}{\partial\bar{\varepsilon}}\{\sigma\}\,d\bar{\varepsilon}\right)$$

$$+ \frac{[C_{ep}]^{-1}[C_e]\{\sigma'\}}{S}\left(\frac{\partial F}{\partial T}\,dT + \frac{\partial F}{\partial\bar{\varepsilon}}\,d\bar{\varepsilon}\right) \qquad (4.17)$$

Substituting (4.14) with $\{\dot{\varepsilon}^c\} = \{d\varepsilon_c\}$ and (4.17) into (4.16) yields the following constitutive relation for the case of thermoelastic–plastic creep deformation:

$$\{d\varepsilon\} = [C_{ep}]^{-1}\{d\sigma\} + [I]\left(\{a\}\,dT + \frac{\partial[C_e]^{-1}}{\partial T}\{\sigma\}\,dT + \frac{\partial[C_e]^{-1}}{\partial \dot{\varepsilon}}\{\sigma\}\,d\dot{\varepsilon}\right)$$

$$+ \frac{[C_{ep}]^{-1}[C_e]\{\sigma'\}}{S}\left(\frac{\partial F}{\partial T}\,dT + \frac{\partial F}{\partial \dot{\varepsilon}}\,d\dot{\varepsilon}\right) + \tfrac{3}{2}(d\bar{\varepsilon}^c/\bar{\sigma})\{\sigma'\} \qquad (4.18)$$

The above equation may be expressed in the usual way of relating the incremental stresses with the incremental strains in a finite, rather than differential, sense for the finite element application:

$$\{\Delta\sigma\} = [C_{ep}]\{\Delta\varepsilon\} - [C_{ep}]\left(\{a\}\,\Delta T + \frac{\partial[C_e]^{-1}}{\partial T}\{\sigma\}\,\Delta T + \frac{\partial[C_e]^{-1}}{\partial \dot{\varepsilon}}\{\sigma\}\,\Delta\dot{\varepsilon}\right.$$

$$\left. + \{\Delta\varepsilon_c\}\right) - \frac{[C_e]\{\sigma'\}}{S}\left(\frac{\partial F}{\partial T}\Delta T + \frac{\partial F}{\partial \dot{\varepsilon}}\Delta\dot{\varepsilon}\right) \qquad (4.19)$$

with $\{\Delta\varepsilon_c\}$ given in (4.15)

The reader is to be cautioned that the determination of H' for the evaluation of the $[C_{ep}]$ matrix by (3.62) and (3.63) should be based on the $\bar{\varepsilon}$ value determined from thermoelastic–plastic strain components excluding creep strain components. Reasons for this exclusion are given after (3.63).

The well-known phenomenon of stress relaxation in a structure during creep deformation is clearly indicated in the above constitutive equation with the inclusion of a negative $\{\Delta\varepsilon_c\}$.

4.4 Finite element formulation of thermoelastic–plastic creep stress analysis

Upon comparing the constitutive equations shown in (4.19) for the thermo-elastic–plastic creep case with those shown in (3.50) for the similar problem but without creep effect, the reader will immediately recognize that the only difference between these two equations is the additional term $\{\Delta\varepsilon_c\}$ in the former equation. Thus, by following the procedure described in Section 3.14, one may readily conclude that the same element equation in (3.100) can be used for the present case with only one minor modification to the incremental thermomechanical load matrix for the creep effect.

The complete element equations have been shown once again to be

$$[K_e]\{\Delta u\} = \{\Delta F\} \qquad (3.100)$$

where
$$[K_e] = \int_v [B(\mathbf{r})]^T [C_{ep}][B(\mathbf{r})] \, dv \tag{3.101a}$$

$\{\Delta F\}$ = incremental thermomechanical load matrix

$$= \{\Delta F_T\} + \{\Delta F_M\} + \{\Delta F_c\} \tag{4.20}$$

$$\{\Delta F_T\} = \int_v [B(\mathbf{r})]^T [C_{ep}]\left(\{a\}\,\Delta T + \frac{\partial [C_e]^{-1}}{\partial T}\{\sigma\}\,\Delta T + \frac{[C_{ep}]^{-1}[C_e]\{\sigma'\}}{S}\frac{\partial F}{\partial T}\,\Delta T\right) dv$$

$$+ \int_v [B(\mathbf{r})]^T [C_{ep}]\left(\frac{\partial [C_e]^{-1}}{\partial \dot\varepsilon}\{\sigma\}\,\Delta\dot\varepsilon + \frac{[C_{ep}]^{-1}[C_e]\{\sigma'\}}{S}\frac{\partial F}{\partial \dot\varepsilon}\,\Delta\dot\varepsilon\right) dv \tag{3.101c}$$

$$\{\Delta F_M\} = \int_v [N(\mathbf{r})]^T\{\Delta f\} \, dv + \int_s [N(\mathbf{r})]^T\{\Delta t\} \, ds \tag{3.101b}$$

The additional incremental load matrix induced by the creep effect can be shown to take the form

$$\{\Delta F_c\} = \int_v [B(\mathbf{r})]^T [C_{ep}]\{\Delta\varepsilon_c\} \, dv \tag{4.21}$$

Comparison of (3.100), (3.101) and (4.20) clearly indicates that the solution of thermoelastic–plastic creep problems differs from that of thermoelastic–plastic analysis only by an additional term called the "pseudo-creep load matrix" in the overall nodal force matrix. All expressions in Section 3.15 for the TEPSA code can be used for the creep effect except the unloading criteria, which will be discussed in a later section.

The inclusion of thermal effects and plasticity in the analysis makes the formulation in (4.20) different from that shown in Greenbaum and Rubinstein[10]. The latter dealt with only elastic creep problems.

4.5 Integration schemes

Since creep is regarded as a time-dependent plasticity, it is in fact a transient problem. The creep strain is accumulated over a small time increment, Δt following nonlinear $\dot\varepsilon$–t relations depicted in Figure 3.4. As in the transient thermal analysis in Section 2.8, a proper time integration scheme is essential in order to achieve both convergence (accuracy) and stability of the solution.

There are several such schemes available for this purpose. Among these schemes are the Euler forward explicit scheme[10], the fourth-order Runge-Kutta formula[11,12] and the Taylor series[13]. The Taylor series scheme has been adopted in the TEPSAC code for its easy implementation and relatively better computational efficiency.

The Taylor series scheme can be outlined as follows:

$$d\varepsilon_c = \dot\varepsilon^c \, dt \Big/ \left(1 - \frac{\Delta t}{2}\frac{\partial \dot\varepsilon^c}{\partial \varepsilon_c}\right) \tag{4.22}$$

119

For materials which obey Norton's law,

$$\dot{\bar{\varepsilon}}^c = K_c \bar{\sigma}^n f(T)$$

The maximum time increment allowed in the computation becomes

$$\Delta t_m = (2/B)K_c f(T)n\bar{\sigma}^{(n-1)} \tag{4.23}$$

where $B = d\bar{\sigma}/d\bar{\varepsilon}_c > 0$.

There is, of course, another limit on the size of time steps which governs the computational stability. A limit which applies to von Mises type visco-plastic material with Euler integration scheme is adopted:

$$\Delta t_c = 4(1 + v)/3EK_c f(T)n\bar{\sigma}^{(n-1)} \tag{4.24}$$

where E, v are the modulus of elasticity and Poisson's ratio.

An automatic time step generation scheme based on the lower value of Δt_m in (4.23) and Δt_c in (4.24) was imposed in the TEPSAC code.

4.6 Solution algorithm

As the finite element formulation with creep effect differs from that of thermoelastic–plastic case by an additional "pseudo-creep load matrix" in (4.21), and the selection of time steps governed by the integration schemes in the foregoing section, the solution algorithm described in Section 3.16 may be adopted here with a few minor modifications as outlined below:

(1) Initialize $\{\varepsilon\}$ and $\{\dot{\varepsilon}\}$ for all elements in the model.
(2) Compute $\{T\}$, $\{u\}$, $\{\varepsilon\}$ and $\{\sigma\}$ due to initial loads (i.e. before creep deformation begins) by the usual thermoelastic analysis.
(3) Set the time increment Δt to be the smaller of Δt_m from the following:

$$\Delta t_m = (2/B)K_c f(T)n\bar{\sigma}^{(n-1)} \tag{4.23}$$

or $\qquad \Delta t_m = 4(1 + v)/3EK_c f(T)n\bar{\sigma}^{(n-1)} \tag{4.24}$

(4) Determine $\{\Delta\varepsilon_c\}$ from $\{\Delta\varepsilon_c\} = \frac{3}{2}(d\bar{\varepsilon}_c/\bar{\sigma})\{\sigma'\}$ as described in (4.15) with $d\bar{\varepsilon}_c$ computed from

$$d\bar{\varepsilon}_c = \dot{\bar{\varepsilon}}^c \Delta t \bigg/ \left(1 - \frac{\Delta t}{2}\frac{\Delta\dot{\bar{\varepsilon}}^c}{\Delta\bar{\varepsilon}_c}\right) \tag{4.22}$$

with the specified $\dot{\bar{\varepsilon}}^c$ and $\bar{\sigma}$ at time t.
(5) Formulate $[K_e]$ from (3.101a),
$\qquad \{\Delta F_M\}$ from (3.101b),
$\qquad \{\Delta F_T\}$ from (3.101c)
$\qquad$ and $\{\Delta F_c\}$ from (4.21).

(6) Assemble the overall stiffness matrix $[K]$ and construct the overall structural equilibrium equations following (3.102):

$$[K]\{\Delta u\} = \{\Delta R\}$$

(7) Modify $\{\Delta R\}$ then $[K]$ for applicable boundary conditions.
(8) Solve for $\{\Delta u\}$ by Gaussian elimination technique and then

$$\{u\} = \{u\} + \{\Delta u\}$$

(9) Compute $\{\Delta \varepsilon\}$ from (3.95).
(10) Compute $\{\Delta \sigma\}$ from (4.19).
(11) Compute $\{\sigma\} = \{\sigma\} + \{\Delta \sigma\}$
$$\{\varepsilon\} = \{\varepsilon\} + \{\Delta \varepsilon\}$$
(12) Update nodal coordinates by letting

$$r = r + \Delta u_r, \qquad z = z + \Delta u_z$$

(13) Repeat Step (3) to find next Δt.
(14) Repeat the same procedure from Step (4).

4.7 Code verification

The formulation and solution algorithm have been incorporated into one version of the TEPSA code, and was renamed as TEPSAC (stands for TEPSA-creep). The accuracy of TEPSAC code has been verified against a case of a thick-walled pressure tube with a flat end enclosure as described in Greenbaum and Rubinstein[10]. An isotropic hardening rule was used in the present example and the effective creep strain rate was assumed to follow the following expression:

$$\dot{\varepsilon}^c = 21.146 \times 10^{-16}\bar{\sigma}^{3.61}$$

where $\dot{\varepsilon}$ = the effective creep strain rate, and $\bar{\sigma}$ = the effective stress. The internal pressure is 445 psi and the Young's modulus of the material is 20×10^6 psi with Poisson's ratio 0.3.

The idealization of the structure for one quarter of the pressure tube is shown in Figure 4.1. A total of 248 elements and 246 nodes were used in the analysis.

Figure 4.2 shows the effective stress contours at time $t = 0$ and the contours for $t = 3$ hours are given in Figure 4.3.

In Figure 4.4 is presented a comparison of the effective stresses at the cylinder-end closure junction for both the inside and the outside surfaces (element 72 and 160). It is apparent that the effective stresses at the outside surface agree well with those given by Greenbaum and Rubinstein[10]. Discrepancies of results, however, are observed at the inside surface. These discrepancies are mainly due to the difference in the finite element models

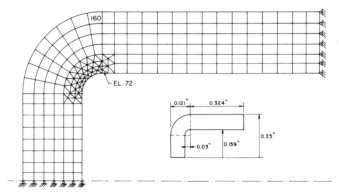

Figure 4.1 Finite element idealization of one-quarter of a pressure tube.

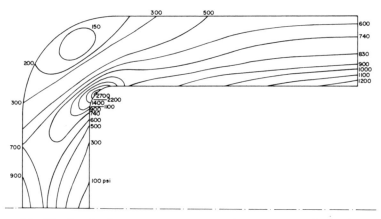

Figure 4.2 Elastic–plastic creep stress contours in a pressure tube at time $t = 0$.

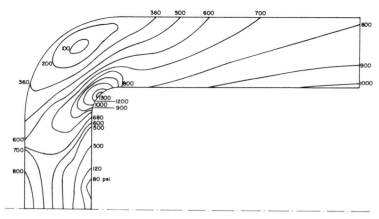

Figure 4.3 Elastic–plastic creep stress contours in a pressure tube at time $t = 3$ hours.

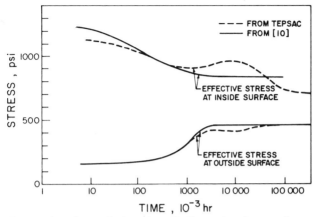

Figure 4.4   Stress relaxation at the junction of a flat-end enclosure of a pressure tube.

used in the analyses. In the present example, the elements in the inner corner are finer than those used by Greenbaum and Rubinstein, which resulted in a higher stress concentration as well as slower convergence to the steady-state creep condition.

4.8   Closing remarks

Creep behavior of many engineering materials in the form of time-dependent deformation under constant static load has been recognized by engineers for over a century, but major efforts in trying to understand its real nature and the development of effective analytical methods did not begin until World War I. A great many publications in both microscopic and macroscopic aspects of creep behavior of common engineering materials have been made available since World War II. However, the majority of the published work has been primarily of an empirical nature. More effort in deriving fundamental theories of creep mechanics is needed.

The formulation of thermoelastic–plastic creep presented in this chapter is largely based on many theories established for the time-invariant plasticity theory, but with little experimental evidence to prove their validity. However, it has the merit of being relatively simple and can be readily adapted to the well-established finite element analysis for thermoelastic–plastic stress analysis, as in the case of TEPSAC code formulation.

Another less developed area, in the author's view, is the lack of unified and rational treatment of creep unloading. Most constitutive laws presented in Section 4.2.1 are valid only for the loading process. No general expression for unloading, such as depicted in Figure 3.4, is available for finite element analysis. Most cyclic creep analyses were formulated on the hypotheses of

123

elastic unloading, as in the case of plasticity theory. The subject of cyclic creep plays a major role in the low cycle fatigue analysis[14] which, in the author's opinion, is one of a number of challenging tasks that confront researchers in solid mechanics.

References

1   Lagneborg, R. 1978. Creep deformation mechanisms. Ch. 1, in *Creep of engineering materials and structures*. G. Bernasconi and G. Piatti (eds.). London: Applied Science Publishers.
2   Mellor, M. 1980. Mechanical properties of polycrystalline ice. In *Physics and mechanics of ice*. P. Tryde (ed.). Springer-Verlag, pp. 217–45.
3   Liu, Y. J. and T.-R. Hsu 1982. *On residual stresses/strains in pressure vessels induced by cyclic thermomechanical loads*. ASME Paper 82-pvp-27.
4   Bailey, R. W. 1935. The utilization of creep test data in engineering design. Advance Paper, *Proc. Inst. Mech. Engng.* 136 pp.
5   Odquist, F. K. G. 1981. Historical survey of the development of creep mechanics from its beginnings in the last century to 1970. In *Creep in structures*. A. R. S. Ponter and D. R. Hayhurst (eds.). Proc. 3rd Symp. Int. Union Theoret. & Appl. Mech., Springer-Verlag. pp. 1–12.
6   Penny, R. K. and D. L. Marriott 1971. *Design for creep*. New York: McGraw-Hill.
7   Norton, F. H. 1929. The creep of steel at high temperature. New York: McGraw-Hill.
8   Dorn, J. E. 1955. Some fundamental experiments on high temperature creep. *J. Mech. & Phys. Solids* **3**, 85–116.
9   Pizzo, P. P. 1979. Rate equations for elevated temperature creep. *J. Engng. Mat. Technol., ASME Trans.* **101**, 396–402.
10 Greenbaum, G. A. and M. F. Rubinstein 1968. Creep analysis of axisymmetric bodies using finite elements. *Nucl. Engng. & Des.* **7**, 379–97.
11 Donea, J. 1978. The application of computer methods to creep analysis, Ch. 12, in *Creep of engineering materials and structures*. G. Bernasconi and G. Piatti (eds.). London: Applied Science Publishers.
12 Donea, J. and S. Giuliani 1973. Creep analysis of transversely isotropic bodies subject to time-dependent loading. *Nucl. Engng. & Des.* **24**, 410–19.
13 Shih, C. F. and H. G. Delorenzi 1977. A stable computational scheme for still time-dependent constitutive equations. *Proc. 4th SMiRT Conf.*, San Francisco, paper No. L2/2.
14 Del Puglia, A. and E. Manfredi 1978. High temperature low-cycle fatigue damage. Ch. 9, in *Creep of engineering materials and structures*. G. Bernasconi and G. Piatti (eds.). London: Applied Science Publishers.

5

ELASTIC–PLASTIC
STRESS ANALYSIS WITH
FOURIER SERIES †

5.1 Introduction

The previous three chapters have presented the reader with the theoretical basis for a combined thermomechanical stress analysis of three-dimensional axisymmetric (r, z, θ) and two-dimensional planar (r, z) structures. The reader will find, after reading Chapter 10, that the TEPSAC code developed on this basis can in fact solve a variety of advanced structural problems. However, all engineering problems are strictly of a three-dimensional nature. Exact solutions therefore require three-dimensional analyses. It is correct to assume that the finite element method, in principle, can handle such analyses much easier than classical methods. However, the cost of solutions increases substantially with each additional dimension in the solution, leaving aside the limitation of computer size. Engineers are thus compelled to idealize the real three-dimensional problems by one- or two-dimensional cases such as described in the foregoing chapters. It is then possible to solve these problems with available resources and facilities.

There are, however, occasions where the geometry of the structure, but not the loading or boundary conditions, can reasonably be assumed to be either planar or axisymmetric. A box bridge structure shown in Figure 5.1a typifies a plane strain with the load variation along the axis normal to the plane. Figure 5.1b illustrates another common problem of a heavy horizontal pressure vessel which has an axisymmetrical geometry but is subject to a significant load variation along the θ-direction due to the gravitational effect. It is not difficult for one to envisage the inadequacy of using conventional two-dimensional analyses, as used in the TEPSA code for the solution of these problems. One alternative approach to the full three-dimensional analysis is to modify the existing two-dimensional solution with a Fourier series approximation for the load variation and the associated structural responses in the third direction.

† Part of this chapter was previously published and reprinted by permission of the Council of the Institution of Mechanical Engineers from *Proceedings of the 4th International Conference on Pressure Vessel Technology*, **2**, May 19–23, 1980 in London.

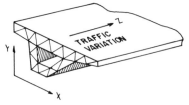

(a) Box-bridge structure, plane strain $(x-y)$ with load varying in z direction

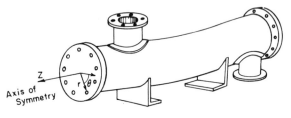

(b) Sag of a long heavy pressure vessel, axisymmetric structure with load varying along θ direction due to gravity

Figure 5.1 Two-dimensional structures with three-dimensional loadings.

The idea of using the Fourier series for the above purpose is not new. Wilson and Nickell used this method for heat conduction analysis[1]. A description of elastic stress analysis of planar and axisymmetric structures is given in Zienkiewicz[2]. In this chapter the extension of this technique to elastic–plastic analysis is presented. Detailed information can be found in Wu[3] and Wu and Hsu[4]. Although the formulations presented here are mostly based on the combination of zeroth and first modes of a Fourier series, no restriction is imposed on the use of higher-order modes. The method presented here, in principle, can be used for highly complicated load variations in the θ direction in axisymmetric structures by a high number of mode mixtures. However, computational effort and machine capability will eventually make such an exercise less economical and less effective.

5.2 Element equation for elastic axisymmetric solids subject to nonaxisymmetric loadings

Consider first an axisymmetric solid as shown in Figure 5.2. The solid is subject to nonaxisymmetric loading components $\{f\}$ for body forces and $\{t\}$ for the surface tractions. Both these loadings may be expanded in a Fourier series of order n as

$$\{f\} = \begin{Bmatrix} f^r \\ f^z \\ f^\theta \end{Bmatrix}_{(r,z,\theta)} = \sum_n \begin{Bmatrix} f_n^r(r,z)\cos n\theta \\ f_n^z(r,z)\cos n\theta \\ f_n^\theta(r,z)\sin n\theta \end{Bmatrix} \tag{5.1a}$$

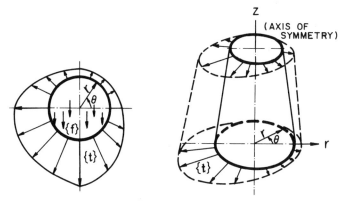

Figure 5.2 Axisymmetric structure subject to nonaxisymmetric load.

$$\{t\} = \begin{Bmatrix} t^r \\ t^z \\ t^\theta \end{Bmatrix}_{(r,z,\theta)} = \sum_n \begin{Bmatrix} t^r_n(r,z)\cos n\theta \\ t^z_n(r,z)\cos n\theta \\ t^\theta_n(r,z)\sin n\theta \end{Bmatrix} \qquad (5.1b)$$

As an example, the surface traction on a heavy pipe subject to a line load q lb/in as shown in Figure 5.3 can be expressed in a Fourier series as follows:

$$\{t\} = \begin{Bmatrix} t^r \\ t^z \\ t^\theta \end{Bmatrix}_{(r,z,\theta)} = \tfrac{1}{2}\begin{Bmatrix} q/(\pi b) \\ q/(\pi b) \\ q/(\pi b) \end{Bmatrix} + \sum_{n=1}^{\infty}(-1)^n\frac{q}{\pi b}\begin{Bmatrix} \cos n\theta \\ \cos n\theta \\ \sin n\theta \end{Bmatrix}$$

Since the loadings are nonaxisymmetric, although the structural geometry is, the displacement solution to this problem is obviously no longer independent of the circumferential coordinate θ, and so the circumferential displacement component U_θ associated with the angular direction θ must be

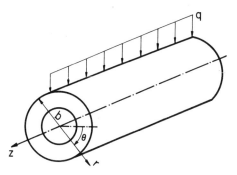

Figure 5.3 A heavy pipe subject to a line pressure load.

included in the computation. The displacement components involved in this case are thus assumed to be:

$$\{U\}_{(r,z,\theta)} = \begin{Bmatrix} U_r \\ U_z \\ U_\theta \end{Bmatrix}_{(r,z,\theta)} = \sum_n \begin{Bmatrix} U_n^r(r,z)\cos n\theta \\ U_n^z(r,z)\cos n\theta \\ U_n^\theta(r,z)\sin n\theta \end{Bmatrix} \tag{5.2}$$

Accordingly, the strain–displacement relationship has to include terms associated with the U_θ components:

$$\{\varepsilon\} = \begin{Bmatrix} \varepsilon_{rr} \\ \varepsilon_{zz} \\ \varepsilon_{\theta\theta} \\ \varepsilon_{rz} \\ \varepsilon_{r\theta} \\ \varepsilon_{z\theta} \end{Bmatrix} = \begin{Bmatrix} \dfrac{\partial U_r}{\partial r} \\[2mm] \dfrac{\partial U_z}{\partial z} \\[2mm] \dfrac{1}{r}\dfrac{\partial U_\theta}{\partial \theta} + \dfrac{U_r}{r} \\[2mm] \dfrac{\partial U_r}{\partial z} + \dfrac{\partial U_z}{\partial r} \\[2mm] \dfrac{1}{r}\dfrac{\partial U_r}{\partial \theta} + \dfrac{\partial U_\theta}{\partial r} - \dfrac{U_\theta}{r} \\[2mm] \dfrac{\partial U_\theta}{\partial z} + \dfrac{1}{r}\dfrac{\partial U_z}{\partial \theta} \end{Bmatrix} \tag{5.3}$$

The substitution of (5.2) into (5.3) leads to the following expression:

$$\{\varepsilon\} = \sum_n \begin{Bmatrix} \varepsilon_n^{rr}(r,z)\cos n\theta \\ \varepsilon_n^{zz}(r,z)\cos n\theta \\ \varepsilon_n^{\theta\theta}(r,z)\cos n\theta \\ \varepsilon_n^{rz}(r,z)\cos n\theta \\ \varepsilon_n^{r\theta}(r,z)\sin n\theta \\ \varepsilon_n^{z\theta}(r,z)\sin n\theta \end{Bmatrix} \tag{5.4}$$

in which

$$
\begin{Bmatrix}
\varepsilon_n^{rr} \\[4pt]
\varepsilon_n^{zz} \\[4pt]
\varepsilon_n^{\theta\theta} \\[4pt]
\varepsilon_n^{rz} \\[4pt]
\varepsilon_n^{r\theta} \\[4pt]
\varepsilon_n^{z\theta}
\end{Bmatrix}_{(r,z)}
= \{\varepsilon_n\} =
\begin{Bmatrix}
\dfrac{\partial U_n^r}{\partial r} \\[10pt]
\dfrac{\partial U_n^z}{\partial z} \\[10pt]
\dfrac{n}{r} U_n^\theta + \dfrac{1}{r} U_n^r \\[10pt]
\dfrac{\partial U_n^r}{\partial z} + \dfrac{\partial U_n^z}{\partial r} \\[10pt]
-\dfrac{n}{r} U_n^r + \dfrac{\partial U_n^\theta}{\partial r} - \dfrac{1}{r} U_n^\theta \\[10pt]
\dfrac{\partial U_n^\theta}{\partial z} - \dfrac{n}{r} U_n^z
\end{Bmatrix}
\qquad (5.5)
$$

The stress–strain relation for an elastic solid has been given in (3.8):

$$
\{\sigma\} = [C_e]\{\varepsilon\}
$$

The following expression applies for the present case:

$$
\{\varepsilon\} = \sum_n [C_e]
\begin{Bmatrix}
\varepsilon_n^{rr} \cos n\theta \\[4pt]
\varepsilon_n^{zz} \cos n\theta \\[4pt]
\varepsilon_n^{\theta\theta} \cos n\theta \\[4pt]
\varepsilon_n^{rz} \cos n\theta \\[4pt]
\varepsilon_n^{r\theta} \sin n\theta \\[4pt]
\varepsilon_n^{z\theta} \sin n\theta
\end{Bmatrix}
\qquad (5.6)
$$

in which the stress component can be expressed as

$$
\{\sigma\} = \sum_n
\begin{Bmatrix}
\sigma_n^{rr}(r, z) \cos n\theta \\[4pt]
\sigma_n^{zz}(r, z) \cos n\theta \\[4pt]
\sigma_n^{\theta\theta}(r, z) \cos n\theta \\[4pt]
\sigma_n^{rz}(r, z) \cos n\theta \\[4pt]
\sigma_n^{r\theta}(r, z) \sin n\theta \\[4pt]
\sigma_n^{z\theta}(r, z) \sin n\theta
\end{Bmatrix}
\qquad (5.7)
$$

with $\{\sigma_n\} = [C_e]\{\varepsilon_n\}$.

The derivation of the element equation first requires an expression of the element potential energy as described in Section 3.14:

$$
\pi = U + W_p = \int_v (\tfrac{1}{2}\{\varepsilon\}^T\{\sigma\} - \{U\}^T\{f\})\, dv - \int_s \{U\}^T\{t\}\, ds
$$

129

By substituting (5.1), (5.2), (5.4) and (5.6) into the above expressions, we obtain

$$
\pi = \int_r \int_z \int_0^{2\pi} \left[\left[\frac{1}{2} \sum_m \left\{ \begin{array}{c} \varepsilon_m^{rr}\cos m\theta \\ \varepsilon_m^{zz}\cos m\theta \\ \varepsilon_m^{\theta\theta}\cos m\theta \\ \varepsilon_m^{rz}\cos m\theta \\ \varepsilon_m^{r\theta}\sin m\theta \\ \varepsilon_m^{z\theta}\sin m\theta \end{array} \right\} \right]^T \left[\sum_n \left\{ \begin{array}{c} \sigma_n^{rr}\cos n\theta \\ \sigma_n^{zz}\cos n\theta \\ \sigma_n^{\theta\theta}\cos n\theta \\ \sigma_n^{rz}\cos n\theta \\ \sigma_n^{r\theta}\sin n\theta \\ \sigma_n^{z\theta}\sin n\theta \end{array} \right\} \right]
$$

$$
- \left(\sum_m \left\{ \begin{array}{c} U_m^r\cos m\theta \\ U_m^z\cos m\theta \\ U_m^\theta\sin m\theta \end{array} \right\} \right)^T \left(\sum_n \left\{ \begin{array}{c} f_n^r\cos n\theta \\ f_n^z\cos n\theta \\ f_n^\theta\sin n\theta \end{array} \right\} \right) \right] d\theta\, r\, dr\, dz
$$

$$
- \int_l \int_0^{2\pi} \left(\sum_m \left\{ \begin{array}{c} U_m^r\cos m\theta \\ U_m^z\cos m\theta \\ U_m^\theta\sin m\theta \end{array} \right\} \right)^T \left(\sum_n \left\{ \begin{array}{c} t_n^r\cos n\theta \\ t_n^z\cos n\theta \\ t_n^\theta\sin n\theta \end{array} \right\} \right) r\, d\theta\, dl
$$

$$
\pi = \int_r \int_z \int_0^{2\pi} \left[\sum_n \sum_m \frac{1}{2} \left\{ \begin{array}{c} \varepsilon_m^{rr}\cos m\theta \\ \varepsilon_m^{zz}\cos m\theta \\ \varepsilon_m^{\theta\theta}\cos m\theta \\ \varepsilon_m^{rz}\cos m\theta \\ \varepsilon_m^{r\theta}\sin m\theta \\ \varepsilon_m^{z\theta}\sin m\theta \end{array} \right\}^T [C_e] \sum_n \left\{ \begin{array}{c} \sigma_n^{rr}\cos n\theta \\ \sigma_n^{zz}\cos n\theta \\ \sigma_n^{\theta\theta}\cos n\theta \\ \sigma_n^{rz}\cos n\theta \\ \sigma_n^{r\theta}\sin n\theta \\ \sigma_n^{z\theta}\sin n\theta \end{array} \right\}
$$

$$
- \sum_m \left\{ \begin{array}{c} U_m^r\cos m\theta \\ U_m^z\cos m\theta \\ U_m^\theta\sin m\theta \end{array} \right\}^T \left(\begin{array}{c} f_n^r\cos n\theta \\ f_n^z\cos n\theta \\ f_n^\theta\sin n\theta \end{array} \right) \right] d\theta\, r\, dr\, dz
$$

$$
- \int_l \int_0^{2\pi} \left(\sum_n \sum_m \left\{ \begin{array}{c} U_m^r\cos m\theta \\ U_m^z\cos m\theta \\ U_m^\theta\sin m\theta \end{array} \right\}^T \left(\begin{array}{c} t_n^r\cos n\theta \\ t_n^z\cos n\theta \\ t_n^\theta\sin n\theta \end{array} \right) \right) r\, d\theta\, dl \qquad (5.8)
$$

The following expressions were used in the above derivations:

$$dv = r \, dr \, dz \, d\theta$$
$$ds = r \, dl \, d\theta$$

Making use of the following orthogonality properties of the trigonometric functions:

$$\int_0^{2\pi} (\cos m\theta)(\cos n\theta) \, d\theta = \begin{cases} 2\pi & m = n = 0 \\ \pi & m = n \neq 0 \\ 0 & m \neq n \end{cases}$$

$$\int_0^{2\pi} (\sin m\theta)(\sin n\theta) \, d\theta = \begin{cases} \pi & m = n \neq 0 \\ 0 & m \neq n \text{ and } m = n = 0 \end{cases}$$

and

$$\int_0^{2\pi} (\cos m\theta)(\sin n\theta) \, d\theta = 0 \qquad \text{for all } m \text{ and } n$$

a simplified expression of π in (5.8) can be obtained:

$$\pi = \eta \left(\sum_n \int_r \int_z (\tfrac{1}{2}\{\varepsilon_n\}^{\mathrm{T}}[C_e]\{\varepsilon_n\} - \{U_n\}^{\mathrm{T}}\{f_n\}) \right) r \, dr \, dz$$

$$- \int_l \{U_n\}^{\mathrm{T}}\{t_n\} r \, dl = \eta \sum_n \pi_n \tag{5.9}$$

in which

$$\eta = \begin{cases} 2\pi & n = 0 \\ \pi & n \neq 0 \end{cases}$$

and

$$\pi_n = \int_r \int_z (\tfrac{1}{2}\{\varepsilon_n\}^{\mathrm{T}}[C_e]\{\varepsilon_n\} - \{U_n\}^{\mathrm{T}}\{f_n\}) r \, dr \, dz - \int_l \{U_n\}^{\mathrm{T}}\{t_n\} r \, dl$$

It should be noted that the simplification in (5.9) is possible only if $\{\varepsilon_n\}$, $\{U_n\}$, $\{f_n\}$, $\{t_n\}$ and $[C_e]$ are independent of the θ-coordinate.

Physically, (5.9) can be interpreted as the decoupling action of a mixed Fourier mode problem. The sum of the potential energy is the sum of N subfunctional π_n. The same minimization process as illustrated in Section 3.14 can be used to arrive at the following set of N stiffness equations:

$$[K_n]\{u_n\} = \{F_n\} \tag{5.10}$$

with $n = 1, 2, \ldots, N$.

Each of the equations in (5.10) may be solved to obtain the displacement solution of each individual mode n. The element displacement components can be computed by summing up all $\{U_n\}$ following (5.2),

131

5.3 Stiffness matrix for elastic solids subject to nonaxisymmetric loadings

The derivation of the $[K_n]$ matrix in (5.10) follows a similar procedure to that described in Section 3.15 for the case of axisymmetric loadings. Only a summary of the equations is given here.

Assuming that the element displacement components $\{U_n\}$ for the mode n are given by

$$\{U_n\}_{(r,z)} = \begin{Bmatrix} U_n^r \\ U_n^z \\ U_n^\theta \end{Bmatrix} = \begin{bmatrix} 1 & r & z & 0 & 0 & 0 & 0 & 0 & 0 \\ 0 & 0 & 0 & 1 & r & z & 0 & 0 & 0 \\ 0 & 0 & 0 & 0 & 0 & 0 & 1 & r & z \end{bmatrix} \begin{Bmatrix} b_{1n} \\ b_{2n} \\ b_{3n} \\ b_{4n} \\ b_{5n} \\ b_{6n} \\ b_{7n} \\ b_{8n} \\ b_{9n} \end{Bmatrix} = [R]\{b_n\}$$

(5.11)

As for (3.104), the nodal displacement components may be related to $\{b_n\}$ by the nodal coordinates:

$$\begin{Bmatrix} u_{nI}^r & u_{nI}^z & u_{nI}^\theta \\ u_{nJ}^r & u_{nJ}^z & u_{nJ}^\theta \\ u_{nK}^r & u_{nK}^z & u_{nK}^\theta \end{Bmatrix} = \begin{bmatrix} 1 & r_i & z_i \\ 1 & r_j & z_j \\ 1 & r_k & z_k \end{bmatrix} \begin{Bmatrix} b_{1n} & b_{4n} & b_{7n} \\ b_{2n} & b_{5n} & b_{8n} \\ b_{3n} & b_{6n} & b_{9n} \end{Bmatrix}$$

(5.12a)

or

$$\{U_n\} = [A]\{b_n\}$$

(5.12b)

and

$$\{b_n\} = [A]^{-1}\{U_n\} = [h]\{U_n\}$$

(5.13)

132

where

$$[h] = \frac{1}{\beta}\begin{bmatrix} r_j z_k - r_k z_j & 0 & 0 & r_k z_i - r_i z_k & 0 & 0 & r_i z_j - r_j z_i & 0 & 0 \\ z_j - z_k & 0 & 0 & z_k - z_i & 0 & 0 & z_i - z_j & 0 & 0 \\ r_k - r_j & 0 & 0 & r_i - r_k & 0 & 0 & r_j - r_i & 0 & 0 \\ 0 & r_j z_k - r_k z_j & 0 & 0 & r_k z_i - r_i z_k & 0 & 0 & r_i z_j - r_j z_i & 0 \\ 0 & z_j - z_k & 0 & 0 & z_k - z_i & 0 & 0 & z_i - z_j & 0 \\ 0 & r_k - r_j & 0 & 0 & r_i - r_k & 0 & 0 & r_j - r_i & 0 \\ 0 & 0 & r_j z_k - r_k z_j & 0 & 0 & r_k z_i - r_i z_k & 0 & 0 & r_i z_j - r_j z_i \\ 0 & 0 & z_j - z_k & 0 & 0 & z_k - z_i & 0 & 0 & z_i - z_j \\ 0 & 0 & r_k - r_j & 0 & 0 & r_i - r_k & 0 & 0 & r_j - r_i \end{bmatrix}$$

$$(5.14)$$

in which $\beta = r_i(z_j - z_k) + r_j(z_k - z_i) + r_k(z_i - z_j)$.

The interpolation function $[N]$ following (3.107) is

$$[N] = [R(r, z)][h]$$

By substituting (5.11) into (5.4), we obtain the strain–displacement relation:

$$\{\varepsilon_n\} = \begin{Bmatrix} \varepsilon_n^{rr} \\ \varepsilon_n^{zz} \\ \varepsilon_n^{\theta\theta} \\ \varepsilon_n^{rz} \\ \varepsilon_n^{r\theta} \\ \varepsilon_n^{z\theta} \end{Bmatrix} = \begin{bmatrix} 0 & 1 & 0 & 0 & 0 & 0 & 0 & 0 & 0 \\ 0 & 0 & 0 & 0 & 0 & 1 & 0 & 0 & 0 \\ 1/r & 1 & z/r & 0 & 0 & 0 & n/r & n & nz/r \\ 0 & 0 & 1 & 0 & 1 & 0 & 0 & 0 & 0 \\ -n/r & -n & -nz/r & 0 & 0 & 0 & -1/r & 0 & -z/r \\ 0 & 0 & 0 & -n/r & -n & -nz/r & 0 & 0 & 1 \end{bmatrix} \begin{Bmatrix} b_{1n} \\ b_{2n} \\ b_{3n} \\ b_{4n} \\ b_{5n} \\ b_{6n} \\ b_{7n} \\ b_{8n} \\ b_{9n} \end{Bmatrix}$$

133

or

$$\{\varepsilon_n\} = [G_n(r, z)]\{b_n\}$$

The matrix $[B]$ may now be formulated by substituting (5.13) into the above expression, resulting in

$$[B_n] = [G_n(r, z)][h] \tag{5.15}$$

The stiffness matrix $[K_n]$ can thus be formulated analogously to (3.115):

$$\begin{aligned}
[K_n] &= \int_v [B_n]^T [C_e][B_n] \, dv \\
&= \int_v ([G_n(r, z)][h])^T [C_e]([G_n(r, z)][h]) \, dv \tag{5.16a} \\
&= [h]^T \left(\int_v [G_n(r, z)]^T [C_e][G_n(r, z)] \, dv \right)[h]
\end{aligned}$$

The integrand in (5.16), after expansion, takes the following form:

$$[G_n(r, z)]^T [C_e][G_n(r, z)] = \frac{E}{(1 + v)(1 - 2v)} \times$$

$$
\begin{bmatrix}
\frac{n^2(1 - 2v) + 2(1 - v)}{2r^2} & \frac{n^2(1 - 2v) + 2}{2r} & \frac{z(n^2(1 - 2v) + 2(1 - v))}{2r^2} & 0 & 0 \\[2ex]
& 2 + \frac{n^2(1 - 2v)}{2} & \frac{z(n^2(1 - 2v) + 2)}{2r} & 0 & 0 \\[2ex]
& & \frac{z^2(n^2(1 - 2v) + 2(1 - v))}{2r^2} + \frac{(1 - 2v)}{2} & 0 & \frac{(1 - 2v)}{2} \\[2ex]
& & & \frac{n^2(1 - 2v)}{2r^2} & \frac{n^2(1 - 2v)}{2r} \\[2ex]
& & & & \frac{(n^2 + 1)(1 - 2v)}{2} \\
\end{bmatrix}
$$

SYM

134

$$
\begin{array}{cccc}
\dfrac{v}{r} & \dfrac{n(3-4v)}{2r^2} & \dfrac{n(1-v)}{r} & \dfrac{nz(3-4v)}{2r^2} \\[2ex]
2v & \dfrac{n(3-2v)}{2r} & n & \dfrac{nz(3-2v)}{2r} \\[2ex]
\dfrac{vz}{r} & \dfrac{nz(3-4v)}{2r^2} & \dfrac{nz(1-v)}{r} & \dfrac{nz^2(3-4v)}{2r^2} \\[3ex]
\dfrac{n^2z(1-2v)}{2r^2} & 0 & 0 & \dfrac{-n(1-2v)}{2r} \\[2ex]
\dfrac{n^2z(1-2v)}{2r} & 0 & 0 & \dfrac{-n(1-2v)}{2} \\[2ex]
\dfrac{n^2z^2(1-2v)}{2r^2}+(1-v) & \dfrac{nv}{r} & nv & \dfrac{nz(4v-1)}{2r} \\[3ex]
& \dfrac{2n^2(1-v)+(1-2v)}{2r^2} & \dfrac{n^2(1-v)}{r} & \dfrac{z(2n^2(1-v)+(1-2v))}{2r^2} \\[2ex]
& & n^2(1-v) & \dfrac{n^2z(1-v)}{r} \\[2ex]
& & & \dfrac{z^2(2n^2(1-v)+(1-2v))}{2r^2}+\dfrac{(1-2v)}{2}
\end{array}
$$

(5.16b)

5.4 Elastic–plastic stress analysis of axisymmetric solids subject to nonaxisymmetric loadings

As described in Section 5.2, the incremental body forces, surface tractions and displacements can be expressed in terms of the Fourier series:

$$
\{\Delta f\} = \begin{Bmatrix} \Delta f_r \\ \Delta f_z \\ \Delta f_\theta \end{Bmatrix}_{(r,z,\theta)} = \sum_n \begin{Bmatrix} \Delta f_n^r(r,z)\cos n\theta \\ \Delta f_n^z(r,z)\cos n\theta \\ \Delta f_n^\theta(r,z)\sin n\theta \end{Bmatrix} \tag{5.17a}
$$

$$
\{\Delta t\} = \begin{Bmatrix} \Delta t_r \\ \Delta t_z \\ \Delta t_\theta \end{Bmatrix}_{(r,z,\theta)} = \sum_n \begin{Bmatrix} \Delta t_n^r(r,z)\cos n\theta \\ \Delta t_n^z(r,z)\cos n\theta \\ \Delta t_n^\theta(r,z)\sin n\theta \end{Bmatrix} \tag{5.17b}
$$

$$
\{\Delta U\} = \begin{Bmatrix} \Delta U_r \\ \Delta U_z \\ \Delta U_\theta \end{Bmatrix}_{(r,z,\theta)} = \sum_n \begin{Bmatrix} \Delta U_n^r(r,z)\cos n\theta \\ \Delta U_n^z(r,z)\cos n\theta \\ \Delta U_n^\theta(r,z)\sin n\theta \end{Bmatrix} \tag{5.18}
$$

135

Accordingly, the strain–displacement relationship corresponding to (5.3) and (5.4) can be obtained:

$$\{\Delta\varepsilon\} = \begin{Bmatrix} \Delta\varepsilon^{rr} \\ \Delta\varepsilon^{zz} \\ \Delta\varepsilon^{\theta\theta} \\ \Delta\varepsilon^{rz} \\ \Delta\varepsilon^{r\theta} \\ \Delta\varepsilon^{z\theta} \end{Bmatrix} = \sum_n \begin{Bmatrix} \Delta\varepsilon_n^{rr}\cos n\theta \\ \Delta\varepsilon_n^{zz}\cos n\theta \\ \Delta\varepsilon_n^{\theta\theta}\cos n\theta \\ \Delta\varepsilon_n^{rz}\cos n\theta \\ \Delta\varepsilon_n^{r\theta}\sin n\theta \\ \Delta\varepsilon_n^{z\theta}\sin n\theta \end{Bmatrix} \tag{5.19a}$$

$$\{\Delta\varepsilon_n\} = \begin{Bmatrix} \Delta\varepsilon_n^{rr} \\[2ex] \Delta\varepsilon_n^{zz} \\[2ex] \Delta\varepsilon_n^{\theta\theta} \\[2ex] \Delta\varepsilon_n^{rz} \\[2ex] \Delta\varepsilon_n^{r\theta} \\[2ex] \Delta\varepsilon_n^{z\theta} \end{Bmatrix} = \begin{Bmatrix} \dfrac{\partial(\Delta u_n^r)}{\partial r} \\[2ex] \dfrac{\partial(\Delta u_n^z)}{\partial z} \\[2ex] \dfrac{\partial(\Delta u_n^\theta)}{r\,\partial\theta} + \dfrac{(\Delta u_n^r)}{r} \\[2ex] \dfrac{\partial(\Delta u_n^r)}{\partial z} + \dfrac{\partial(\Delta u_n^z)}{\partial r} \\[2ex] \dfrac{\partial(\Delta u_n^r)}{r\,\partial\theta} + \dfrac{\partial(\Delta u_n^\theta)}{\partial r} - \dfrac{(\Delta u_n^\theta)}{r} \\[2ex] \dfrac{\partial(\Delta u_n^\theta)}{\partial z} + \dfrac{\partial(\Delta u_n^z)}{r\,\partial\theta} \end{Bmatrix} \tag{5.19b}$$

The stress–strain relationship, with no temperature and strain-rate-dependent material properties, can be expressed following (3.50):

$$\{\Delta\sigma\} = \begin{Bmatrix} \Delta\sigma_{rr} \\ \Delta\sigma_{zz} \\ \Delta\sigma_{\theta\theta} \\ \Delta\sigma_{rz} \\ \Delta\sigma_{r\theta} \\ \Delta\sigma_{z\theta} \end{Bmatrix} = [C_{ep}]\{\Delta\varepsilon\} \tag{5.20}$$

$$= [C_{ep}] \sum_n \begin{Bmatrix} \Delta\varepsilon_n^{rr}\cos n\theta \\ \Delta\varepsilon_n^{zz}\cos n\theta \\ \Delta\varepsilon_n^{\theta\theta}\cos n\theta \\ \Delta\varepsilon_n^{rz}\cos n\theta \\ \Delta\varepsilon_n^{r\theta}\sin n\theta \\ \Delta\varepsilon_n^{z\theta}\sin n\theta \end{Bmatrix} = \sum_n [C_{ep}] \begin{Bmatrix} \Delta\varepsilon_n^{rr}\cos n\theta \\ \Delta\varepsilon_n^{zz}\cos n\theta \\ \Delta\varepsilon_n^{\theta\theta}\cos n\theta \\ \Delta\varepsilon_n^{rz}\cos n\theta \\ \Delta\varepsilon_n^{r\theta}\sin n\theta \\ \Delta\varepsilon_n^{z\theta}\sin n\theta \end{Bmatrix}$$

in which $[C_{ep}]$ is in a form with parameters of material properties $[C_e]$, H', and state of stresses $\bar{\sigma}$, $\{\sigma'\}$ as shown in (3.48).

5.5 Derivation of element equation

An assumption has to be made, as in Chapter 3, that there exists a functional which is stationary during an incremental load of the solid undergoing plastic deformation. The functional in this case is identical to the potential energy of the solid:

$$\pi = \int_v (\tfrac{1}{2}\{\Delta\varepsilon\}^T\{\Delta\sigma\} - \{\Delta U\}^T\{\Delta f\})\, dv - \int_s \{\Delta U\}^T\{\Delta t\}\, ds$$

and $\partial\pi/\partial\{\Delta U\} = 0$.

Thus by substituting (5.17)–(5.20) into the above expression, we derive the relation

$$\pi = \int_r \int_z \int_0^{2\pi} \sum_m \sum_n \left[\tfrac{1}{2} \begin{Bmatrix} \Delta\varepsilon_m^{rr}\cos m\theta \\ \Delta\varepsilon_m^{zz}\cos m\theta \\ \Delta\varepsilon_m^{\theta\theta}\cos m\theta \\ \Delta\varepsilon_m^{rz}\cos m\theta \\ \Delta\varepsilon_m^{r\theta}\sin m\theta \\ \Delta\varepsilon_m^{z\theta}\sin m\theta \end{Bmatrix}^T [C_{ep}] \begin{Bmatrix} \Delta\varepsilon_n^{rr}\cos n\theta \\ \Delta\varepsilon_n^{zz}\cos n\theta \\ \Delta\varepsilon_n^{\theta\theta}\cos n\theta \\ \Delta\varepsilon_n^{rz}\cos n\theta \\ \Delta\varepsilon_n^{r\theta}\sin n\theta \\ \Delta\varepsilon_n^{z\theta}\sin n\theta \end{Bmatrix} \right.$$

$$\left. - \begin{Bmatrix} \Delta U_m^r\cos m\theta \\ \Delta U_m^z\cos m\theta \\ \Delta U_m^\theta\sin m\theta \end{Bmatrix}^T \begin{Bmatrix} \Delta f_n^r\cos n\theta \\ \Delta f_n^z\cos n\theta \\ \Delta f_n^\theta\sin n\theta \end{Bmatrix} \right] d\theta\, r\, dr\, dz$$

$$- \int_l \int_0^{2\pi} \left(\sum_m \sum_n \begin{Bmatrix} \Delta U_m^r\cos m\theta \\ \Delta U_m^z\cos m\theta \\ \Delta U_m^\theta\sin m\theta \end{Bmatrix}^T \begin{Bmatrix} \Delta t_n^r\cos n\theta \\ \Delta t_n^z\cos n\theta \\ \Delta t_n^\theta\sin n\theta \end{Bmatrix} \right) r\, d\theta\, dl \qquad (5.21)$$

It should be noted that the elastoplastic matrix $[C_{ep}]$ here is dependent of the circumferential coordinate, θ, as the $[C_p]$ matrix involves entries in the form of products of deviatoric stresses $\{\sigma'\}$ and the coefficient of the effective stress $\bar{\sigma}$. All these quantities vary along the circumferential coordinate. The orthogonality characteristics of the trigonometric functions which made it possible to simplify (5.8) in the elastic case into a sum of subfunctionals of separate modes cannot be used here for the elastoplastic case. Separation of various Fourier modes in the formulation has thus become a serious problem in elastoplastic analysis.

5.6 Mode mixing stiffness equations

Due to the complexity of the mode mixing character of the problem described above, the analysis will be presented with the combination of only the zeroth and first mode of the Fourier series. The type of problem which matches this description includes the case of a horizontally mounted heavy pressure vessel as illustrated in Figure 5.1b.

The incremental loading components for this type of combination can be expressed as

$$\{\Delta f\} = \{\Delta f_0\} + \begin{Bmatrix} \Delta f_1^r \cos\theta \\ \Delta f_1^z \cos\theta \\ \Delta f_1^\theta \sin\theta \end{Bmatrix} = \begin{Bmatrix} \Delta f_0^r \\ \Delta f_0^z \\ \Delta f_0^\theta \end{Bmatrix} + \begin{Bmatrix} \Delta f_1^r \cos\theta \\ \Delta f_1^z \cos\theta \\ \Delta f_1^\theta \sin\theta \end{Bmatrix} \quad (5.22a)$$

and

$$\{\Delta t\} = \{\Delta t_0\} + \begin{Bmatrix} \Delta t_1^r \cos\theta \\ \Delta t_1^z \cos\theta \\ \Delta t_1^\theta \sin\theta \end{Bmatrix} = \begin{Bmatrix} \Delta t_0^r \\ \Delta t_0^z \\ \Delta t_0^\theta \end{Bmatrix} + \begin{Bmatrix} \Delta t_1^r \cos\theta \\ \Delta t_1^z \cos\theta \\ \Delta t_1^\theta \sin\theta \end{Bmatrix} \quad (5.22b)$$

in which $\{\Delta f_0\}$, $\{\Delta t_0\}$ and $\{\Delta f_1\}$, $\{\Delta t_1\}$ are the corresponding incremental body forces and surface tractions of mode 0 and mode 1.

The respective incremental element displacement components may be assumed to be

$$\{\Delta U\} = \{\Delta U_0\} + \begin{Bmatrix} \Delta U_1^r \cos\theta \\ \Delta U_1^z \cos\theta \\ \Delta U_1^\theta \sin\theta \end{Bmatrix} = \begin{Bmatrix} \Delta U_0^r \\ \Delta U_0^z \\ \Delta U_0^\theta \end{Bmatrix} + \begin{Bmatrix} \Delta U_1^r \cos\theta \\ \Delta U_1^z \cos\theta \\ \Delta U_1^\theta \sin\theta \end{Bmatrix} \quad (5.23)$$

The nodal displacement components are expressed as

$$\begin{Bmatrix} \{\Delta u_0\} \\ \{\Delta u_1\} \end{Bmatrix} = \begin{bmatrix} 1 & r & z & 0 & 0 & 0 & 0 & 0 & 0 & 0 & 0 & 0 & 0 & 0 & 0 & 0 & 0 & 0 \\ 0 & 0 & 0 & 1 & r & z & 0 & 0 & 0 & 0 & 0 & 0 & 0 & 0 & 0 & 0 & 0 & 0 \\ 0 & 0 & 0 & 0 & 0 & 0 & 1 & r & z & 0 & 0 & 0 & 0 & 0 & 0 & 0 & 0 & 0 \\ 0 & 0 & 0 & 0 & 0 & 0 & 0 & 0 & 0 & 1 & r & z & 0 & 0 & 0 & 0 & 0 & 0 \\ 0 & 0 & 0 & 0 & 0 & 0 & 0 & 0 & 0 & 0 & 0 & 0 & 1 & r & z & 0 & 0 & 0 \\ 0 & 0 & 0 & 0 & 0 & 0 & 0 & 0 & 0 & 0 & 0 & 0 & 0 & 0 & 0 & 1 & r & z \end{bmatrix} \begin{Bmatrix} b_1 \\ b_2 \\ \vdots \\ b_9 \\ b_1' \\ \vdots \\ b_9' \end{Bmatrix}$$

or

$$\{\Delta u\} = [R(r, z)]\{b\} \quad (5.24)$$

It should be noted that the matrix $\{b\}$ is now an 18×1 column vector in this case.

Similar to the derivation for the elastic element equations, the nodal displacement components can also be expressed as

$$\{\Delta u\} = [A]\{b\}$$

Thus, the matrix

$$\{b\} = [A]^{-1}\{\Delta u\} = [h]\{\Delta u\}$$

and

$$\{\Delta u\} = [R(r, z)]\{b\} = [R(r, z)][h]\{\Delta u\} = [N(r, z)]\{\Delta u\} \qquad (5.25)$$

which leads to the following strain-displacement relation:

$$\{\Delta\varepsilon\} = \left\{\begin{array}{c} \Delta\varepsilon^{rr} \\ \Delta\varepsilon^{zz} \\ \Delta\varepsilon^{\theta\theta} \\ \Delta\varepsilon^{rz} \\ \Delta\varepsilon^{r\theta} \\ \Delta\varepsilon^{z\theta} \end{array}\right\} = \left[\begin{array}{cccccccccccc}
0 & 1 & 0 & 0 & 0 & 0 & 0 & 0 & 0 & 0 & \cos\theta & {}^{135}0 \\
0 & 0 & 0 & 0 & 0 & 1 & 0 & 0 & 0 & 0 & 0 & 0 \\
1/r & 1 & z/r & 0 & 0 & 0 & 0 & 0 & \dfrac{\cos\theta}{r} & \cos\theta & \dfrac{z\cos\theta}{r} \\
0 & 0 & 1 & 0 & 1 & 0 & 0 & 0 & 0 & 0 & \cos\theta \\
0 & 0 & 0 & 0 & 0 & 0 & 0 & 0 & \dfrac{-\sin\theta}{r} & -\sin\theta & \dfrac{-z\sin\theta}{r} \\
0 & 0 & 0 & 0 & 0 & 0 & 0 & 0 & 0 & 0 & 0
\end{array}\right.$$

$$\left.\begin{array}{cccccc}
0 & 0 & 0 & 0 & 0 & 0 \\
0 & 0 & \cos\theta & 0 & 0 & 0 \\
0 & 0 & 0 & \dfrac{\cos\theta}{r} & \cos\theta & \dfrac{z\cos\theta}{r} \\
0 & \cos\theta & 0 & 0 & 0 & 0 \\
0 & 0 & 0 & \dfrac{-\sin\theta}{r} & 0 & \dfrac{-z\sin\theta}{r} \\
\dfrac{-\sin\theta}{r} & -\sin\theta & \dfrac{-z\sin\theta}{r} & 0 & 0 & \sin\theta
\end{array}\right] \left\{\begin{array}{c} b_1 \\ b_2 \\ \vdots \\ b_9 \\ b'_1 \\ \vdots \\ b'_9 \end{array}\right\}$$

$$= [G(r, z)]\{b\}$$

$$= [G(r, z)][h]\{\Delta u\} = [B(r, z)]\{\Delta u\} \qquad (5.25a)$$

By following a similar procedure to that outlined in Section 3.14, the element stiffness matrix is given by the integral

$$[K] = \int_z \int_r \int_0^{2\pi} [B(r, z)]^T [C_{ep}(E, v, \sigma(r, z, \theta))][B(r, z)] r \, d\theta \, dr \, dz \quad (5.26)$$

The integrand in the above integral is an 18×18 matrix and its elements are given in Appendix 3 with the following definitions.

(1) $A = \text{COS}\,(\theta)$, $B = \text{SIN}\,(\theta)$
(2) $XI(1) = 1.$
  $XI(2) = 1/r$
  $XI(3) = 1/r^2$
  $XI(4) = z/r$
  $XI(5) = z/r^2$
  $XI(6) = z^2/r^2$
(3) $DS(I, J)$ for $I, J = 1, 2, 3, 4, 5, 6$ are the elastic-plastic matrix $[C_{ep}]$ as shown below:

$$[C_{ep}] = 2G$$

$$
\begin{bmatrix}
\dfrac{1-v}{1-2v} - \dfrac{\sigma_{rr}'^2}{S_0} & & & & & \\[2ex]
\dfrac{v}{1-2v} - \dfrac{\sigma_{rr}'\sigma_{zz}'}{S_0} & \dfrac{1-v}{1-2v} - \dfrac{\sigma_{zz}'^2}{S_0} & & & & \text{SYM} \\[2ex]
\dfrac{v}{1-2v} - \dfrac{\sigma_{rr}'\sigma_{\theta\theta}'}{S_0} & \dfrac{v}{1-2v} - \dfrac{\sigma_{zz}'\sigma_{\theta\theta}'}{S_0} & \dfrac{1-v}{1-2v} - \dfrac{\sigma_{\theta\theta}'^2}{S_0} & & & \\[2ex]
- \dfrac{\sigma_{rr}'\sigma_{z\theta}'}{S_0} & - \dfrac{\sigma_{zz}'\sigma_{z\theta}'}{S_0} & - \dfrac{\sigma_{\theta\theta}'\sigma_{z\theta}'}{S_0} & \dfrac{1}{2} - \dfrac{\sigma_{z\theta}'^2}{S_0} & & \\[2ex]
- \dfrac{\sigma_{rr}'\sigma_{\theta r}'}{S_0} & - \dfrac{\sigma_{zz}'\sigma_{\theta r}'}{S_0} & - \dfrac{\sigma_{\theta\theta}'\sigma_{\theta r}'}{S_0} & - \dfrac{\sigma_{z\theta}'\sigma_{\theta r}'}{S_0} & \dfrac{1}{2} - \dfrac{\sigma_{\theta r}'^2}{S_0} & \\[2ex]
- \dfrac{\sigma_{rr}'\sigma_{rz}'}{S_0} & - \dfrac{\sigma_{zz}'\sigma_{rz}'}{S_0} & - \dfrac{\sigma_{\theta\theta}'\sigma_{rz}'}{S_0} & - \dfrac{\sigma_{z\theta}'\sigma_{rz}'}{S_0} & - \dfrac{\sigma_{\theta r}'\sigma_{rz}'}{S_0} & \dfrac{1}{2} - \dfrac{\sigma_{rz}'^2}{S_0}
\end{bmatrix}
$$

with

$$S_0 = \tfrac{2}{3}\bar{\sigma}^2\left(1 + \frac{H'}{3G}\right)$$

The mixed mode element can thus be expressed in the form

$$\sum_{m=0}^{M} [K^{n,m}]\{\Delta u^m\} = \{\Delta F^n\} \quad (5.27)$$

140

where $m, n = 0, 1, 2, 3, \ldots, M$ are mode numbers of the Fourier expansion; $[K^{n,m}]$ is the element stiffness matrix as shown in (5.26), or in a more general form in:

$$[K^{n,m}] = \int_v [B^n(r, z)]^T [\Omega^n(\theta)]^T [C_{ep}(r, z)][\Omega^m(\theta)][B^m(r, z)]\, dv \quad (5.28)$$

$\{\Delta F^n\}$ is the nodal force matrix, or

$$\{\Delta F^n\} = \int_v [N^n(r, z)]^T \{\Delta f^n\}\, dv + \int_s [N^n(r, z)]^T \{\Delta t^n\}\, ds \quad (5.29)$$

and

$$[\Omega^n(\theta)] = \begin{bmatrix} \cos n\theta & 0 & 0 & 0 & 0 & 0 \\ & \cos n\theta & 0 & 0 & 0 & 0 \\ & & \cos n\theta & 0 & 0 & 0 \\ & & & \cos n\theta & 0 & 0 \\ & \text{SYM} & & & \sin n\theta & 0 \\ & & & & & \sin n\theta \end{bmatrix}$$

$$(5.30)$$

The overall stiffness equations for the entire system can be obtained by summing up all the element stiffness and nodal force matrices shown in (5.28) and (5.29) for all elements following normal finite element analysis procedure as outlined in Section 3.14.

The stiffness matrix in (5.28) for mode number up to M can be expressed as

$$[K^M] = \int_v [W^M(\theta)]^T [B(r, z)]^T [C_{ep}(r, z, \theta)][B(r, z)][W^M(\theta)]\, dv \quad (5.31)$$

in which

$$[W^M(\theta)] = ([\bar\Omega^0(\theta)], [\bar\Omega^1(\theta)], [\bar\Omega^2(\theta)], \ldots, [\bar\Omega^M(\theta)])[I]$$

and $[I]$ is the $6(M + 1) \times 6(M + 1)$ identity matrix.

For problems involving a large number of Fourier modes, the following decoupling scheme may be used as an approximation only.

By rearranging (5.27) into the form

$$[K^{n,n}]\{\Delta u^n\} = \{\Delta F^n\} - \sum_{\substack{m=0 \\ m \neq n}}^{M} [K^{n,m}]\{\Delta u^m\}$$

we may solve the stiffness equations for each Fourier mode by using the values of $\{\Delta u^m\}$ and $[K^{n,m}]$ obtained from the preceding loading step. Mathematically it can be shown as

$$[K^{n,n}]_{i-1}\{\Delta u^n\}_i = \{\Delta F^n\}_i - \sum_{\substack{m=0 \\ m \neq n}}^{M} [K^{n,n}]_{i-1}\{\Delta u^m\}_{i-1} \qquad (5.32)$$

where the subscripts i and $(i-1)$ denote the sequence of loading steps.

The accuracy of the approximate solution of (5.27) on using (5.32) obviously depends on the size of the assigned load increments. Substantial saving of computational effort is to be expected for this method over the exact integration scheme given in (5.31), which may be achieved with a smaller number of modes involved.

5.7 Circumferential integration scheme

Integration of the stiffness matrix given in either of (5.26), (5.27) or (5.28) in the circumferential direction presents some difficulty in classical methods. Gaussian quadrature has been recommended for this purpose. A detailed description of this integration scheme can be found in Wilson and Nickell[1]. Proper selection of the total number of Gauss points, of course, is a compromise of accuracy and computing effort. This scheme works on the following mathematical form:

$$[K] = \int_0^{2\pi} \left(\int_r \int_z [B]^T [C_{ep}][B] r \, dr \, dz \right) d\theta$$

$$= \sum_{k=1}^{N} \left[A_k \int_r \int_z ([B]^T [C_{ep}][B])(\theta_k) r \, dr \, dz \right] \qquad (5.33)$$

in which A_k are the weighting factors of Gaussian quadrature, N is the total number of Gaussian points evaluated, and θ_k are the individual Gauss points along the circumferential direction. The integrand $[B]^T [C_{ep}][B]$ is an 18×18 matrix which has to be formulated term by term before the integration. These terms have been tabulated in Appendix 3.

It is clear that the integration scheme adopted here can be regarded as an approximation process only, not just from the mathematical point of view, but also from the material behavior point of view. Nonlinear properties such as H' and stress-dependent quantities, $\bar{\sigma}$, $\{\sigma'\}$, etc. are evaluated at those selected Gauss points θ_k. The decision on the selection of total number of Gauss points should therefore be made on both mathematical and material behavior grounds.

As for the accuracy of numerical integration by Gaussian quadrature, many references such as Wilson and Nickell[1] are available for assessing the required number of integration points to achieve satisfactory accuracy of integration over a plane triangular element. However, in the present case, integrands of elements of the stiffness matrix (as shown in Appendix 3) are dependent on circumferential coordinate. Integration of these elements over the circumferential coordinate has to be specially discussed. By referring to Appendix 3, it can be seen that these integrands can be classified into three general categories, namely

$$I_1 = a_1/(b_1 \cos^2 \theta + b_2 \cos \theta \sin \theta + b_3 \sin^2 \theta)$$

$$I_2 = (a_1 \cos \theta + a_2 \sin \theta)/(b_1 \cos^2 \theta + b_2 \cos \theta \sin \theta + b_3 \sin^2 \theta)$$

$$I_3 = (a_1 \cos^2 \theta + a_2 \sin^2 \theta + a_3 \sin \theta \cos \theta)/ \\ (b_1 \cos^2 \theta + b_2 \cos \theta \sin \theta + b_3 \sin^2 \theta)$$

where a_1, a_2, a_3, b_1, b_2, b_3 are various constants in the integrand. As shown in Table 5.1, it can be seen that the Gaussian quadrature scheme with *nine* integration points appeared to give reasonably accurate results, e.g. to the order of 10^{-3} for the integrand I_2.

Table 5.1 Accuracy of Gaussian quadrature as a function of Gauss points for the stiffness matrix of modes 0 and 1[†].

No. of Gauss points	I_1	I_2	I_3
7	2.8560	0.2717×10^{-1}	3.5410
9	2.8492	0.8879×10^{-3}	3.4921
12	2.8490	0.4361×10^{-3}	3.5043
16	2.8488	-0.1002×10^{-4}	3.5040
24	2.8488	0.4422×10^{-5}	3.5040
32	2.8488	0.2499×10^{-5}	3.5040

[†] Integration results of I_1, I_2, I_3 by Gaussian quadrature scheme with various numbers of integration points.

5.8 Numerical example

The sagging of a horizontally mounted pipe or pressure vessel is considered to be a typical example of an axisymmetric structure subject to asymmetric loads. The asymmetric loads in the form of combined effects of fluid gravity, internal pressure, and the weight of structural material as well as vessel/support interface conditions can be described by various forms of Fourier expansions[5]. Here, the problem of a simply supported heavy pipe containing high pressure gas is analyzed as a numerical illustration. This example is chosen as the combined gravity and internal gas pressure load can be

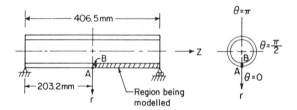

(a) A Simply Supported Heavy Pipe

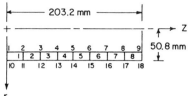

(b) Finite Element Discretization

Figure 5.4 Dimensions and finite element idealization of a horizontally mounted heavy pipe[3,4]. Assumed densities 623, 1380 and 2770 tons/m³.

Table 5.2 Convergence of results by element number.

Load P (ksi)	4 element $\bar{\varepsilon}$ (%)	4 element δ (10^{-3} in)	8 element $\bar{\varepsilon}$ (%)	8 element δ (10^{-3} in)	16 element $\bar{\varepsilon}$ (%)	16 element δ (10^{-3} in)
0	0.002	0.486	0.002	0.531	0.002	0.533
39	0.201	2.149	0.204	3.705	0.203	3.621
39.8	0.273	2.790	0.312	7.482	0.307	6.986
40.7	0.465	3.176	1.14	9.747	0.988	8.923
41.7	0.796	3.311	2.22	10.58	2.02	9.856
43.2	1.026	3.394	3.94	11.31	3.79	10.86

δ = maximum deflection at point A in Figure 5.4.
Density of pipe material: 22.5 lbf/in³, or 623 tons/m³.

described by a simple Fourier expansion involving zero and first modes, i.e. $M = 2$ in the analysis.

The geometry and dimensions of the pipe are shown in Figure 5.4. Also shown in this figure is the finite element idealization. Due to symmetry, only a quarter of the pipe section was modelled. Test runs were made on the finite element models with 4, 8 and 16 elements. The comparison of results in the maximum deflection at point A, (δ) is shown in Table 5.2.

144

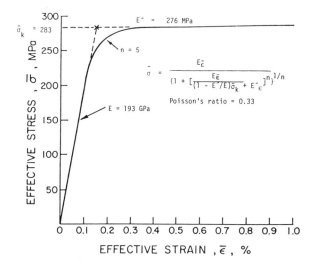

Figure 5.5 Effective stress–strain curve for pipe material.

It was observed that the improvement of the results was marginal in the case of 16 elements over the case of 8 elements. An 8-element model was then used throughout the rest of the analysis.

Three different pipe material densities were assumed in order to assess the effect of gravity load on the mechanical behavior of pipe, which accounts for the mode 1 of the Fourier expansion formulation. Numerical values of these densities are shown in Figure 5.4. The effective stress–strain relation of the pipe material is shown in Figure 5.5.

The polynomial expression given in (3.24) was used to describe this relation with the following numerical values for the parameters:

modulus of elasticity, $E = 193\,\text{GPa}$

plastic modulus, $E' = 276\,\text{MPa}$

$\bar{\sigma}_K$ = stress at the intersections of lines drawn from E and $E' = 283\,\text{MPa}$

stress power, $n = 5$

The initial yield strength in tension and the Poisson ratio of the material are respectively assumed to be 276 MPa and 0.33.

Since only the zeroth and first mode were required in the Fourier expansions, the element stiffness matrix was calculated by exact integration as shown in (5.28). The expressions of the integrands and the result of integrations for the element stiffness matrix can be found in Appendix 3. Nine integration points were used along the circumference of all toroidal elements in the model.

Figure 5.6 shows the maximum sag of the pipe at point A in Figure 5.4 with increasing internal gas pressure at three different material densities. As can be observed, a drastic increase of sagging occurs after the initiation of plastic deformation in the pipe. Similar variations were observed for the maximum

145

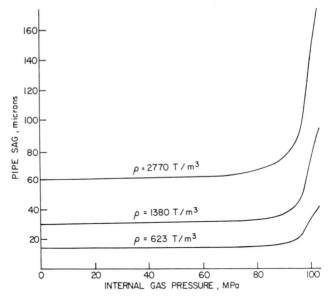

Figure 5.6   Elastoplastic sag of a pipe[3,4].

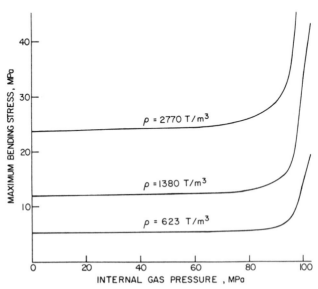

Figure 5.7   Maximum bending stress in a pipe[3,4].

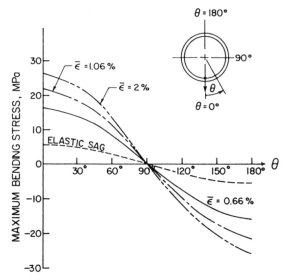

Figure 5.8 Variation of bending stresses along the circumference[3,4].

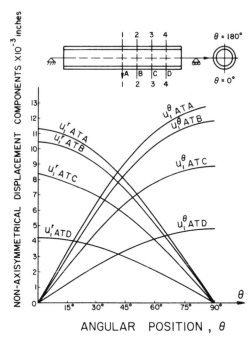

Figure 5.9 Asymmetric displacement components in a pipe[3].

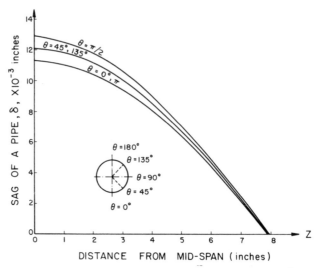

Figure 5.10 Sag of a pipe along the axial direction[3].

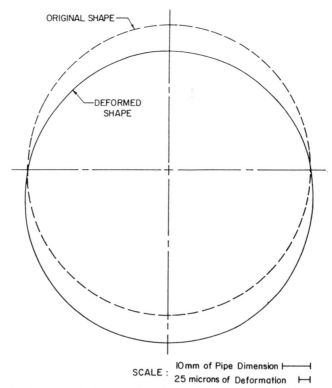

Figure 5.11 Asymmetric deformation of a pipe cross section at mid-span with a 2% maximum effective strain[3,4].

bending (longitudinal) stress in the pipe (point B in Figure 5.4) as illustrated in Figure 5.7. The variation of the bending stress along the circumference of the same toroidal element is depicted in Figure 5.8.

The mode 1 component of the pipe deformation (mode 0 represents the axisymmetric deformation due to internal pressure loading) at various cross sections in the pipe for the case of $\varepsilon_{max} = 3.94\%$ is illustrated in Figure 5.9, and the sag at various points along the circumference is depicted in Figure 5.10. Finally, the deformed cross section of the pipe in comparison to its original shape is shown in Figure 5.11.

5.9 Discussion of the numerical example

The special feature of the present analysis in the code was verified with analytical solutions derived on the elastic sag of a heavy rod by Pearson[6]. The results derived from the present analysis, Pearson's formulation, and the simple beam theory are summarized in Table 5.3. It is worth noting that while the agreement of results between the present method and Pearson's analysis is excellent, the popular simple beam theory consistently underestimated the sag by 7%.

Table 5.3 Code verification on the sag of a heavy rod by its own weight.

ρ	Present method		Pearson[6]		Simple beam theory	
	δ	σ	δ	σ	δ	σ
623	18.8	8.8	18.8	8.7	17.4	8.7
1380	41.7	19.5	41.9	19.4	38.7	19.3
2770	83.4	39.0	83.8	38.7	77.4	38.6

ρ = density, ton/m^3; δ = maximum deflection, microns; σ = maximum bending stress, MPa.

The main advantage of using the combined finite element analysis and the Fourier expansion method over the general three-dimensional finite element analysis is the substantial saving in the amount of input data, computing effort and computer storage. Table 5.4 indicates that the present method can achieve a saving of about five times in the amount of input data, 20 times in the required computer storage and two times in the required computer effort on the basis of nine elements[†] along the circumference in an equivalent three-dimensional finite element model. These savings may be greater for mode numbers up to about five. The advantages of the present method in computing effort may become less obvious for cases involving

[†] Corresponding to the nine Gauss points used in the numerical integration.

Table 5.4 Computational effort for the sag of a heavy pipe.

	Present method	Equivalent three-dimensional finite element method
Total nodes	18	162
Total elements	8	72
Size of stiffness matrix	108×108	486×486
Integration points	443 232	909 792

mode number higher than five. However, the simplicity in data preparation and less data storage requirements remain attractive. For computations involving moderate mode numbers, the mode separation scheme described in (5.32) may prove to be more efficient than a general three-dimensional finite element analysis.

The maximum bending stress associated with the elastoplastic sag of the pipe, as shown in Figure 5.7, is as high as four times that estimated by the elastic analysis. It is therefore important, from the design safety point of view, that elastic–plastic analysis be used when dealing with pipelines of heavy materials or with long unsupported span.

5.10 Summary

Based on the numerical example presented above, a brief conclusion may be drawn regarding the use of the Fourier expansion approximation method for the elastic–plastic analysis of a special type of three-dimensional problem.

The introduction of Fourier expansion to the two-dimensional finite element analysis such as the TEPSA code has been demonstrated to be an effective method for the prediction of elastic–plastic sag of heavy pipes. The inadequacies of the simple beam theory, which has been a popular approximation for this purpose, are apparent as indicated in the numerical illustration.

Although the Fourier expansion approximation is considered to be an effective method for the solutions of asymmetric loadings on axisymmetric structures, its accuracy remains to be verified by experiment. The idea of evaluating the θ-direction dependent material properties by the integration of such properties at the selected Gauss points along the circumference requires further deliberation and verification as well. Only then can the merit of this method be fully utilized for practical engineering applications.

References

1 Wilson, E. L. and R. E. Nickell 1966. Application of the finite element method to heat conduction analysis, *Nucl. Engng. & Des.* **4,** 276–86.
2 Zienkiewicz, O. C. 1971. *The finite element method in engineering science.* New York: McGraw-Hill. Ch. 13.
3 Wu, R. Y. 1977. *On the application of Fourier expansion to elasto-plastic finite element stress analysis.* M.Sc. Thesis, Dept. of Mech. Engng., Univ. of Manitoba, Winnipeg, Manitoba, Canada.
4 Wu, R. Y. and T.-R. Hsu 1980. On elasto-plastic sag of pipes by two-dimensional finite element analysis, *Proc. 4th Int. Conf. Press. Vess. Technol.* **2,** 237–42.
5 Duthie, G. and A. S. Tooth 1977. The analysis of horizontal cylindrical vessels supported by saddles welded to the vessel—a comparison of theory and experiment. *Proc. 3rd Int. Conf. Press. Vess. Technol.*, Part 1, Tokyo. 25–38.
6 Pearson, K. 1889. On the flexure of heavy beams subjected to continuous systems of load. *Q. J. Pure & Appl. Math.* **24** and **31,** 63.

6

ELASTODYNAMIC STRESS ANALYSIS WITH THERMAL EFFECTS

6.1 Introduction

There are generally two types of engineering analyses which fit into the definition of "dynamic stress analysis". The first type primarily deals with the modal and frequency analysis of a vibrating solid whereas the second type studies the stress wave propagation in structures induced by impact or other forms of rapidly applied loads, e.g. by explosions. Both analyses can be handled by a similar mathematical model, i.e. the equation of motion, and both are essential elements of modern engineering analysis. This chapter, however, is devoted to the study of the mechanical aspects of stress wave propagation in elastic media only.

Like the subject of creep deformation of solids described in Chapter 4, the phenomenon of stress wave propagation in solids has been recognized by scientists and engineers as long as a century ago by such well-known persons as Stokes, Poisson, Rayleigh. However, systematic mathematical treatment of this subject was not formulated until early this century, e.g. in Chapter 8 of Love[1] and in Lapwood[2]. A more complete special volume on this subject was available as recently as 1963[3].

The subject of thermoelastodynamic stress analysis gained its prominence in the 1950s, presumably because of the rapid development of the aerospace industry. Thermoelastic equations with dynamic effect, such as those shown in Boley and Weiner[4] and Nowacki[5] or as described in (3.10) and (3.11) with the inclusion of inertia forces, were used to assess the propagation of stress waves in elastic media due to thermal shocks[6,7]. Results indicated that the stress waves induced by thermal forces alone with inertia effect was substantial under the circumstances described in these papers, but further studies[4,8] indicated that this inertia effect could indeed be neglected in most practical applications without introducing significant error.

The finite element formulations presented in this chapter will be constructed on the assumptions that the inertia term can be neglected in the thermal analysis which will be performed separately from the mechanical analysis. The inertia (or dynamic) effect, however, is included in the mechanical analysis which also includes temperature-dependent material properties. The theory, as so far formulated, can in fact be regarded as an elastodynamic analysis with material nonlinearity due to thermal effect. Like those finite element formulations presented in the preceding two chapters, the elastodynamic analysis derived here can be readily adapted to the TEPSAD[†] code for the dynamic stress analysis on structures of both axisymmetric and planar geometries.

6.2 Theoretical background

The theory of stress wave propagation in elastic media is a well-developed subject. Detailed description of the mathematical formulation can be found in several treatises, e.g. in Kolsky[3] and Fung[9]. Only a brief outline of key equations will be reviewed here.

As usual, there are a number of assumptions to be made in the derivation of the finite element formulations for the TEPSAD code. These are:

(1) although the variation of the mechanical properties of the material with respect to temperature is included in the formulation, the temperature field is solved separately by the method described in Chapter 2;
(2) damping by the material is neglected;
(3) a complete wave reflection from the boundary of a finite solid is assumed;
(4) a lumped mass matrix is used instead of the consistent mass matrix as for other similar analyses;
(5) the dynamic behavior of the material is treated as an equivalent static problem as a series of points in time.

Almost all classical dynamic analyses start with the equation of motion of the solid. A general expression of this equation is similar to that shown in (3.10) with the addition of the inertia force:

$$\sigma_{ij,j} + X_i = \rho \ddot{u}_i \tag{6.1}$$

or, in a different form, but similar to (3.11):

$$G u_{i,kk} + (\lambda + G) u_{k,ki} + X_i = \rho \ddot{u}_i \tag{6.2}$$

The equations of motion as shown above, in a way, are modified versions of (3.10) and (3.11) and hence can be regarded as the "dynamic equilibrium equations". The definition of variables shown above can be found after these equations.

Assumption (5) above implies the necessity of using numerical time integration over the discretized equations of motion given in (6.1) or (6.2). Since most stress waves in solid media have high frequencies, it is more efficient to use the direct time integration scheme[10] as will be further elaborated in a later section.

[†] Stands for TEPSA-Dynamic.

6.3 Hamilton's variational principle

In Chapter 3, we have demonstrated the principle of minimization of the potential energy stored in a deformed solid under static loads in order to satisfy the equilibrium condition. The counterpart of this principle in the dynamic loading situation is Hamilton's principle according to the argument presented in [9]. Indeed, by referring to a solid of volume v and surface s, the variation of the equations of motion in (6.1) or (6.2) leads to the following important relation[9] for the dynamic equilibrium condition:

$$\delta \pi = \delta \int_{t_1}^{t_2} (K + P)\, dt = 0 \tag{6.3}$$

where

π = a functional used in the variational process

K = kinetic energy of the deforming solid

$$= \frac{1}{2} \int_v \rho \frac{\partial u_i}{\partial t} \frac{\partial u_i}{\partial t}\, dv \tag{6.4}$$

ρ = mass density of the material

P = potential energy stored in the solid

$$= U - W \tag{6.5}$$

U = strain energy

$$= \frac{1}{2} \int_v \sigma_{ij} \varepsilon_{ij}\, dv \tag{6.6a}$$

W = work done to the solid by external forces

$$= \int_v X_i u_i\, dv + \int_s \tau_i u_i\, ds \tag{6.6b}$$

τ_i = components of surface tractions

6.4 Finite element formulation

The finite element formulation for mechanical vibration analyses has been well documented, e.g. in Zienkiewicz[11], Bathe and Wilson[12] and Desai and Abel[13]. The same treatment for the propagation of stress waves in elastic-plastic solids can be found in Biffle[14] and Scarth[15]. The following derivation, however, is focused on the element equation in a discretized elastic solid with thermal effect.

For the case of dynamic loading of a solid involving substantial temperature and strain rate changes, variation of the material properties with these changes, as illustrated in Figures 3.7 & 8, must be taken into consideration in the computation. The analysis becomes nonlinear and the incremental deformation approach based on piecewise linearity has been adopted as the basis for the finite element formulation.

Following the constitutive equation for elastic solids with nonlinear material characteristic as given in (3.15), the incremental stresses and strains during a small load (or time in the case of dynamic loading) can be related by

$$\{\Delta\sigma\} = [C_e]\{\Delta\varepsilon'\} \tag{6.7}$$

where

$$\{\Delta\varepsilon'\} = \text{increment of total strain components}$$

$$= \{\Delta\varepsilon\} - \{\varepsilon^0\} \tag{6.7a}$$

$$\{\varepsilon^0\} = \{a\}\,\Delta T + \left(\frac{\partial[C_e]^{-1}}{\partial T}\,\Delta T + \frac{\partial[C_e]^{-1}}{\partial\dot\varepsilon}\,\Delta\dot\varepsilon\right)\{\sigma\} \tag{6.7b}$$

The incremental potential and kinetic energies can be formulated for an element of volume v and surface s from (6.4) to (6 6):

$$\Delta P = \int_v \tfrac{1}{2}\{\Delta\varepsilon'\}^{\mathrm{T}}\{\Delta\sigma\}\,dv - \int_s \{\Delta U\}^{\mathrm{T}}\{\Delta\tau\}\,ds \tag{6.8a}$$

and

$$\Delta K = \int_v \tfrac{1}{2}\rho\{\Delta\dot U\}^{\mathrm{T}}\{\Delta\dot U\}\,dv \tag{6.8b}$$

where $\{\Delta\varepsilon'\}$, $\{\Delta\sigma\}$, $\{\Delta\tau\}$ and $\{\Delta U\}$ are the corresponding incremental total strains, stresses, surface tractions and displacements in the element.

Body forces have been neglected in the above formulation.

As piecewise linearity of the σ vs ε relationship is assumed during the small load increment, i.e. $\Delta t = t_2 - t_1$, (6.3) may be expressed in an incremental sense to give:

$$\delta(\Delta\pi) = \delta\int_{t_1}^{t_2} (\Delta K + \Delta P)\,dt = 0$$

By substituting ΔP and ΔK in (6.8a, b) into the above equation, the dynamic equilibrium condition is achieved:

$$\delta\int_{t_1}^{t_2}\left(\tfrac{1}{2}\int_v \rho\{\Delta\dot U\}^{\mathrm{T}}\{\Delta\dot U\}\,dv + \tfrac{1}{2}\int_v \{\Delta\varepsilon'\}^{\mathrm{T}}\{\Delta\sigma\}\,dv\right.$$

$$\left. - \int_s \{\Delta U\}^{\mathrm{T}}\{\Delta\tau\}\,ds\right)dt = 0 \tag{6.9}$$

155

The discretization scheme as described in Chapter 1 and (3.94) and (3.95) can be used here to relate the element displacements and strains to the respective nodal displacement increments as follows:

$$\{\Delta U\} = [N]\{\Delta u\}$$

and

$$\{\Delta \varepsilon'\} = [B]\{\Delta u\}$$

where the interpolation function $[N]$ and the matrix $[B]$ have been expressed in (3.108) and (3.111) respectively.

The expression of (6.9) in terms of $\{\Delta u\}$ can be achieved by replacing $\{\Delta \varepsilon'\}$ and $\{\Delta U\}$ by the above relations, together with the constitutive equation given in (6.7):

$$\delta \int_{t_1}^{t_2} \left(\frac{1}{2} \int_v \rho \{\Delta \ddot{u}\}^T [N]^T [N] \{\Delta \ddot{u}\} \, dv + \frac{1}{2} \int_v \{\Delta u\}^T [B]^T [C_e][B]\{\Delta u\} \, dv \right.$$

$$- \int_v \{\Delta u\}^T [B]^T [C_e]\{\varepsilon^0\} \, dv + \frac{1}{2} \int_v \{\varepsilon^0\}^T [C_e]\{\varepsilon^0\} \, dv$$

$$\left. - \int_s \{\Delta u\}^T [N]^T \{\Delta \tau\} \, ds \right) dt = 0$$

Since $\{\Delta u\}$, $[C_e]$ and $\{\varepsilon^0\}$ are independent of the local coordinates but a function of time t, whereas the functions $[N]$ and $[B]$ are not dependent on time t, the above variational process yields the following element equations of dynamic equilibrium after integrating the first term in the above equation by parts.

$$[M_e]\{\Delta \ddot{u}\} + [K_e]\{\Delta u\} = \{\Delta F_e\} \tag{6.10}$$

where

$[M_e]$ = element mass matrix

$$= \int_v \rho [N]^T [N] \, dv \tag{6.11a}$$

$[K_e]$ = element stiffness matrix

$$= \int_v [B]^T [C_e][B] \, dv \tag{6.11b}$$

$\{\Delta F_e\}$ = element incremental force matrix

$$= \int_s [N]^T \{\Delta \tau\} \, ds$$

$$+ \int_v [B]^T [C_e] \left(\{a\} \Delta T + \frac{\partial [C_e]^{-1}}{\partial T} \{\sigma\} \Delta T + \frac{\partial [C_e]^{-1}}{\partial \dot{\varepsilon}} \{\sigma\} \Delta \dot{\varepsilon} \right) dv \tag{6.11c}$$

Although the element mass matrix $[M_e]$ in (6.11a) is expressed in a consistent form in terms of the continuous interpolation function $[N]$, it was, however, lumped at the nodes in the computation. This "lumping" procedure is similar to that used to determine the heat capacitance matrix in heat conduction analysis.

The overall dynamic equilibrium equation or equation of motion for the entire discretized structure can be expressed in a usual way

$$[M]\{\Delta \ddot{u}\} + [K]\{\Delta u\} = \{\Delta F\} \qquad (6.12)$$

in which

$\qquad [M]$ = overall mass matrix

$$= \sum_{1}^{m} [M_e] \qquad (6.13a)$$

$\qquad [K]$ = overall stiffness matrix

$$= \sum_{1}^{m} [K_e] \qquad (6.13b)$$

$\qquad \{\Delta F\}$ = overall incremental force matrix

$$= \sum_{1}^{m} \{F_e\} \qquad (6.13c)$$

$\qquad m$ = total number of elements in the system

It is worth noting that $[M]$ is a diagonal matrix and $[K]$ is symmetric about the diagonal elements as in the static analysis.

6.5 Direct time integration scheme

There are two commonly used methods of solving the global equations of motion in (6.12) by numerical means. These are the normal mode method and direct time integration techniques. Descriptions of both of these methods are available in Clough[10], Zienkiewicz[11], Bathe and Wilson[12] and Clough and Penzien[16]. Because of the high frequencies often encountered in the stress wave propagation, the direct time integration technique, called the Newmark-β method[17], has been adopted in the present analysis. It provides unconditionally stable solutions and has been used successfully in dealing with similar problems by other researchers[14, 15, 18, 19]

The state of the equations of motion in (6.12) at time t may be expressed as

$$[M]\{\Delta \ddot{u}_t\} + [K]\{\Delta u_t\} = \{\Delta F_t\} \qquad (6.14)$$

where $\Delta u_t = u_{t+\Delta t} - u_t$ and so on.

The Newmark-β method approximates the displacements and velocities at a time $t + \Delta t$ by the following recurrence relations:

$$\{u_{t+\Delta t}\} = \{u_t\} + \Delta t\{\dot{u}_t\} + \Delta t^2[(\tfrac{1}{2} - \beta)\{\ddot{u}_t\} + \beta\{\ddot{u}_{t+\Delta t}\}] \qquad (6.15a)$$

$$\{\dot{u}_{t+\Delta t}\} = \{\dot{u}_t\} + \Delta t[(1 - \gamma)\{\ddot{u}_t\} + \gamma\{\ddot{u}_{t+\Delta t}\}] \qquad (6.15b)$$

where the parameters β and γ dictate the assumed variation of the acceleration, e.g. $\beta = \tfrac{1}{6}$ and $\gamma = 0.5$ corresponds to a case of linear acceleration, whereas a combination of $\beta = 0.25$ and $\gamma = 0.5$ results in a constant acceleration and an unconditionally stable solution.

By rearranging (6.15a) one may express the following quantities in an incremental form:

$$\{\Delta u_t\} = \Delta t\{\dot{u}_t\} + \frac{\Delta t^2}{2}\{\ddot{u}_t\} + \Delta t^2\beta\{\Delta \ddot{u}_t\} \qquad (6.16a)$$

and

$$\{\Delta \ddot{u}_t\} = \frac{1}{\beta}\left(\frac{1}{\Delta t^2}\{\Delta u_t\} - \frac{1}{\Delta t}\{\dot{u}_t\} - \frac{1}{2}\{\ddot{u}_t\}\right) \qquad (6.16b)$$

Upon substituting (6.16a, b) into (6.14) and rearranging the terms, a familiar form of the discretized equilibrium equation for the finite element analysis is obtained:

$$[K^*]\{\Delta u_t\} = \{\Delta F_t^*\} \qquad (6.17)$$

where

$$[K^*] = \text{equivalent stiffness matrix}$$

$$= [K] + \frac{1}{\beta \Delta t^2}[M] \qquad (6.18a)$$

$$\{\Delta F_t^*\} = \text{equivalent force matrix}$$

$$= \{\Delta F_t\} + [M]\left(\frac{1}{\beta \Delta t}\{\dot{u}_t\} + \frac{1}{2\beta}\{\ddot{u}_t\}\right) \qquad (6.18b)$$

The matrices $[K]$, $\{M\}$ and $\{\Delta F_t\}$ are given in (6.13a, b, c).

The above derivation has effectively converted a nonlinear second-order differential equation (6.12) into a set of linear algebraic equations at discrete instants (6.17). The usual finite element analysis algorithms as presented in the foregoing chapters can therefore be applied for this type of problem.

6.6 Solution algorithm

As (6.17) is the discretized form of the equation of motion which is a second-order differential equation at time t, two initial conditions are required to start the time-integration algorithm. Two such common initial conditions are

initial nodal displacement vector $\{u\}_{t=0} = 0$, or a constant;
initial nodal velocity vector $\{\dot{u}\}_{t=0} = 0$, or a constant.

The initial nodal acceleration vector can be evaluated from the other two initial conditions by the equation of motion, i.e. (6.17), as

$$\{\ddot{u}\}_{t=0} = [M]^{-1}(\{F\}_{t=0} - [K]\{u\}_{t=0})$$ (6.19)

Once the initial conditions are specified, (6.17) can be solved with the progression of proper time steps.

The reader will recognize a fact that the solution algorithm described for the static loading cases in Chapter 3 can be readily used for the solution of (6.17) with, of course, an additional mass matrix included in computations of $[K^*]$ and $\{F_i^*\}$ as shown in (6.18a, b). Once the solution of the nodal incremental displacements become available, the corresponding element incremental strain components may be computed by a similar relationship to that shown in (3.121). The incremental stress components in the elements, of course, will be computed from the constitutive equation given in (6.7). The total stresses and strains can be evaluated from the corresponding incremental values as indicated in (3.124) and (3.125).

The above algorithm has been incorporated into the base TEPSA code for structures of both axisymmetric and plane strain geometries with a modified code name TEPSAD.

6.7 Numerical illustration

TEPSAD was used to assess the stresses and deformation in a thick-walled cylinder subject to sudden application of pressure loading at its inside surface. Physically, this case is similar to the firing of a cannon.

The geometry and the corresponding finite element idealization of this case illustration are described in Figure 6.1. A total of 32 square elements were used in the analysis. The analytical solution of this problem with a step pressure loading function is available in Alzheimer and Forrestal[20]. It is obvious that a step loading function cannot be implemented in a finite element analysis and a ramp-type load with a small finite rising time t_0 had to be used in the present analysis. The step and ramp-type loadings have been illustrated in Figure 6.2 with a maximum constant pressure load of $P_0 = 6.9$ MPa.

159

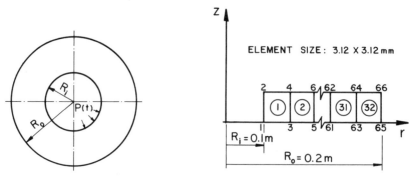

(a) Geometry   (b) F. E. Idealization

Figure 6.1 Geometry and finite element idealization of a thick wall cylinder.

The following material properties were used in the case illustration:

Young's modulus, E: 71 862 MPa
Poisson ratio, v: 0.3
weight density, ρ: 27.2 kg/m^3

The numerical values of $\beta = 0.25$ and $\gamma = 0.5$ for Newmark parameters were used in the TEPSAD code for unconditional stable solutions.

Since the dilatational wave velocity, C, in this material is $C = (E/\rho_m)^{1/2} = 5144$ m/s, in which ρ_m is the mass density of the material, and the length across one element, ΔL, was 3.12 mm as shown in Figure 6.1b, the maximum allowable time step was

$$\Delta t_{max} = \Delta L/C = 0.607 \ \mu s$$

The total time duration for the pressurization was assumed to be 116.6 μs, which is equivalent to the time required for the wave to pass the cylinder wall six times. As shown in Figure 6.2, four ramp rates were used with the rising times to be $t_0 = 0.607$, 1.944, 9.719 and 19.44 μs. The effect of these ramp rates on the numerical solutions will be observed. The time step sizes were

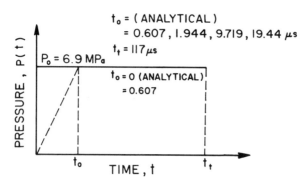

Figure 6.2 Loading history with various ramp rates.

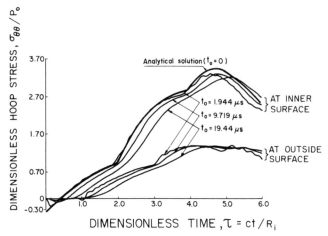

Figure 6.3 Dynamic hoop stress at surfaces with various loading ramps.

chosen to be $\Delta t = 0.305$, 0.607 and 0.91 μs, which correspond to $0.5\,\Delta t_{max}$, Δt_{max} and $1.5\,\Delta t_{max}$.

Figure 6.3 shows the normalized hoop stresses at both inner and outside surfaces of the cylinder using three different ramp rising times in the TEPSAD analysis. The analytical solution from Alzheimer and Forrestal[20], with step loading function corresponding to zero rise time (or $t_0 = 0$) at the inner surface, has been included. As can be observed from this figure, the numerical solution deviates further from the analytical solution with slower assumed rising times as expected. Another interesting phenomenon to observe from this figure is that an instantaneous compressive hoop stress was generated at the inner surface of the cylinder by the step loading as in the case of analytical solution, whereas the occurrence of this negative hoop

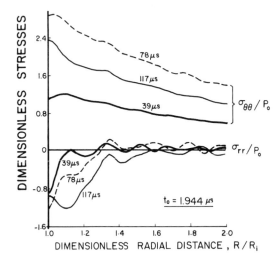

Figure 6.4 Distribution of dynamic stresses in the cylinder

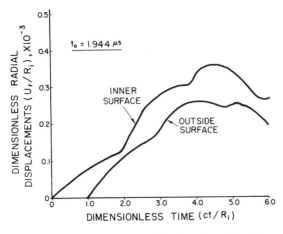

Figure 6.5 Radial deformation of the cylinder surfaces.

stress not only was delayed in time but also showed reduction in magnitudes with longer rise times in the numerical solutions. This negative hoop stress at the inner surface of the cylinder due to internal pressure is unique for the dynamic loadings only. It occurs because of the delayed response to the load by the material away from the inner surface, resulting in a hoop constraint beneath this surface, which in turn provides the compressive effect for the part of the material already under the influence of the internal pressure loading. The variations of both radial and hoop stress components in the cylinder at three selected time steps are illustrated in Figure 6.4.

Figure 6.5 shows the radial displacements at both the inner and outer surfaces of the cylinder at various instants. Results in both Figures 6.4 & 5 were produced by a ramp rise time of 1.944 μs. Numerical results indicated little difference with the sizes of time increment used in the computation.

References

1   Love, A. E. H. 1944. *A treatise on the mathematical theory of elasticity*, 4th edn. New York: Dover.
2   Lapwood, E. R. 1949. The disturbance due to a line source in a semi-infinite elastic medium. *Phil. Trans. R. Soc. Lond.* **242**A, July, 63–100.
3   Kolsky, H. 1963. *Stress waves in solids*. New York: Dover.
4   Boley, B. A. and J. H. Weiner 1960. Theory of thermal stresses. New York: John Wiley, pp. 54–60.
5   Nowacki, W. 1962. *Thermoelasticity*. Reading, Mass.: Addison-Wesley. Chs 1, 4.
6   Danilovskaya, V. Y. 1950. Temperature stresses in an elastic semi-space due to a sudden heating of its boundary (in Russian). *Prikl. Mal. Mekh.* **14**, 316–18.
7   Mura, T. 1956. *Dynamical thermal stresses due to thermal shocks*. Research Report No. 8, Meiji University, Japan.
8   Sternberg, E. and J. G. Chakravorty 1958. *On inertia effects in a transient thermoeleastic problem*. Technical Report No. 2, Contract Nonr-562(25), Brown University.

9 Fung, Y. C. 1967. *Foundations of solid mechanics.* Englewood Cliffs, NJ: Prentice-Hall. Ch. 11.

10 Clough, R. W. 1971. Analysis of structural vibrations and dynamic response. In *Recent advances in matrix methods of structural analysis and design.* R. H. Gallagher, Y. Yamada and J. T. Oden (eds.). University of Alabama Press.

11 Zienkiewicz, O. C. 1971. *The finite element method in engineering science.* New York: McGraw-Hill.

12 Bathe, K. and E. L. Wilson 1976. *Numerical methods in finite element analysis.* Englewood Cliffs, NJ: Prentice-Hall.

13 Desai, C. S. and J. F. Abel, 1972. *Introduction to the finite element method.* New York: Van Nostrand Reinhold.

14 Biffle, J. H. 1973. *Finite element analysis for wave propagation in elastic–plastic solids.* PhD dissertation, Division of Engineering Mechanics, The University of Texas at Austin, Report No. 73-4.

15 Scarth, D. A. 1982. *The elastodynamic analysis of cracks under axisymmetric conditions by a path-independent integral.* M.Sc. Thesis, Department of Mechanical Engineering, University of Manitoba, Winnipeg, Canada.

16 Clough, R. M. and J. Penzien 1975. *Dynamics of structures.* New York: McGraw-Hill.

17 Newmark, N. M. 1959. A method of computation for structural dynamics. *Proc. Am. Soc. Civil Engrs.* **85,** EM3, 67–94.

18 Verner, E. A. and E. B. Becker 1973. Finite element stress formulation for wave propagation. *Int. J. Num. Meth. Engng.* **7,** 441–59.

19 Bathe, K. J. and E. L. Wilson 1973. Stability and accuracy analysis of direct integration methods. *Earthquake Engineering and Structural Dynamics* **1,** 283–91.

20 Alzheimer, W. E. and M. J. Forrestal 1968. *Elastic waves in thick-walled cylinders.* Rep. No. SC-TM-68-646, Sandia Laboratories.

163

7

THERMOFRACTURE MECHANICS

PART 1

REVIEW OF FRACTURE MECHANICS CONCEPT[†]

7.1 Introduction

The mechanisms of fracture which may be encountered in engineering structures can be classified into two general groups. The first category is termed "brittle fracture", which occurs in brittle materials such as glass, or in mild steel at very low temperature. Brittle fracture may also occur in most other engineering materials under very high loading rates or under the "plane strain" conditions encountered in heavy sectioned structural parts where the dimensions of the original defect are small compared to the characteristic dimensions of the part. This type of fracture is associated with relatively low fracture energy (i.e. the input energy required to propagate the crack) and small plastic deformation prior to and during crack extension. The second type of fracture falls into the general category of "ductile fracture" or high energy fracture[‡], and usually occurs in non-brittle materials under "plane stress" conditions. For example, thin-walled tubes and shell structures composed of materials with high ductility would be expected to undergo large plastic deformation prior to and during a rupture process.

Research in the field of fracture mechanics was initially concerned with investigating brittle fracture problems, since these types of failures are of more disastrous consequence and easier to analyze than ductile fracture cases. However, because of the clear evidence of thermally affected fracture characteristics of most materials as well as the growing demand for fracture prevention in machine components operating in high temperature environment, the subject of thermofracture has attracted much needed attention

[†] A substantial portion of this part was condensed from Kim, Y. J., 1981. *Stable crack growth in ductile materials—a finite element approach.* Ph.D. thesis, University of Manitoba.

[‡] The term "high energy" refers to the high input energy required to propagate the crack (usually dissipated as heat), and should not be confused with the high time rate of energy input as occurs in dynamic fracture.

164

from prominent researchers in recent years. Parts 2 and 3 of this chapter will deal with problems involving strong thermal effects.

In this part, the concept of linear elastic fracture mechanics (LEFM) is briefly described, and the current level of development of elastic–plastic fracture mechanics is also discussed. A review of the literature on the application of the finite element method to fracture mechanics problems is presented at the end of this chapter.

7.2 Linear elastic fracture mechanics

The linear elastic fracture mechanics approach to evaluating stresses and displacements associated with each fracture mode follows the Griffith–Irwin theory[1,2]. In this approach, the general stress field near a crack tip can be expressed as the superposition of stress fields due to the three modes of fracture, each mode associated with a kinematic movement of two crack surfaces relative to each other. These deformation modes, illustrated in Figure 7.1, are denoted as the opening mode, the edge sliding mode and the tearing mode.

The opening mode (mode I) is associated with local displacements in which the crack surfaces move apart in a direction perpendicular to these surfaces (symmetric with respect to the x–y and z–x planes).

The edge sliding mode (mode II) is characterized by displacements in which the crack surfaces slide over one another and remain perpendicular to the leading edge of the crack (symmetric with respect to the x–y plane and skew-symmetric with respect to the z–x plane).

The tearing mode (mode III) is defined by the crack surfaces sliding with respect to one another parallel to the leading edge of the crack (skew-symmetric with respect to the x–y and x–z planes).

Griffith's theory[3], first proposed in 1921, was based on the assumption that incipient fracture in ideally brittle materials occurs when the magnitude of the elastic energy supplied to the crack tip during an incremental increase in crack length exceeds the magnitude of the energy required to create the new crack surface during the same incremental increase in crack length. This

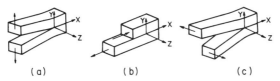

(a)   (b)   (c)

Figure 7.1   Three modes of fracture: (a) mode I; (b) mode II; (c) mode III.

strain energy release rate G, or the elastic energy made available per unit extension of the crack area, is

$$G = -d\pi/dA \tag{7.1}$$

where π is the potential energy of the structure and A is the cracked area.

In 1956, Irwin[4] developed the analytical basis of the elastic crack tip stress field theory, which in turn was the starting point of modern fracture mechanics. In his theory the stress intensity factor[†], K, is extracted from the solutions for stresses and displacements near the crack tip and is a combination of applied load P, crack length c and specimen configuration.

$$K = Pf(c, \text{geometry}) \tag{7.2}$$

K is said to be the controlling parameter of a crack tip field, because stresses and displacements are proportional to this factor. In general, the stress and displacement fields can be expressed mathematically as follows:

$$\sigma_{ij} = \frac{K}{\sqrt{r}} f_{ij}(\theta) + \cdots \tag{7.3a}$$

$$u_i = K\sqrt{r}\, g_i(\theta) + \cdots \tag{7.3b}$$

where (r, θ) is a polar coordinate system at the crack tip shown in Figure 7.2. The truncated terms of (7.3) are the terms with higher order in r, and for small radius of r (i.e. very close to the crack tip), only the first term is significant. Crack extension will occur when the intensity of the stress field in the close vicinity of the crack tip reaches a critical value. This means that fracture must be expected to occur when K reaches a critical value, K_c.

Irwin et al.[5] also showed that there exists a unique relationship between K and G as follows:

$$G = K^2/E_0 \tag{7.4}$$

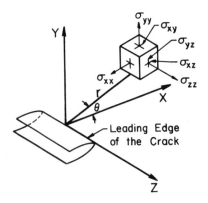

Figure 7.2 Coordinate systems at crack tip.

[†] K_I, K_{II} and K_{III} are defined as Mode I, Mode II and Mode III stress intensity factors, respectively.

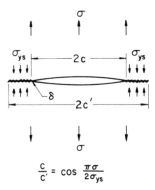

$$\frac{c}{c'} = \cos \frac{\pi \sigma}{2\sigma_{ys}}$$

Figure 7.3 The Dugdale model.

in which $E_0 = E$ (plane stress) $= E/(1 - v^2)$ (plain strain), where E is a Young's modulus and v is the Poisson ratio. Thus the consistency of the two theories, Griffith theory and Irwin theory, was apparent.

Due to the singular nature of (7.3a), a plastic zone is always formed at the crack tip where the stress field exceeds the yield strength of the material. Since (7.3) is based on the theory of elasticity, K has significance only when the geometry of the crack of the remaining ligament exceeds the plastic zone size by a factor of about 50[1], a criterion which is met by the plane strain condition. In order to use an experimentally determined plane strain fracture toughness K_{IC} value as a fracture criterion, the American Society for Testing and Materials (ASTM) specifies the following thickness requirement.

$$B > 2.5K_{IC}^2/\sigma_y^2 \tag{7.5}$$

where B is the thickness of the specimen, and σ_y is the yield strength of the material.

7.3 Elastic–plastic fracture mechanics

Most large complex engineering structures such as airplane fuselage, ships, pipelines, etc. have such small wall thickness that plane stress conditions prevail. Also, these structures are usually composed of ductile materials. Thus for many structural applications, the linear elastic analysis used to calculate the stress intensity factor K is invalidated by the formation of large plastic zones around the crack tip. Currently, much effort is being devoted to the development of elastic–plastic fracture mechanics analysis as an extension of the LEFM[1,2]. Among the various techniques the following approaches are most popular:

(1) plastic zone corrections;
(2) crack opening displacement;
(3) J integral;
(4) crack growth resistance R curve.

The above techniques provide considerable promise for appraising structural integrity in terms of an allowable loading or crack size. The engineering significance of each technique is reviewed in the following sections.

7.3.1 Plastic zone corrections

The first attempt at extending fracture mechanics beyond the LEFM limits involved a correction to the crack length to account for the effect of the plastic zone while continuing to use the LEFM approach. This procedure, as proposed by Irwin[4], involved moving the crack tip to the center of the plastic zone characterized by a distance r_y, i.e.

$$c \rightarrow c + r_y \tag{7.6}$$

The distance r_y is evaluated as

$$r_y = \frac{1}{2\pi} \frac{K_c^2}{\sigma_y^2} \quad \text{(plane stress)} \tag{7.7a}$$

$$= \frac{1}{6\pi} \frac{K_{IC}^2}{\sigma_y^2} \quad \text{(plane strain)} \tag{7.7b}$$

Although Irwin's plastic zone correction gives consistent results for small scale yielding, the limits of its applicability are uncertain.

A more rigorous correction for the plastic zone size was proposed by Dugdale[6]. He assumed that yielding occurs in a thin strip-like zone at the crack tip, extending the crack by a distance $c'-c$ (Figure 7.3). The stresses in this yielded zone are considered to be a continuous distribution of point loads, which act to restrain the crack from opening. An expression for the restraining stress intensity factor can then be obtained by integrating from c to c' with the appropriate Westergaard stress function:

$$K = 2\sigma_y(c/\pi)^{1/2} \cos^{-1}(c/c') \tag{7.8}$$

The size of the plastic zone is obtained by equating the restraining stress intensity factor in (7.8) with the K value for the opening of the crack, $K = \sigma\sqrt{(\pi c')}$. Hence

$$r_y = \frac{\pi}{16} \frac{K^2}{\sigma_y^2} \tag{7.9}$$

The plastic zone size calculated by (7.9) is about 20% bigger than that calculated by (7.7) for the plane stress case.

7.3.2 Crack opening displacement

Wells[7] proposed that the fracture behavior in the vicinity of a crack could be characterized by the opening of the crack faces—namely the crack opening displacement (COD), as shown in Figure 7.3. Furthermore, he showed that the COD concept was analogous to the G value, so that the COD value could be related to the plane-strain fracture toughness K_{IC}. Since COD measurement can be made when there is considerable plastic flow around the crack tip, this technique gives useful information for elastic–plastic fracture analysis.

An extension of the Dugdale analysis yields an expression for the crack opening displacement normal to the crack plane at the crack tip, δ:

$$\delta = \frac{8}{\pi} \frac{\sigma_y}{E} c \ln \left[\sec \left(\frac{\pi}{2} \frac{\sigma}{\sigma_y} \right) \right] \tag{7.10}$$

which for the cases $\sigma/\sigma_y \ll 1$ reduces to

$$\delta = \frac{K_I^2}{\sigma_y E} \tag{7.11}$$

Equation (7.11) implies that at the onset of crack instability, where K_I reaches K_{IC}, the COD value reaches a critical value δ_C. Under plane strain conditions, unstable fracture will occur upon crack initiation, and thus $\delta_i = \delta_C$. Like K_{IC}, a δ_C value under plane strain conditions is a material property, and is independent of specimen geometry. However, under plane stress conditions, stable crack growth can occur after the COD value reaching δ_i. The COD value rises to a value designated as δ_{max} ($> \delta_i$), upon which unstable crack extension occurs. Under plane stress conditions, values of δ_i and δ_{max} are dependent on specimen geometry.

7.3.3 The J integral

Rice[8] proposed a path-independent contour integral, the J integral, for a two-dimensional deformation field, evaluated over the contour Γ in a counter clockwise direction, as illustrated in Figure 7.4.

$$J = \int_{\Gamma} \left(w \, dy - t_i \frac{\partial u_i}{\partial x} \, ds \right) \tag{7.12}$$

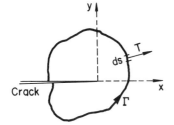

Figure 7.4 Contour for J integral.

where w is the strain energy density function, t_i is the surface traction vector $(= \sigma_{ij} n_j)$, u_i is the displacement vector, and ds is a differential element of an arc length along Γ.

The J integral, derived under the assumption of nonlinear elastic material behavior, is defined as the elastic energy release rate (per unit crack extension) to the crack tip. Since Rice[8] proved J to be path independent, one may evaluate J remote from the crack tip, where a well-defined elastic field prevails, and use this value of J to represent the energy release rate to the crack tip, that is, paths adjacent to or remote from the crack tip produce the same result. This scheme is suitable in the case of small scale yielding, for which the concept of path independence of energy release rate to the crack tip is assumed to remain valid.

Just as K was found to describe the elastic crack tip stress field in the LEFM approach, Hutchinson[9] and Rice and Rosengren[10] showed that the stress-strain field at the crack tip may be expressed as

$$\sigma_{ij} = \left(\frac{J}{r}\right)^{1/(n+1)} f_{ij}(\theta, n) + \cdots$$

$$\varepsilon_{ij} = \left(\frac{J}{r}\right)^{n/(n+1)} g_{ij}(\theta, n) + \cdots$$

(7.13)

where r and θ are polar coordinates centered at the crack tip and n is the power hardening coefficient in the assumed uniaxial stress–strain law, which is of the form $\varepsilon \propto \sigma^n$. For the linear elastic case, $n = 1$, (7.13) reduces to LEFM equations with $J = K^2/E_0$.

Just as the plastic zone size governs the validity of LEFM, so the size of the intensely nonlinear zone restricts the application of the J-integral approach. According to Paris[11], if the analysis using J is to be relevant, the size of this zone, I, must satisfy

$$I = 2\frac{J}{\sigma_y} \ll \text{planar dimensions}$$

(7.14)

Further, if a plane strain behavior is to be maintained, the thickness B must satisfy

$$B \geqslant 25\frac{J}{\sigma_y}$$

(7.15)

This restriction is an order of magnitude less severe than the corresponding LEFM requirement in (7.5).

The original concept of the J integral was constructed on an assumption that the material behaves elastically although its constitutive relation needs to be a nonlinear one[12]. The underlying meaning of this condition is the preclusion of any local unloading in the material during the global loading process. Extension of this concept to the plastically deforming solid, however, became available shortly after its inception. Rice and Rosengren proved that this parameter could be used to characterize the crack tip in a solid undergoing a small scale plastic yielding[10]. Basic formulations for the J_k parameters for mixed mode fracture problems were also presented by Budiansky and Rice[13]. During the past decade, the J integral has become the most frequently used parameter for the elastic–plastic fracture of solids[14]. Landes and Begley[15] and Harper and Ellison[16] even went as far as to extend this parameter for creep fracture analyses.

The seemingly unlimited expansion of the influence of the J integral in non-brittle type of fracture analyses has indeed caused concern among many researchers about the ultimate bounds of its many unique characteristics.

In a presentation made in 1978, Rice[17] summarized published results and offered the following comments:

(1) *Computational theory.* The deformation plasticity theory may be a more accurate theory than the incremental isotropic hardening flow theory as described in Chapter 3. This argument appears to be contradictory to common belief as described in Lin[18].

(2) *Path-independence.* The J parameter is essentially path-independent with paths greater than a radial distance of approximately 4 times the crack tip opening displacement from the tip of the crack[19].

(3) *Validity of J during stable crack growth.* A concept has been established that J must increase during a stable crack growth. Paris *et al.*[20] proposed that $\Delta J / \Delta c$ be used as a measure of instability of a growing crack. It is not clear though whether the associated progressive local unloading in the material which gives way for new crack surfaces would invalidate the use of J during such a situation.

Despite the potential controversial issues as described above, the J parameter remains a viable tool for characterizing elastic–plastic fracture of solids. Continuing research effort in establishing its applicability in various aspects of plastic deformation of solid structures is warranted.

7.3.4 Crack growth resistance R curve

The COD and J integral methods described previously relate their values at crack initiation to K_{IC} under plane strain conditions. These values may not be applied when determining the fracture toughness under plane stress conditions. The plane stress fracture toughness, K_C, is generally 2–10 times larger than K_{IC} and varies with specimen thickness.

Representation of the fracture toughness of thin sheet materials by a resistance curve has been attempted by a number of researchers[21-23], and is still under development. The concept of the crack growth resistance R curve is based on the observation that during the fracture process of most

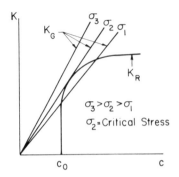

Figure 7.5 A typical crack growth resistance R curve.

sheet materials the unstable fracture is always preceded by a certain amount of stable crack growth under a monotonically rising load.

Figure 7.5 illustrates a typical R curve with the crack length as the abscissa and the crack growth resistance as the ordinate. The curve labeled K_R is the R curve, determined from experiment, with the stress intensity factor K_R at any crack length c being that required to propagate the crack from length c_0 to c. The first stage of the R curve is a vertical line representing a "no crack growth" situation. The point where the R curve deviates from a vertical line is the initiation of "stable crack growth". As the crack extends, the crack growth resistance also increases. This increased crack growth resistance is due to the increased size of the plastic zone and work-hardened material ahead of the crack tip, which increases the amount of work required to move the crack tip. Finally, a stage is reached on the R curve where no increased crack growth resistance accompanies crack growth. At this stage, the effects of increased crack length and decreased ligament size have overcome the effect of work hardening ahead of the crack tip.

The fracture criterion in the R curve concept is that for a given initial crack length c_0, failure will occur when the stress intensity factor from the applied loading, known as the crack driving force K_G equals K_R. Note that this condition must be fulfilled at the same instantaneous crack length c. Thus a family of K_G curves is drawn at various applied stress levels σ_1, σ_2, etc. The K_G curve that fulfills the fracture criterion is tangent to the K_R curve at the critical point. The stress corresponding to this K_G curve is the critical failure stress for an initial crack length c_0. Instability occurs at this stress level as the point of tangency of the K_G curve always exceeds tangency of the K_R curve. Further crack extension is spontaneous as

$$\frac{\partial K_G}{\partial c} \geqslant \frac{\partial K_R}{\partial c} \tag{7.16}$$

Analytical studies on stable crack growth[24, 25] have indicated that the strain field ahead of an extending crack is governed by a logarithmic singularity rather than the stronger inverse square root singularity as in the stationary case. This weaker singularity is due to the extension of the crack into the zone where the material is less plastically deformed, thus preventing complete refocusing of the strain field at the tip of the extended crack. This reduction in strain concentration accompanying the crack tip may be one of the main reasons for stable crack growth.

7.4 Application of the finite element method to fracture mechanics

Due to mathematical complexities, only a few crack problems, encompassing simplified boundary conditions, have been solved analytically. The finite element method, on the other hand, has been widely used to solve a large variety of fracture problems. This well-known numerical technique is capable of performing an elastic or elastic–plastic stress analysis to just about any two- or three-dimensional crack problem. Stress intensity factors under complicated geometry and loading conditions have been obtained using the linear elastic finite element method. Stress analysis, the determination of fracture criteria and the prediction of fracture instability have been attempted using the elastic–plastic finite element method.

7.4.1 Computation of stress intensity factors

Kobayashi et al.[26] and Chan et al.[27] were the first to use the finite element method to calculate the stress intensity factor. Once the numerical values of the nodal displacements and the element stresses near the crack tip are obtained, the K value can be calculated using (7.3) at several points. The K value at the crack tip is evaluated by extrapolating these values to the crack tip and by disregarding the first few points very close to the crack tip.

Watwood[28] calculated the K value by computing the strain energy release rate G. The total strain energy of the structure, as calculated for a given crack length, may be lacking in accuracy for a certain finite element mesh. However, there is a cancellation of these errors when taking the difference in strain energy for two different crack lengths, so that reasonable accuracy can be obtained with a relatively coarse mesh.

A number of special singular elements (Wilson[29], Byskov[30], Hilton and Sih[31], Tracey[32]) have been presented to accommodate the singularity at the crack tip. These elements are the result of incorporating both the eigenfunction expansions for the crack tip field conditions and the finite element method. The theoretical background of this approach is based on the mathematical properties of the two numerical techniques employed, i.e. the asymptotic expansion becomes increasingly more accurate as one approaches the crack tip, while the finite element method is very accurate everywhere except near the crack tip.

Another class of special singular elements is the quarter point element, used in conjunction with the standard isoparametric element. The Jacobian transformation from physical to isoparametric coordinates will produce spatial derivatives (i.e. strains) which are singular at the crack tip, if nodal points along the sides of the element are positioned in a certain way. Henshell[33] and Barsoum[34, 35] perceived that by moving the middle nodal point of a quadratic isoparametric element to the quarter point closest to the crack tip, the strain singularity is achieved.

7.4.2 Elastic–plastic analysis of a stationary crack

Swedlow et al.[36–38] and Marcal and King[39] pioneered the application of elastic–plastic finite element analysis to ductile crack problems. They analyzed edge- and center-cracked plates under plane stress and plane strain conditions, and reasonable agreement was obtained between the numerical and experimental results. Since then the numerical accuracy of these predictions has been improved through the better description of elastic–plastic material behavior and improved computing capability.

Miyamoto et al.[40] presented results for two- and three-dimensional analyses of cracked plates and also showed results for cyclic loading, although not accurate enough for useful application. A more accurate analysis has been performed by Larsson and Carlsson[41] by which crack tip plastic zones are assumed to respond only to an elastic outer field of the singular stress field.

Precise stress analyses at the crack tip have been attempted by Rice and his co-workers[42–46]. Solutions for small scale yielding of plane strain cases have been obtained for nonhardening materials[42, 43] and also for hardening materials[44]. These investigations have been further extended to the analysis of large crack tip geometry change by using the finite strain theory[45].

Wells[47] and Turner and Cheung[48] calculated the COD value by the finite element method and showed that the crack initiation may be characterized by the critical value of COD.

Sumpter and Turner[49], Parks[50] and Miyamoto and Kageyama[51] showed that the finite element method can be used to calculate the path-independent J integral and the crack initiation can be characterized by the critical value of the J integral.

7.4.3 Elastic–plastic analysis of a growing crack

Kobayashi et al.[52] and Anderson[53] proposed a nodal force relaxation technique simulating crack growth in the finite element model. Hsu and Bertels[54] also proposed the breakable element concept which can model the crack growth. Since then the finite element method has been used to investigate stable crack growth behavior for plane stress[55–60] and plane strain[61–67] cases under monotonically increasing loading condition.

Sorensen[63, 66] investigated stable crack growth of anti-plane shear and plane strain cases under an arbitrary loading history. His results showed that while the stress distribution ahead of a growing crack was found to be nearly the same as that ahead of a stationary crack, the strain values were lower for a growing crack.

Shih *et al.*[64] and Kanninen *et al.*[65] investigated several fracture criteria characterizing stable crack growth by using the experimentally obtained applied load vs stable crack growth curve as input information. The fracture criteria examined in their studies include the J integral, its rate of change during crack growth dJ/dc, the crack tip opening angle (CTOA), the energy release rate G and the crack tip force F.

d'Escatha and Devaux[67] evaluated stable crack growth based on microscopic development, i.e. void nucleation, void growth and coalescence. The purpose of this model was to predict the fracture properties of a material during the process of crack initiation, stable crack growth and crack instability. Various parameters used to correlate stable crack growth were evaluated by this model, including CTOA, the J integral and F.

Lee and Liebowitz[61] used the applied load–crack growth curve as input information, to produce a linear relationship between the plastic energy and crack growth.

Newman[68] performed finite element analyses using one of the fracture criteria to determine the applied load–crack growth behavior and instability for a given specimen geometry. He studied the effects of various parameters such as mesh size, strain hardening and critical strain on finite element fracture predictions for both monotonic and cyclic loading conditions. While some interesting observations were made, no attempt was made to correlate the predictions with actual material behavior.

Varanasi[56] used ultimate tensile strength as a fracture criterion to predict unstable crack growth. Belie and Reddy[60] used critical strain as a fracture criterion and adopted the "zero modulus–unload reload" scheme as a technique for crack growth modelling. Even though some comparisons were made between experimental results and numerical results in these two papers, the accuracy was very crude due to their unrealistic computational method.

The elastic–plastic finite element method was also used to investigate crack growth under cyclic loading conditions with simplified assumptions[54, 68–70].

PART 2

THERMOELASTIC–PLASTIC FRACTURE ANALYSIS

7.5 Introduction

The principal characteristics of the two distinct mechanisms of fracture, namely brittle fracture and ductile fracture, have been described in Part 1. Brittle fracture is often catastrophic. It can be handled satisfactorily by the theory of LEFM. Ductile fracture of solids, on the other hand, usually exhibits a sizeable stable crack growth preceding the ultimate breakdown of the structure. The unique blunting phenomenon associated with substantial plastic deformation ahead of the crack tip as described in Paris *et al.*[20] and Shih *et al.*[64] cannot be observed in solids fractured in "brittle" fashion. Elastic–plastic analysis is considered to be a necessary means to assess this type of fracture situation.

Although the consequence of these two types of fracture appears to be radically different, the boundary between the two is by no means rigid. A brittle solid may fracture like a ductile material at an elevated temperature environment. On the other hand, a ductile material such as Zircalloy can be embrittled by excessive hydrogen diffusion, a phenomenon which is a well-known fact in the nuclear industry. Another important environmental effect on the transition of brittle–ductile fracture of solids is the corrosion factor. However, of these three typical environmental effects, the thermal factor, i.e. temperature, is the prime concern to engineers. Elaborate discussion of its effect on the material's behavior has been well documented, e.g. in Rolfe and Barsom[71]. This part will present the reader with the concept of using the thermoelastic–plastic finite element analysis theory described in Chapter 3 for such applications as the determination of the J integral and crack extension by a special method involving the use of the "breakable" element algorithm. The finite element algorithm adopted in Chapter 3 takes into consideration of material properties variation with temperature and strain rates, which is another important factor in the fracture mechanism[71]. It provides an additional versatility of being able to handle combined thermal and mechanical load, which is a desirable feature when dealing with fracture analysis on structures involving significant temperature gradients.

7.6 Fracture criteria

As mentioned in the foregoing section, stable crack growth preceding fracture instability in ductile materials may occur when the plastic zone size at the crack tip is large as compared to the thickness of the material. The problem is of particular theoretical interest since the plastic zone behind the extending crack tip is unloaded to an elastic state, while that portion of the plastic zone ahead of the crack tip expands. The variation of the crack tip parameters during this crack extension process may provide useful information which may lead to realistic fracture criteria in the presence of large scale yielding.

The continuum equations described in Chapter 3 governing the stress-strain field near a growing crack tip may not be expected to account for the microstructural phenomenon of separation inherent in the fracture process. The same is true in the finite element modelling of a growing crack tip, but

176

for this analysis the difficulty is usually obviated through the use of fracture criteria based on some macroscopic field quantity. Extensive research of this point has identified nine possible requirements[64] which must be fulfilled by a candidate fracture criterion. However, the question of what is an acceptable fracture criterion still remains open to discussion.

Schaeffer et al.[72], Gavigan et al.[73] and Evans et al.[74] attempted to measure the strain field around the crack tip. Their results indicated that the strains at a small distance away from the crack tip are independent of geometry and, to a lesser extent, of material. As a consequence of this it is postulated that a rupture surface exists as an extension of the von Mises yield surface[75]. Fracture is assumed to occur when the state of strain at the crack tip has reached the rupture surface; the corresponding value of the effective strain defined in (3.22) is used as a fracture criterion. This value is calculated by numerically modelling a stable crack growth experiment. The effective strain corresponding to the maximum applied load with no crack extension is designated as the effective rupture strain, $\bar{\varepsilon}_{\text{rup}}$.

7.7   *J* integral with thermal effect

Despite the undisputable popularity of the *J* integral concept, particularly for application in the elastic–plastic fracture analyses, its validity and the inherent path-independence when used with the presence of thermal effect have not been widely discussed. The inadequacy of using the *J* integral in (7.12) for cracked structure subject to a nonuniform temperature field appeared to be first mentioned by Blackburn et al.[76]. Shortly afterward, Ainsworth et al.[77] proposed a modified form of the *J* integral to accommodate the thermal effect in nonlinear elastic solids. A similar form was also proposed by Kishimoto et al.[78] from a different approach. The same problem was investigated by Wilson and Yu[79] without making any reference to the previous work described in Ainsworth et al.[77] and their final expressions for the *J* integral are different from those presented there and in Kishimoto et al.[78]. However, the two sets of expressions can be shown to give the same results.

The following derivation is based on the thermoelastic–plastic theory described in Chapter 3, which could be readily incorporated into the TEPSA code for numerical evaluations. The reader will find that the expression of the *J* integral derived by this approach is indeed identical to those given in Ainsworth et al.[77] and Kishimoto et al.[78].

Consider a crack situated in a plane of a solid defined by a coordinate system (x_1, x_2) as illustrated in Figure 7.6. Assuming that the tip of the crack is the only place where the mathematical stress singularity exists, then the potential energy release in the region enclosed by the complete contour, i.e.

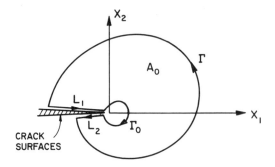

CRACK
SURFACES

Figure 7.6 A close contour on a plane with a line crack.

$\Gamma + L_1 + \Gamma_0 + L_2$ per unit volume, can be expressed as:

$$\pi = \oint (U\, dx_2 - t_i u_{i,1}\, ds) = 0 \qquad (7.17)$$

where U = strain energy density, t_i = surface traction to the contour, u_i = displacement components, and ds = segment of contour length.

Equation (7.17) is valid only for the case of mode I fracture with no presence of temperature field in the solid.

For the case where a temperature field, $T(x_1, x_2)$, exists in the cracked solid, the expression for the strain energy density is modified according to (3.12b) to take the form:

$$U = U(\varepsilon_{ij}, T) = \tfrac{1}{2}\sigma_{ij}(\varepsilon_{ij} - aT\delta_{ij}) \qquad (3.12b)$$

with a replacement of ΔT by T for the convenience of subsequent derivation.

The expression for thermoelastic stress becomes

$$\sigma_{ij} = \lambda\varepsilon_{kk}\,\delta_{ij} + 2\mu\varepsilon_{ij} - \frac{E}{1-2v}\,aT\delta_{ij} \qquad (3.12a)$$

The reader may prove that

$$\partial U/\partial\varepsilon_{ij} = \sigma_{ij} \qquad (7.18)$$

to justify the definitions given in (3.12a) and (3.12b).

The line integral for the potential energy release in (7.17) may be converted into an area integration via Green's theorem to be[†]

---

† Green's theorem is formulated to convert a surface integral to a volume integral. Here, the conversion is done between a line integral and area integral for a plane with uniform thickness. In such case, a contour line is equivalent to a surface, whereas an area enclosed by the contour is equivalent to a volume.

$$\pi = \oint \left(U \, dx_2 - t_i \frac{\partial u_i}{\partial x_1} \, ds \right)$$

$$= \oint \left(U \, dx_2 - \sigma_{ij} n_j \frac{\partial u_i}{\partial x_1} \, ds \right)$$

$$= \int_{A_0} \left[\frac{\partial U}{\partial x_1} - \left(\sigma_{ij} n_j \frac{\partial u_i}{\partial x_1} \right)_{,j} \right] dA \qquad (7.19)$$

where n_j is the outward normal unit vector to the contour.

Since $U = U(\varepsilon_{ij}, T)$, as shown in (3.12b), and ε_{ij} and T vary in (x_1, x_2) plane, the following chain rule of partial differentiation applies:

$$\frac{\partial U}{\partial x_1} = \frac{\partial U}{\partial \varepsilon_{ij}} \frac{\partial \varepsilon_{ij}}{\partial x_1} + \frac{\partial U}{\partial T} \frac{\partial T}{\partial x_1}$$

The first term in the above expression can be expressed to be $\sigma_{ij} \partial \varepsilon_{ij}/\partial x_1$ following the relation given in (7.18). The derivative $\partial U/\partial T$ in the second term can be shown to be $-\sigma_{ij} a \delta_{ij}$ or $-a\sigma_{ii}$. The term $\partial U/\partial x_1$ in (7.19) thus takes the form

$$\frac{\partial U}{\partial x_1} = \sigma_{ij} \frac{\partial \varepsilon_{ij}}{\partial x_1} - a\sigma_{ii} \frac{\partial T}{\partial x_1} \qquad (7.20)$$

The second term in (7.19) can be shown to have the form

$$- \sigma_{ij} \frac{\partial \varepsilon_{ij}}{\partial x_1} \qquad (7.21)$$

with the aid of $\sigma_{ij,j} = 0$ and $\frac{1}{2}(u_{i,j} + u_{j,i}) = \varepsilon_{ij}$ given in (3.10) and (3.6) respectively.

Substitution of (7.20) and (7.21) into (7.19) leads to a new relationship:

$$\pi = \oint \left(U \, dx_2 - t_i \frac{\partial u_i}{\partial x_1} \, ds \right) = - \int_{A_0} a\sigma_{ii} \frac{\partial T}{\partial x_1} \, dA \qquad (7.22)$$

which shows that the potential energy release in the region enclosed by the contour is no longer zero with the presence of a thermal field. It is obviously in contradiction to the case of purely mechanical loads as shown in (7.17).

Expand (7.22) by introducing all segments of the contour to give

$$\int_{\Gamma + L_1 + \Gamma_0 + L_2} \left(U \, dx_2 - t_i \frac{\partial u_i}{\partial x_1} \, ds \right) + \int_{A_0} a\sigma_{ii} \frac{\partial T}{\partial x_1} \, dA = 0$$

179

The segments L_1 and L_2 which are close to the stress-free crack surfaces make no contribution to the potential energy release, and so the above expression can be reduced to the form

$$\int_{\Gamma}\left(U\,dx_2 - t_i\frac{\partial u_i}{\partial x_1}\,ds\right) + \int_{A_0} a\sigma_{ii}\frac{\partial T}{\partial x_1}\,dA = \int_{\Gamma_0}\left(U\,dx_2 - t_i\frac{\partial u_i}{\partial x_1}\,ds\right)$$

with Γ_0 now taking the same direction as Γ.

Since the contour Γ_0 surrounding the crack tip is an arbitrary path of integration which can be taken to be infinitesimally small to characterize the behavior of the crack tip, it fits into the definition of the J integral described in Rice[12]. The proper expression of the J integral involving temperature field can thus be expressed by the above equality to yield

$$J = \int_{\Gamma}\left(U\,dx_2 - t_i\frac{\partial u_i}{\partial x_1}\,ds\right) + \int_{A_0} a\sigma_{ii}\frac{\partial T}{\partial x_1}\,dA \qquad (7.23)$$

which is identical to that given in Ainsworth et al.[77] and Kishimoto et al.[78]

It is readily seen from (7.23) that for cases where no temperature field is involved or an isothermal condition prevails, the J integral given in the above expression reduces to the usual form as given in (7.12).

The derivation of (7.23) was apparently based on linear thermoelastic theory†. However, this expression may be extended to nonlinear elastic or even plastic ranges by replacing the total quantities in (7.23) with respective incremental values as described in Chapter 3, although some concern has been expressed by some researchers about the accuracy of using the incremental plasticity theory as described in Section 7.4.

Similar forms of the J integral for the cases involving Mode I and II fracture were derived by Chen and Chen[80] by extending the work of Wilson and Yu[79].

Numerical evaluation of (7.23) by finite element algorithm first requires the discretization of the solid into the form

$$J = \sum_{m=1}^{M}\left[W_m\,\Delta y_m - \left(\frac{t_m^x\,\Delta u_m}{\Delta x} + \frac{t_m^y\,\Delta v_m}{\Delta x}\right)\Delta s_m\right]$$

$$+ a\sum_{n=1}^{N}(\sigma_{xx,n} + \sigma_{yy,n} + \sigma_{zz,n})T_n A_n/\Delta x \qquad (7.24)$$

where M = total number of elements along the assigned path of integration Γ;

m = subscript designating the mth element along Γ;

N = total number of elements on and enclosed by Γ;

---

† One clear violation of the nonlinear theory is the invalid use of the relationship given in (7.18).

n = subscript designating the nth element in the region bounded by Γ;

W_m = strain energy in element m;

Δy_m = component of the segment of the path Δs in the y-direction;

Δs = path length across the element m;

t_m^x, t_m^y = respective surface tractions acting on Δs in the x- and y-directions;

$\Delta u_m, \Delta v_m$ = respective element displacement components in the x- and y-directions;

$\sigma_{xx,n}, \ldots$ = stress component in element n in the directions $x, \ldots$;

T_n = temperature in element n;

A_n = area of element n.

Equation (7.24) has been modified further to be adapted to the thermoelastic–plastic algorithm described in Chapter 3 in the form

$$
\begin{aligned}
J = \sum_{m=1}^{M} \Bigg\{ & \sum_{t} \tfrac{1}{2}(\{\sigma\}_{m,t} + \{\sigma\}_{m,t-\Delta t})(\{\varepsilon\}_{m,t} - \{\varepsilon\}_{m,t-\Delta t})\, \Delta z_m \\
& - [(\sigma_{rr,m}\cos\theta_m + \sigma_{rz,m}\sin\theta_m)\varepsilon_{rr,m} \\
& + (\sigma_{rz,m}\cos\theta_m + \sigma_{zz,m}\sin\theta_m)(u_{z,t} - u_{z,t-\Delta t})_m/\Delta r]\, \Delta s_m \Bigg\} \\
& + \sum_{n=1}^{N} a(\sigma_{rr,n} + \sigma_{zz,n} + \sigma_{\theta\theta,n})T_n A_n/\Delta r
\end{aligned}
\tag{7.25}
$$

Notation in (7.24) and (7.25) is described in Figure 7.7. The time t and its increment Δt in a thermoelastic–plastic stress analysis, of course, are fictitious. They are used for the designation of loading steps.

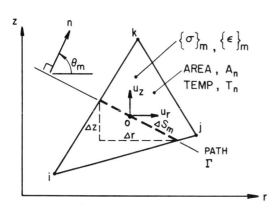

Figure 7.7 *J* integral by the finite element method.

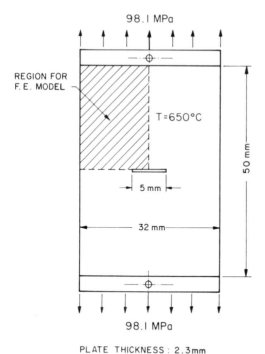

98.1 MPa

REGION FOR
F. E. MODEL

T=650°C

50 mm

5 mm

32 mm

98.1 MPa

PLATE THICKNESS : 2.3mm

Figure 7.8 Geometry and dimensions of a center-cracked plate.

7.8 Numerical illustrations of *J* integrals with thermal effect

The following numerical illustrations will demonstrate the thermal effects on the evaluation of the *J* integral. The suitability of the *J* integral for thermo-fracture analysis may be assessed from the results of these examples.

Figure 7.8 shows the geometry and dimensions of a center-cracked plate used for all these numerical illustrations. Temperature-dependent material properties given in Figure 3.24 were used in the analysis. A special version of the TEPSA code incorporating the algorithm given in (7.25) was used to determine the numerical values of *J* integrals subject to various thermal conditions.

7.8.1 *J integral in uniform temperature field*

The cracked plate is assumed to have a uniform temperature distribution everywhere. However, two conditions were used:

Case 1. The plate was initially at 21°C but was heated to 650°C with a simultaneous mechanical load increment to a maximum of 98.1 MPa.

Case 2. The plate was uniformly heated to a temperature of 650°C followed by the application of mechanical load to the same maximum value.

182

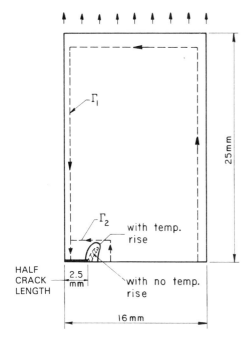

Figure 7.9 Contours for numerical evaluation of the *J* integral in isothermal conditions.

Two paths, Γ_1 and Γ_2, were chosen for the *J* integral. As can be seen from Figure 7.9, Γ_2 was much closer to the crack than Γ_1.

Table 7.1 shows the numerical values of the *J* integral determined by integration along Γ_1 and Γ_2 for the two cases at 5 selected load levels. Analytical results computed from the Dugdale approximate formula[81] are tabulated as the right-hand column for the comparison of results. As can be observed from this table, the maximum deviation of *J* values determined by different paths is 14.74%. Another point worthy of note is that more pronounced path sensitivity resulted for the Case 1 condition in which the effect

Table 7.1 *J* integral in a center-cracked plate with uniform temperature field.

	Case 1 (with temp. variation: 21–650°C)			Case 2 (at constant temp.: 650°C)			Analytical results (Dugdale approximation)
	J (kPa mm)		Deviation	*J* (kPa mm)		Deviation	
Load (MPa)	Path Γ_1	Path Γ_2	(%)	Path Γ_1	Path Γ_2	(%)	*J* (kPa mm)
35.0	67.2600	77.1760	14.74	69.3840	70.8360	2.09	70.4349
50.0	154.7510	162.3040	4.88	144.6735	147.7362	2.12	147.6970
65.0	284.4360	282.5085	−0.68	253.3642	258.7788	2.14	259.6639
80.0	452.5068	438.3666	−3.12	406.9274	415.6193	2.16	415.7092
98.1	729.3461	693.5849	−4.90	687.3081	696.0703	1.27	688.8769

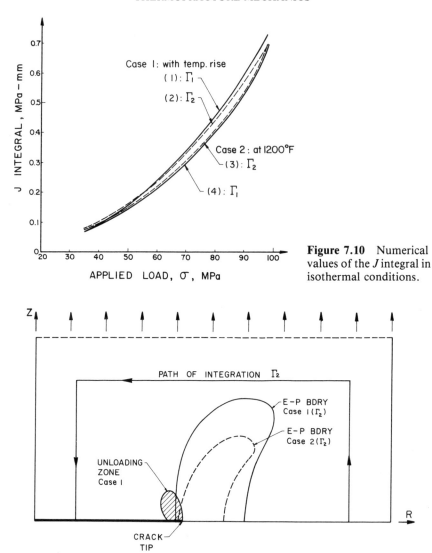

Figure 7.10 Numerical values of the J integral in isothermal conditions.

Figure 7.11 Plastic zones for case 1: simultaneous application of thermal and mechanical loads; case 2: preheated condition.

of temperature-dependent material properties in the forms of the associated strain components described in (3.45a–c) were included in the evaluation process.

Figure 7.10 illustrates the variation of the J parameter with respect to applied mechanical loads, whereas the comparison of the spreading of plastic zones for the two cases is indicated in Figure 7.9 with an enlarged view and unloaded region depicted in Figure 7.11.

7.8.2 J integral in nonuniform temperature field

We shall now examine the effect of thermal stresses induced by the non-uniform temperature distribution introduced to the same center-cracked plate with the geometry described in Figure 7.8. The length of the plate, however, was doubled. As will be seen in later illustrations, three paths were selected for the J integrations.

A temperature variation following the function

$$T(x_1) = \Delta T \left(\frac{x_1}{W}\right)^2 \qquad (7.26)$$

was assumed to exist in the cracked plate, in which x_1 is the coordinate along the crack with its origin at the center of the crack, W is the half width of the plate (or 16 mm in this case) and ΔT is the maximum temperature differential between the center of the crack ($x_1 = 0$) to the edge of the plate ($x_1 = W$). The temperature at the center of the crack was assumed to be 800°F (427°C).

The above temperature variation function is identical to what was used in Ainsworth et al.[77].

The value ΔT in these case studies was determined by the formula:

$$\Delta T = 1.8 \sigma_y / E a$$

where σ_y = initial yield strength of the material at 800°F = 290 MPa, E, a = are the respective Young's modulus and the linear thermal expansion coefficient with numerical values of 198 GPa and $1.0 \times 10^{-5}/°F$, and $\Delta T = 263.7°F$ was thus used for all case studies.

The following two cases were examined:

Case 3. A temperature field following (7.26) was first applied and followed immediately by a mechanical load ramp (Figure 7.12). Numerical values of the J integral were determined using constant material properties at 800°F.

Table 7.2 shows the numerical values of the J integral at various load levels obtained from three paths.

Graphical representation of the above variation is illustrated in Figure 7.13. Normalized J values are used in the plot with a normalizing factor, J_e, which can be expressed as[77]:

$$J_e = E a^2 (\Delta T)^2 (\pi c) [\tfrac{1}{3} - (c/W)^2/2]^2$$

A linear thermal expansion coefficient $a = 10 \times 10^{-6}/°F$ was used.

Case 4. With identical loading conditions as in the previous case, except that all properties were to vary with temperature following Figure 3.24, and associated strain components given in (3.45a–c) as well as the temperature-dependent plastic potential function described in the last term of (3.50) were included in the numerical evaluation.

185

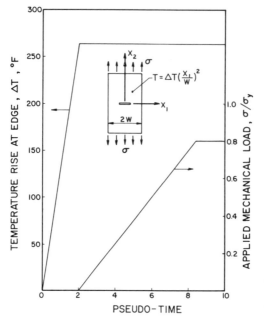

Figure 7.12 Variation of thermal and mechanical loads with a nonuniform temperature field.

Table 7.2 J integral in a center-cracked plate with nonuniform temperature field but with constant material properties.

Pseudo-time	Maximum temperature difference, ΔT (°F)	Mechanical load σ/σ_y	J integral (kPa mm)			Deviation $(\Gamma_1-\Gamma_3)/\Gamma_1$ (%)
			Γ_1	Γ_2	Γ_3	
1	128	0	263.32[†]	269.53[†]	266.24[†]	
2	263.68	0	1238.10	1266.10	1232.06	
3	263.68	0.1	1824.08	1862.23	1765.72	3.2
4	263.68	0.2	2785.11	2838.55	2550.00	8.4
5	263.68	0.3	3718.77	3772.97	3227.85	13.2
6	263.68	0.4	5226.38	5215.50	4265.48	18.4
7	263.68	0.5	7267.02	7115.34	5654.15	22.2
8	263.68	0.6	9860.41	9499.88	7390.84	25.0
9	263.68	0.7	13 117.57	12 547.77	9679.33	26.2
10	263.68	0.8	17 913.25	17 089.96	13 218.55	26.2

† Approximated analytical value by Dugdale[81] is 262.74.

Significant difference in the results was obtained, as will be observed from Table 7.3. Again, the graphical representation of this variation can be found in Figure 7.14.

The plastic zones in the cracked plate at various loading levels for both Cases 3 and 4 have been shown in Figures 7.15 and 7.16 respectively. These contours were determined based on the Γ_1 path in each case. A very peculiar

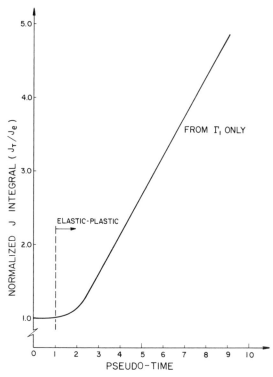

Figure 7.13 Normalized J integral using constant material properties.

Table 7.3 J integral in a center-cracked plate with nonuniform temperature and temperature-dependent properties.

Pseudo-time	Maximum temperature difference, ΔT (°F)	Mechanical load σ/σ_y	J integral (kPa mm)			Deviation $(\Gamma_1-\Gamma_3)/\Gamma_1$ (%)
			Γ_1	Γ_2	Γ_3	
1	128.00	0	245.07	253.84	251.76	
2	263.68	0	991.65	1038.71	1023.86	
3	263.68	0.1	1574.04	1647.79	1591.55	−1.11
4	263.68	0.2	2342.39	2451.54	2274.57	2.90
5	263.68	0.3	3437.02	3583.17	3129.18	8.90
6	263.68	0.4	5012.57	5152.03	4294.43	14.30
7	263.68	0.5	7214.27	7290.57	5900.05	18.20
8	263.68	0.6	10 312.80	10 335.60	8228.73	20.20
9	263.68	0.7	14 896.69	14 915.48	11 851.41	20.40
10	263.68	0.8	54 595.80	56 294.88	51 162.98	6.20

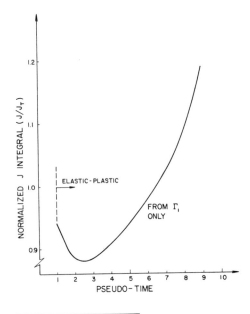

Figure 7.14 Normalized J integral using temperature-dependent material properties.

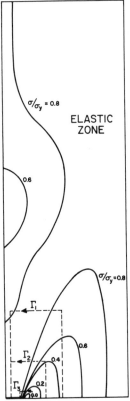

Figure 7.15 Spread of plastic zones with a nonuniform temperature field and constant material properties.

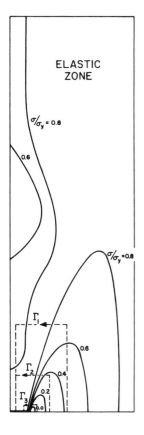

Figure 7.16 Spread of plastic zones with a nonuniform temperature field and temperature-dependent material properties.

phenomenon can be observed from these illustrations about the substantial size of the plastic zones along the center line of the plate in addition to those near the tip of the crack.

These excessive plastic regions were no doubt induced by the thermal stresses as a result of the nonuniform thermal field. This phenomenon can partially explain the significant difference between the J values determined by different paths at higher load levels (see last column in Tables 7.2 and 7.3). A similar conclusion that the J integral becomes path-dependent was drawn by Ainsworth et al.[77].

7.9 The "breakable element"

The "breakable element" concept is based on the successive reduction of the stiffness matrix of the crack tip element during a simulated crack growth. One stiffness reduction scheme that has been used in the past to treat crack propagation in concrete structures[82] consisted of setting Young's modulus

189

to zero. Another procedure[83] used a zero Young's modulus for the crack tip element to achieve the singularity present in elastic crack analysis.

In 1976, Hsu and Bertels[54] proposed the stiffness reduction scheme to model crack growth, referred to as the "breakable element". However, in their case study breakable elements would skip during crack growth. After a similar attempt to model the crack growth by a simple reduction of the stiffness matrix[84], the authors found that the computational accuracy can be further improved by creating a pseudo-nodal point in the breakable element used in the stiffness reduction scheme. This pseudo-nodal point moves through the breakable element as the crack tip extends, producing smoother crack growth[85].

7.9.1 Implementation of breakable element algorithm

The breakable element concept as described above can be readily incorporated into the TEPSA code. The major steps of implementation can be outlined as follows:

Step 1. Position the breakable elements along the expected path of crack extension as illustrated in Figure 7.17. In this particular case, breakable elements are positioned along the expected crack path (Figure 7.17a). The following steps are then employed to simulate the crack growth in the finite element model at any given load step.

Step 2. Upon completion of the thermoelastic–plastic stress analysis at a given load step, the effective strains in the breakable elements are extrapolated as a smooth curve toward the crack tip using a least squares curve fitting technique[86]. The best numerical results were obtained by extrapolating the average effective strains at the centroids of the first four elements immediately in front of the crack tip. The distribution of the effective strain ahead of the crack tip can be expressed as:

$$\bar{\varepsilon}(x) = a_1 x^3 + a_2 x^2 + a_3 x + a_4 \qquad (7.27)$$

where x denotes the distance from the crack tip along the crack path, and a_1, a_2, a_3 and a_4 are constants derived from the least squares analysis of the average strains at the element centroids (x_1, x_2, x_3, x_4). If the extrapolated strain at the crack tip, $\bar{\varepsilon}_{ext}(\bar{\varepsilon}(0))$, does not exceed the assigned fracture criterion, $\bar{\varepsilon}_{rup}$, then the stress analysis proceeds with the application of the next load increment.

Step 3. If $\bar{\varepsilon}_{ext}$ exceeds $\bar{\varepsilon}_{rup}$, the crack growth process begins. Figure 7.17b illustrates schematically the start of crack growth at load step i. The strains in the proportional length Δx of the breakable element adjacent to the crack tip (crack tip element) are estimated to have exceeded $\bar{\varepsilon}_{rup}$. Thus the portion Δx of the element is deemed to have fractured and to have become incapable of carrying any load. The amount of crack extension, Δx, is evaluated by solving for the value of x in (7.27) at which $\bar{\varepsilon}(x) = \bar{\varepsilon}_{rup}$.

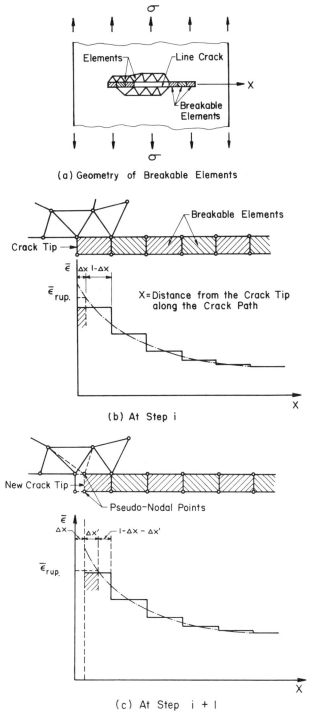

(a) Geometry of Breakable Elements

(b) At Step i

(c) At Step i + I

Figure 7.17 Schematic description of breakable element algorithm.

Step 4. For the next load step $i + 1$, the $[K_e]$ matrix of the crack tip element is evaluated with a proportionally reduced volume. The original nodal points at the crack tip are shifted by the amount Δx to the positions of the pseudo-nodal points as illustrated in Figure 7.17c, which specify the current location of the crack tip. This nodal point shift changes the $[B]$ matrix, which in turn reduces the $[K]$ matrix in (3.101a) to

$$[K'] = \int_{v'} [B']^T [C][B'] \, dv' \qquad (7.28)$$

where $[K']$ is the reduced $[K]$ matrix, $[B']$ is the $[B]$ matrix after the nodal point shift, and v' is the reduced volume of the crack tip element. Obviously the nodal point shift in the $[B]$ matrix results in larger strain and stress increments in (3.110) and (3.96) respectively. At this load step an additional portion, $\Delta x'$, may be found to exceed $\bar{\varepsilon}_{rup}$. This is dealt with similarly as outlined in Step 3.

Step 5. The shifting to the new pseudo-nodal points at each load step continues so long as $\bar{\varepsilon}_{ext}$ exceeds $\bar{\varepsilon}_{rup}$. When the aspect ratio of the remaining ligament of the crack tip element reaches a critical value, say 8, the pseudo-nodal points are considered to have reached the next nodal points, thereby maintaining numerical stability. In this manner, the whole element is gradually broken and simultaneously its stiffness progressively reduced to and then maintained as a zero stiffness element.

7.9.2 *Nodal force relaxation*

Once the crack front has passed through a crack tip element, a nodal force relaxation procedure is implemented in order to redistribute the stress field previously supported by the crack tip element before the element breaks. The following sequence of implementation has been developed for this purpose:

(1) The nodal reaction force of the "broken" crack tip element (say element number 1 in Figure 7.18) is calculated from the accumulated stress before the element breaks:

$$\{F\} = \int_{v} [B]^T \{\sigma\} \, dv \qquad (7.29)$$

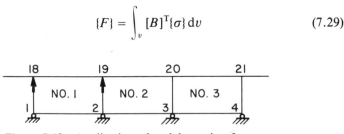

Figure 7.18 Application of nodal reaction forces.

(2) Since relaxation occurs in the direction normal to the crack growth path, only the loading direction components of the reaction forces at nodes 18 and 19 as illustrated in Figure 7.18 are applied over five[†] equal incremental loading steps. A stress analysis on the whole structure is performed with these nodal reaction forces, while the external load remains constant. The increments of displacements and stresses so derived are added to the accumulated displacements and stresses of the structure.

(3) The amount of crack extension due to the nodal force relaxation is determined after the fifth relaxation step using the extrapolation scheme outlined in the previous section. If the extended crack tip exceeds the next nodal point (i.e. node 3 in Figure 7.18), continuous nodal force relaxation of the subsequent element (i.e. element No. 2 in Figure 7.18) must also be carried out. Otherwise, a further load increment is necessary in order to increase the strain distribution ahead of the crack tip.

In Step 2, the stress analysis for the entire structure requires routines which change the element stiffness according to the incremental theory of plasticity, since the relaxation technique results in simultaneous unloading of the newly created crack surface behind the crack tip and loading of the region ahead of it. Thus the $[K_e]$ matrix is calculated using the $[C_e]$ and $[C_{ep}]$ matrices for the unloading and loading elements respectively.

The determination of element loading or unloading is done iteratively using a trial application of Step 2. Initially all elements are assumed to continue loading. Equation (3.102) is solved and the incremental stresses due to the trial load step are added in the usual manner. For each element the effective stress ($\bar{\sigma}_2$) for the trial load step is compared with the effective stress ($\bar{\sigma}_1$) prior to the application of the trial step. The loading and unloading elements are determined for $\bar{\sigma}_2 \geqslant \bar{\sigma}_1$ and $\bar{\sigma}_2 < \bar{\sigma}_1$ respectively (Figure 7.19). The stress state prior to the trial step is then re-established and the crack is extended as outlined above with the loading and unloading elements now identified.

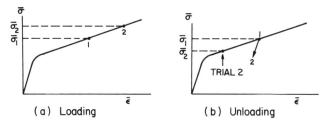

Figure 7.19 Determination of loading or unloading in an element.

† It was found that increasing the number of loading steps beyond five provided only a marginal improvement in the results for a significant increase in the computational expense.

7.9.2 *"Breakable elements" for mixed mode crack growth*

An attempt was made by Kim[87] to apply the "breakable" element concept to cases involving combined Mode I and II fracture. Prediction of crack growth in this type of problems requires not only a realistic fracture criterion, but also the proper positioning of the breakable elements along the crack growth path which has to be determined first.

There are three common schemes proposed by various researchers for the prediction of the direction for crack propagation in brittle solids:

(1) *Maximum tangential stress criterion.* Erdogan and Sih[88] have analyzed the crack extension in a plate with a central crack inclined at angle to the direction of the axially applied uniform stress. They assumed that the crack would propagate in the radial direction at which the tangential stress at the crack tip is maximum. Physically this means that the crack will propagate in the direction at which the stress component tending to open the crack tip is a maximum.

(2) *Minimum strain energy density criterion*[89]. The theoretical basis for the formulation of this criterion is that the crack will spread in the radial direction along which the strain energy density is a minimum, and the onset of fracture is determined when a critical value of the minimum strain energy density is reached.

(3) *Maximum energy release rate criterion.* Palaniswamy and Knauss[90] proposed that a crack will propagate in the radial direction along which the elastic energy release per unit crack extension is a minimum and the onset of fracture is determined by some critical value of the energy release rate. Hellen[91] applied this criterion in a step-by-step procedure to study the problem of fatigue crack growth with the aid of the finite element method. The numerically predicted crack path showed good agreement with the experimental results.

(4) *Minimum effective strain criterion for crack growth in ductile solids.* The fact that significant plastic deformation near the tip of a crack normally precedes its growth in a ductile solid requires appropriate modification of the above criteria developed for brittle solids. A criterion based on a minimum effective strain was proposed by the author and his graduate student[87]. This criterion was based on two hypotheses which can be shown to be consistent with those already mentioned above: that a crack grows in the radial direction along which the effective strain attains a minimum value and that the onset of fracture is determined when a critical value of the minimum effective strain is reached. The first statement in the above hypotheses is justified by considering, following the work of Sih[92], that a crack grows in the direction of least distortion energy (or plastic work). The description of the plastic deformation of solids in Chapter 3 clearly indicates that the effective strain $\bar{\varepsilon}$ is proportional to this distortion energy. The second

statement in the hypotheses suggests the use of the effective rupture strain $\bar{\varepsilon}_{rup}$ as the fracture criterion, as for the Mode I fracture analysis. A clear advantage of using this criterion is apparent: the determination of $\bar{\varepsilon}$ in a cracked structure can be used to predict not only the critical load at which crack growth will occur, but also the direction of the growth.

Finite element analysis was performed on a thin aluminum plate containing a slant line crack at $45°$ from the direction of the unidirectional load. Detailed description of the analysis and experimental investigation can be found in Kim[87]. Analytical results indicated crack growth paths to be bounded at $-53°$ by the maximum tangential stress criterion and at $-51°$ as predicted by the minimum strain energy density criterion, with the results obtained by the minimum effective strain criterion lying between these two bonds. All these values fell well within close ranges measured by experiments.

Once the crack growth path in a mixed mode case is determined by using either of the above four proposed criteria (preference is obviously given to the fourth one), breakable elements can then be positioned along this path. The rate of the crack growth can be determined by using the general "breakable" element algorithm.

7.10   Numerical illustrations of stable crack growth

The breakable element algorithm was used to predict the stable growth of a line crack in a large thin aluminum plate[85, 87]. A fracture criterion of $\bar{\varepsilon}_{rup} = 0.12$ was used for both Mode I and mixed Mode I and II cases. This value was established following the procedure outlined in [84]. The predicted growth of the crack under monotonically increasing mechanical loads by the finite element analysis showed excellent correlation with those observed by experiments. The unique advantages of using the breakable elements for the detailed description of stresses, strains and crack profiles at any given loading stage have been fully demonstrated in these references.

The following numerical illustrations will demonstrate the application of the same breakable element algorithm for the prediction of slow crack growth in a solid in a thermal environment. The geometry and dimensions of the center-cracked plate are identical to those already described in Figure 7.8. Again, the temperature-dependent material properties illustrated in Figure 3.24 are used in the finite element analysis. Since the purpose of these case studies is to illustrate the algorithm of the breakable elements and also the significance of the thermal effect, a hypothetical $\bar{\varepsilon}_{rup} = 0.03$ was used as the fracture criterion. Also, a much coarser size of breakable elements was adopted in the finite element idealization which is shown in Figure 7.20. A total of 11 such elements were placed along the crack growth path. Each

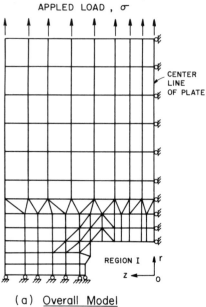

(a) Overall Model

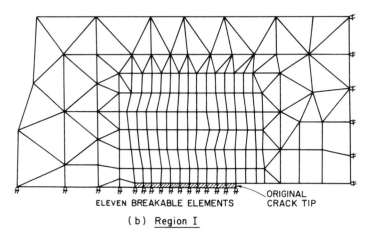

(b) Region I

Figure 7.20 Finite element idealization of a center-cracked plate.

element was 0.2 mm long, which is about 1/12th of the half crack length, which was one order of magnitude higher than that used in Kim and Hsu[85] and Kim[87]. A total of 315 elements and 285 nodes were used in the model with a refined mesh near the line crack as shown in Figure 7.20b. The low $\bar{\varepsilon}_{rup}$ value and large breakable elements were necessary in order to economize the computational effort and time.

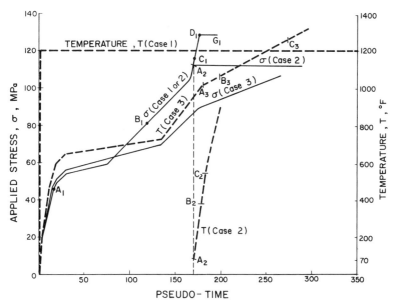

Figure 7.21 Variation of temperature and mechanical loads in a center-cracked plate.

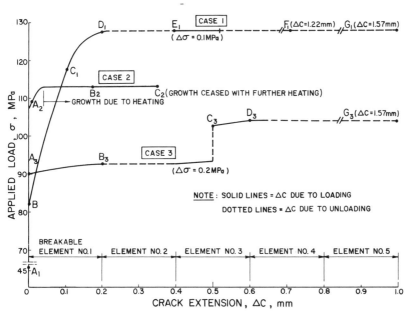

Figure 7.22 Crack growth in a center-cracked plate in various thermal environments.

Three cases were considered. These were:

Case 1. The plate was preheated to 1200°F (650°C) followed by a monotonically increasing mechanical load acting normal to the crack.

Case 2. The mechanical load was applied to the plate first and then followed by a linear temperature rise in the plate.

Case 3. Both mechanical and thermal loads were applied simultaneously to the plate.

Loading rates for the above three cases are depicted in Figure 7.21.

Crack growth for all three cases has been illustrated in Figure 7.22. With reference to Figure 7.21, Case 1 represents the growth of a crack due to the mechanical loading in a preheated plate with substantially less strength. It is thus not surprising that the growth is initiated at a significantly lower

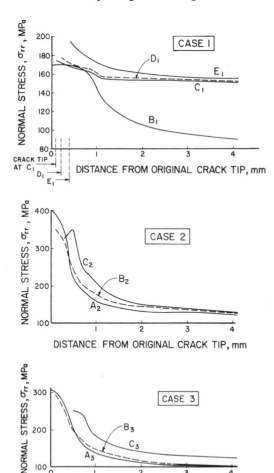

Figure 7.23 Distribution of normal stress along the path of crack growth.

loading level of 82 MPa (Point B_1) than that in Case 2 in which mechanical load was applied to the cracked plate at room temperature. The crack growth initiation load in this case is 107 MPa (Point A_2). Case 3, being the case with simultaneous application of thermal and mechanical loads, indicates a crack growth initiation load of 90 MPa (Point A_3) which falls between the above two load levels.

The patterns of crack growth for these three cases are also radically different. As material was regarded to be "soft" at elevated temperature in Case 1, a significant amount of stable growth (0.2 mm as observed in Figure 7.22) was allowed before reaching the instability point, D_1, beyond which no further load was needed to maintain further growth. In contrast to Case 1

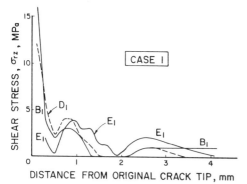

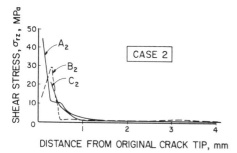

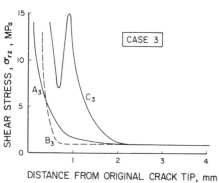

Figure 7.24 Distribution of shear stress along the path of crack growth.

199

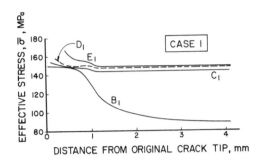

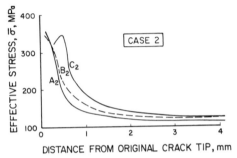

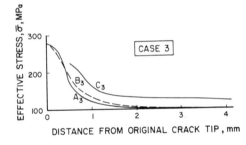

Figure 7.25 Variation of effective stress along the path of crack growth.

was a much less stable crack growth for Case 2 in which only 0.04 mm growth was observed. Case 3 presents a rather interesting growth pattern. A rather low instability load was computed (load at B_3). However, the unstable growth was arrested after a substantial extension of 0.5 mm and further growth was revitalized after an additional 10 MPa load was introduced. The ultimate instability occurred at D_3. Another interesting point worth noting is that a significant growth of the crack due to heating alone took place in Case 2 (point C_2). Further growth could not be observed with additional heating.

The distribution of normal stress and shear stress components along the path of crack growth for all three cases at selected growing stages have been shown in Figures 7.23 & 24 respectively. Large fluctuations were observed in the shear stress distributions due to numerical discrepancies.

Figures 7.25 & 26 show the effective stress and strain distribution at the same locations as for Figures 7.23 & 24. Again, fluctuations of effective

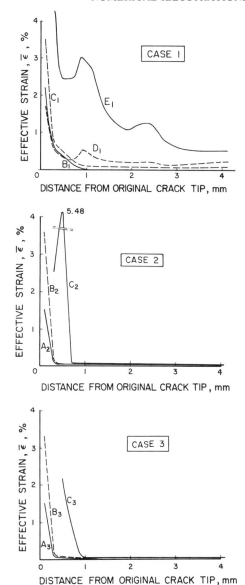

Figure 7.26 Variation of effective strain along the path of crack growth.

strain distribution can be observed in the breakable elements after passing the instability points.

The spreading and kinetics of plastic zones in a growing crack have been illustrated in Figure 7.27 and the profiles of the crack at various stages of growth are shown in Figure 7.28. The unique blunting phenomenon preceding the crack extension in a ductile solid has been depicted by the breakable element algorithm.

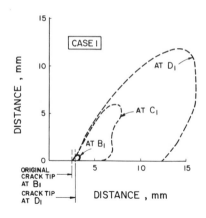

Figure 7.27 Plastic zones ahead of a growing crack for Case 1.

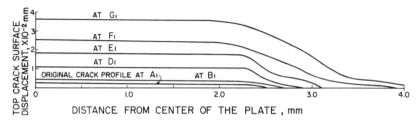

Figure 7.28 Profiles of the crack at various stages of growth for Case 1.

Among many unique advantages of the breakable elements algorithm as described in Refs. 85 and 87, a significant point which deserves special attention is its ability to accommodate much larger mesh size than that permitted by other finite element algorithms for a similar purpose. In the above numerical illustrations, unusually large breakable elements were used. Although the crack growth appeared to be smooth for Cases 1 and 2, as demonstrated in Figure 7.22, a visible discontinuity was observed in Case 3. Such a discontinuity could have been avoided had more and smaller breakable elements been used. The excessive size of the breakable elements used in the case studies was also responsible for the fluctuations in the distributions of stresses and strains in Figures 7.24–26 as those quantities are assumed to be constant in each individual element as described in Chapter 3.

Case 1 and Case 2 were designed to illustrate a very important fact that exists in thermomechanical analysis. As pointed out by Argyris and Kelsey[93], the work done on a structure, and hence the stored strain energy, depend on the sequence of application of mechanical loads and temperature. As readily seen from the foregoing descriptions, the thermal and mechanical loads applied to the same center-cracked plate were in the reversed sequences in Cases 1 and 2. As a result, substantially different crack growths were observed.

PART 3

T<small>HERMOELASTIC–PLASTIC CREEP FRACTURE ANALYSIS</small>[†]

7.11 Literature review

The growing concern about the possible damage of structural components by creep at high temperature has been discussed in Chapter 4. Here we will take a close look at recent developments in analytical models to deal with this situation. Of particular interest is the growth of a crack primarily induced by the creep deformation of the material. This subject, which is usually referred to as creep fracture, is a critical element in the design of high temperature components as minute flaws or defects in the material can grow without additional loading until the moment of catastrophic failure, which leaves little or no time for the operator to take necessary preventive actions.

It is therefore understandable that considerable effort has been made by many prominent researchers in recent years attempting to characterize the creep crack growth behavior in solids. As can be expected, early work has been closely related to the LEFM approach, as described in Section 7.2. Some forms of power functions of the stress intensity factor of a notched solid were proposed to characterize the creep crack growth rates[95–99]. During the same period, another model based on the net section stress parameter was proposed for this purpose[100–102] as well as models based on the rates of change of COD and CTOD (crack tip opening displacement)[103–105]. The inherently nonlinear nature of the problem was eventually incorporated[15, 16, 106] in a power line integral C^*. Although the last concept differs in the physical interpretation, the mathematical formulation appears to be the direct time differentiation of the J integral given in (7.12) for the elastic-plastic fracture analysis. Various fracture criteria have also been proposed for the analytical models. Among these are the critical local strain criterion[101], the energy dissipation[107], the generalized creep damage hypothesis[108], the Dugdale model[109], the critical plastic zone size[110] and the critical COD[111]. A complete detailed review of these criteria has been included in Fu[112].

The various approaches described above indeed appear to be both inconsistent as well as somewhat confusing. Some researchers, e.g. Fu[112] and Riedel and Rice[113] have reached the qualitative conclusion that the stress intensity factor concept can be used to characterize the crack growth if the creep zone is small in comparison to the crack length and the size of the solid. The power line integral C^* is considered to be a viable loading parameter which can determine the near-tip stress field. The net section stress is

---

[†] The main portion of Part 3 has been published in *Journal of Engineering Fracture Mechanics*. Permission to reproduce this material was granted by Pergamon Press Inc.

regarded as a good approximate criterion for the case involving a large inelastic zone in the solid.

7.12   Generalized creep fracture model

It is apparent from the description in the foregoing section that there is no single parameter that can describe the growth of a crack in a solid under general thermomechanical loading conditions. The concept of the C_g^* integral was formulated in an attempt to achieve just this purpose[94].

The generalized thermomechanical creep crack growth model, which includes the C_g^* integral and the tearing modulus of the material, was constructed on the theory of energy balance and the energy dissipation in the fracture process region. Like the J and C^* integrals, the C_g^* integral is determined by a contour integration with the potential energy induced by the general thermomechanical loadings computed from the TEPSAC code.

The derivation of this generalized model is based on the following assumptions:

(1)   The material is assumed to be initially isotropic and homogeneous. A single through crack is involved in a solid of uniform thickness as illustrated in Figure 7.29;

(2)   Because of the large rate of void initiation and growth and linkage taking place at the tip of the crack during creep deformation, a small fracture process zone such as the ones described by Broberg[114, 115] is assumed to exist in that region as shown in Figure 7.29. The continuum mechanics theory ceases to apply within this region.

(3)   An energy balance is always maintained between all energies produced by the thermoelastic–plastic creep loading in the solid excluding the fracture process zone and that dissipated into that zone. The latter region is regarded as an energy sink in the system.

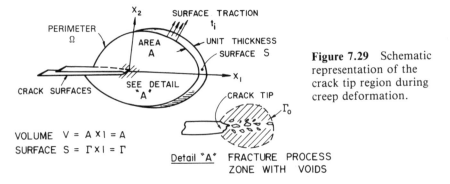

Figure 7.29 Schematic representation of the crack tip region during creep deformation.

7.12.1 Creep crack growth model

As the C_g^* integral, in a true sense, is a generalized form of the C^* integral[15, 16], it may be defined as

$$C_g^* = d\dot{\pi}/dc \qquad (7.30)$$

where $\dot{\pi}$ = potential energy rate function including all inelastic strain energies, and c = crack length.

Since the metallurgical interpretation of creep deformation of solids is one of continuous initiation and growth of voids in the region with high magnitude of stresses and/or temperature, as illustrated in Figure 7.29, the crack extension can be treated as the result of the linkage of these growing voids. It is therefore easy to envisage that creep crack growth is indeed a discontinuous process and to have a reason to assume that the extension of a crack due to creep occurs when the accumulation of the creep damage energy reaches a critical value in the fracture process zone. Mathematical translation of this concept yields the form

$$E_c = \int_0^t C_g^* \, dt \qquad (7.31)$$

where E_c is the aforementioned critical creep damage energy corresponding to the creep crack growth.

Numerically E_c may be expressed in terms of an equivalent J integral:

$$E_c = \xi J_e \qquad (7.32)$$

The term J_e is the equivalent J integral for the case involving all forms of elastic and inelastic deformation.

The variation of E_c in (7.32) with respect to the growing crack during a time increment dt can be derived by simply differentiating this equation to give

$$C_g^* = \frac{dE_c}{dc} = \xi \frac{dc}{dt} \frac{dJ_e}{dc} = \xi \dot{c} \frac{dJ_e}{dc} \qquad (7.33)$$

Following the definition of the tearing modulus[20]:

$$T = \frac{E}{\sigma_0^2} \frac{dJ}{dc}$$

and consider J to be a special case of the general term J_e, the above definition can be extended[†]:

$$T = \frac{E}{\sigma_0^2} \frac{dJ_e}{dc}$$

---

† The assumption of $J \doteq J_e$ leading to the $\dot{c}$ model in (7.34) has been shown to give excellent correlation with experimental results[94], although it is generally believed that the rupture modes of creep cracking differ from that of plastic tearing.

for the case of crack growth under general thermoelastic–plastic creep loads. Thus, by substituting the last expression into (7.33), a general crack growth model may be derived:

$$\dot{c} = \frac{EC_g^*}{\sigma_0^2}\bigg|\xi T \qquad (7.34)$$

where C_g^* is a generalized power integral defined in (7.30), E is the modulus of elasticity of the material and $\sigma_0 = (\sigma_y + \sigma_u)/2$ is the flow stress, σ_y and σ_u being the respective initial yield and ultimate tensile strengths of the material. The tearing modulus T is another material property according to Paris et al.[20].

The creep crack growth model given in (7.34) is simple to use provided that the numerical value of C_g^* can be determined.

7.12.2 Expression of the C_g^* integral

The exact expression for the C_g^* integral may be derived from the mathematical definition given in (7.30).

By referring to Figure 7.30, the potential energy in the solid containing a line crack produced by the surface tractions t_i and the body forces f_i can be expressed following the definition given in Section 3.14 that the incremental potential energy, or the rate of change of that energy in the case of a non-stationary process, can be expressed as

$$\dot{\pi} = \dot{U} - \dot{W} = \int_v \sigma_{ij}\dot{\varepsilon}_{ij}\,dv - \int_v f_i\dot{U}_i\,dv - \int_s t_i\dot{U}_i\,ds$$

where $\dot{U}$, $\dot{W}$ are the respective rates of strain energy storage and applied load, $\dot{U}_i$ are the rates of variation of displacement.

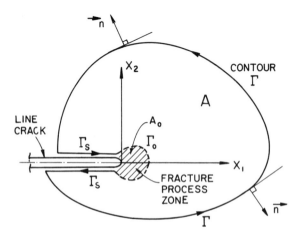

Figure 7.30 Closed contour for the C_g^* integral.

The equilibrium condition requires that $\dot\pi = 0$, or

$$\int_s t_i \dot U_i \, ds - \int_v (\sigma_{ij}\dot\varepsilon_{ij} - f_i \dot U_i)\, dv = 0$$

As the solid has a unit thickness, which implies that $V = A$ and $S = \Omega$, where Ω is the perimeter, the above equation may be expressed as

$$\int_\Omega t_i \dot U_i \, d\Omega - \int_{A+A_0} (\sigma_{ij}\dot\varepsilon_{ij} - f_i \dot U_i)\, dA = 0 \qquad (7.35)$$

By referring to Figures 7.29 and 7.30, we see that the rate of potential energy production in the system due to the action of f_i and t_i is equal to the left-hand side of (7.35):

$$\dot\pi_s = \int_{\Omega-\Gamma_0} t_i \dot U_i \, d\Omega - \int_A (\sigma_{ij}\dot\varepsilon_{ij} - f_i \dot U_i)\, dA$$

whereas the rate of potential energy dissipated into the fracture process zone is

$$\dot\pi_{fp} = \int_{-\Gamma_0} t_i \dot U_i \, d\Gamma_0 + C_g^* \, dc$$

in which $C_g^* \, dc$ represents the rate of energy required to extend the crack by an amount dc.

Following the third assumption of the energy balance concept:

$$\dot\pi_s = \dot\pi_{fp}$$

and realizing that $\Omega = \Gamma + \Gamma_s - \Gamma_0$ and A_0 is negligibly small, the equivalence of the last two equations leads to the following:

$$C_g^* = \frac{d}{dc}\left(\int_{\Gamma+\Gamma_s} t_i \dot U_i \, d\Gamma - \int_A (\sigma_{ij}\dot\varepsilon_{ij} - f_i \dot U_i)\, dA \right)$$

which becomes, after differentiation,

$$C_g^* = \int_{\Gamma+\Gamma_s} t_i \frac{d\dot U_i}{dc}\, d\Gamma - \int_A \left(\sigma_{ij}\frac{d\dot\varepsilon_{ij}}{dc} - f_i \frac{d\dot U_i}{dc} \right) dA \qquad (7.36)$$

For the case under a generalized thermomechanical loading condition, the total strain components in the above equation can be expressed, by the well-known partition theory[116], as

$$\dot\varepsilon_{ij} = \dot\varepsilon_{ij}^e + \dot\varepsilon_{ij}^p + \dot\varepsilon_{ij}^c$$

where $\dot\varepsilon_{ij}^e$, $\dot\varepsilon_{ij}^p$ and $\dot\varepsilon_{ij}^c$ are the respective elastic, plastic and creep strain rate components.

The C_g^* in (7.36) can be expanded to give

$$C_g^* = \int_{\Gamma+\Gamma_s} t_i \frac{d\dot U_i}{dc}\, d\Gamma - \int_A \left[-f_i \frac{d\dot U_i}{dc} + \sigma_{ij}\frac{d\dot\varepsilon_{ij}^e}{dc} + \sigma_{ij}\frac{d\dot\varepsilon_{ij}^c}{dc} + \sigma_{ij}\frac{d\dot\varepsilon_{ij}^p}{dc} \right] dA$$

$$(7.37)$$

The expression of C_g^* during the crack extension can be derived on the basis of (7.37) and a similar procedure described in Kishimoto *et al.*[78] and Liu and Hsu[94]:

$$
\begin{aligned}
C_g^* = \sum_{m=1}^{2} \Bigg\{ & \int_{\Gamma+\Gamma_s} \left(\dot{W}_c n_m - t_i \frac{\partial \dot{U}_i}{\partial x_m} \right) d\Gamma \\
& + \int_A \left[-f_i \frac{\partial \dot{U}_i}{\partial x_m} + \sigma_{ij} \frac{\partial \dot{\varepsilon}_{ij}^*}{\partial x_m} \right] dA \Bigg\} \begin{cases} \cos \theta_0 & \text{for } m = 1 \\ \sin \theta_0 & \text{for } m = 2 \end{cases}
\end{aligned} \quad (7.38)
$$

where $\dot{W}_c$ = creep strain energy, an analogue to creep-elastic theory

$$
= \int_{\varepsilon_{ij}} \sigma_{ij} \, d\dot{\varepsilon}_{ij}^c
$$

$\dot{\varepsilon}_{ij}^*$ = thermoelastic–plastic strain rate components

$$
= \dot{\varepsilon}_{ij}^e + \dot{\varepsilon}_{ij}^p
$$

For a simple Mode I (or opening mode) crack growth, all differential quantities with respect to x_2 should vanish and $\theta_0 = 0$ in (7.38).

The creep crack growth model given in (7.34) with the C_g^* integral expressed in (7.38) was used to correlate experimental results described in Landes and Begley[15], Nikbin *et al.*[106] and Taira *et al.*[117] and showed excellent agreement as described in detail in Liu and Hsu[94].

7.12.3 C_g^* integral for various creep scales

As described earlier in this section, the unique advantage of the C_g^* integral approach is its ability to characterize the crack growth in a solid subject to general thermal and mechanical loads. The expressions in (7.37) and (7.38) were formulated exactly on that basis. The following expressions of the C_g^* integrals are deduced from its general form in (7.38) for various scales of creep loadings. Efforts to use different and often confusing parameters for treating various creep fracture conditions as described in Section 7.11 can be avoided. For the sake of simplicity, only Mode I cases are illustrated below:

(1) *For small scale creep.* This case can be handled by eliminating all in-elastic components in (7.38) as small scale creep fracture can be regarded as the domination of elastic strains everywhere in the solid except very near the crack tip. The resultant C_g^* expression will fit into the definition of elastic energy release G given in (7.1), which can be related to the stress intensity K by (7.4). The use of parameter K for small-scale creep crack growth, of course, has been suggested before[112, 113].

(2) *For extensive scale creep.* The term extensive scale creep means long-term creep in which the creep strain rate dominates the deformation of the solid. For material with little work hardening, the following approximation

$$
\dot{\sigma}_{ij} = \dot{\varepsilon}_{ij}^e = \dot{\varepsilon}_{ij}^p = 0
$$

can be made, reducing (7.38) to

$$C_g^* = \int_{\Gamma + \Gamma^s} \left(\dot{W}_c n_1 - t_i \frac{\partial \dot{U}_i}{\partial x_1} \right) d\Gamma \qquad (7.39)$$

The reader will readily find that the expression of C_g^* in (7.39) is identical to the C^* integral proposed by Landes and Begley[15] for this type of creep fracture analysis.

(3) *For the limiting case of creep.* The physical interpretation of this type of creep fracture is that the fracture process zone is so widely spread that a homogeneous stress and strain distribution is observed in the solid. For a sufficiently large fracture process region, the first term in (7.39) becomes negligibly small in comparison to the second term. The C_g^* integral in (7.38) has been further reduced to give

$$C_g^* = - \int_{\Gamma + \Gamma_s} t_i \frac{\partial \dot{U}_i}{\partial x_1} d\Gamma \qquad (7.40)$$

which can be readily related to the net cross-section stress σ_{net} in the solid with extensive crack growth. The latter parameter, again, has been mentioned[112,113] for the characterization of this type of creep crack growth in a solid.

7.13 Path dependence of the C_g^* integral

As has been demonstrated above, the well-documented C integral proposed by Landes and Begley[15] and expressed in (7.39) can be regarded as a special case of the general C_g^* integral given in (7.38). The C integral, which was extended from a similar concept of the J integral for the time invariant fracture is considered to be independent of path for integration[15,16]. A numerical example presented in the sequel[118] shows that the path independence of the C_g^* integral, on the other hand, was observed.

The example chosen for the power line integrations involving the creep fracture of a thin plate made of 304 stainless steel is shown in Figure 7.8.

Pertinent material properties used in the numerical evaluation are tabulated as follows:

Young's modulus:	140	GPa
shear modulus of elasticity:	53	GPa
Poisson ratio:	0.3	
plastic modulus:	700	MPa
yield strength:	103	MPa

creep law: $\dot{\varepsilon}^c = 1.37 \times 10^{-18} \sigma_n^{7.1}/\text{hr}$
with the nominal stress, $\sigma_n < 176.5\,\text{MPa}$.

Figure 7.31 shows the region for the finite element model with four selected contour regions, i.e. Γ_1 to Γ_4 as indicated.

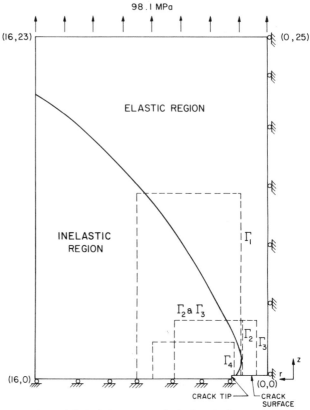

Figure 7.31 Region of finite element analysis for C_g^* integral of a center-cracked plate described in Figure 7.8.

Two sets of finite element models were used:

>*set A:* 154 elements at 0.4 × 1 mm
>*set B:* 201 elements at 0.3 × 0.6 mm

Figure 7.32 shows the numerical values of the C^* integral computed by the TEPSAC code using the four chosen contours. The same values of the C^* integral by contour Γ_2 were superimposed onto the respective C_g^* values in Figure 7.33. This figure also shows the corresponding experimentally determined values of the C integral, i.e. $C_{\exp}^*$ as defined in Liu *et al.*[118]. It is apparent that substantial variation of numerical values of C^* exist in Figure 7.32 with those C^* obtained from contour Γ_4 deviating most from the remaining sets. A possible explanation for such deviation is that the region for the C^* integration encompassed by Γ_4 is entirely within the inelastic zone, which violates the assumption that the C^* integral is valid only for small-scale inelastic deformation of the solid. The C_g^* integral, on the other hand, did not seem to be affected by the size of the inelastic region under the contour

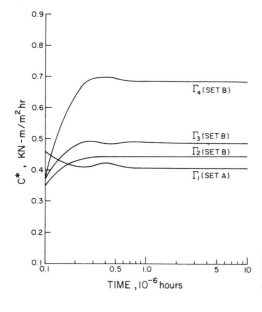

Figure 7.32 Variation of C^* integrals as computed by different contours.

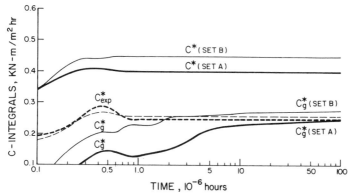

Figure 7.33 Correlation of the C_g^* integral with experimental results.

of integration. It also shows better correlation with the experimental values, as can be seen from Figure 7.33. One obvious shortcoming of the C_g^* integral is its numerical instability at the early stage of the creep deformation, although it eventually converged to a stable solution.

7.14 Creep crack growth simulated by "breakable element" algorithm

An attempt was first made by Hsu and Zhai[119] to use the "breakable element" algorithm described in Section 7.9 to simulate the growth of a line crack in a thin plate subject to combined thermoelastic–plastic creep load.

Formulations of the stress analysis were based on those presented in Chapters 2–4. While a definite fracture criterion $\bar{\varepsilon}_{rup}$ was not available at that time, three arbitrary values at $\bar{\varepsilon}_{rup}$ = 0.03, 0.075 and 0.12 were assumed in the analysis. It has been shown from this parametric study that the use of the "breakable element" algorithm could indeed provide some detailed information on the mechanical behavior of the material near the crack tip during a creep fracturing process as in the cases described in the foregoing sections.

The numerical illustration presented below in some ways can be regarded as an extension to the Case 1 study described in Section 7.10, except that the mechanical loading ceased to increase at 98.1 MPa as illustrated in Figure 7.34. Other conditions such as the geometry of the preheated center-cracked plate (Fig. 7.8), the finite element idealization (Fig. 7.20) as well as the temperature-dependent material properties (Fig. 3.24) and the fracture criterion, $\bar{\varepsilon}_{rup}$ = 3%, have also been adopted.

Upon reaching the pre-set mechanical load at 98.1 MPa, creep deformation began in the cracked plate. The same creep law for the material used in Section 7.13 was used for this case.

Figure 7.35 shows the crack growth vs time into the creep. An initial crack growth due to the applied load up to Point A in Figure 7.34 was produced by the mechanical load up to 98.1 MPa. Further growth to Point B, where complete rupture of the first breakable element took place, was prompted by pure creep deformation in the plate following the breakable element algorithm. The rupture of the breakable element No. 1 required proper

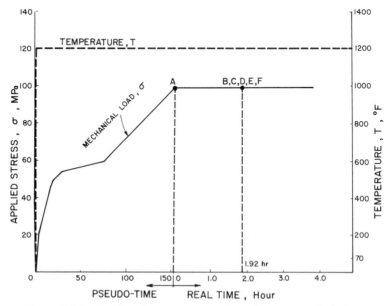

Figure 7.34. Loading history on a preheated center-cracked plate.

212

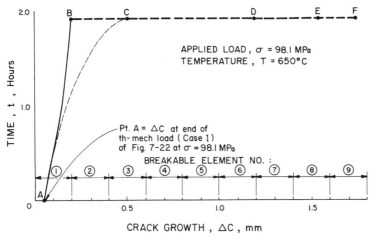

Figure 7.35 Crack growth due to creep deformation.

redistribution of the load in the cracked plate, which was carried out by a nodal force relaxation procedure, as described in Section 7.9.2. This redistribution of the load caused an equivalent further crack growth to Point C. The real crack growth during this period of 1.92 hours would therefore be more likely to follow the continuous curve shown as thin dotted line. An unstable growth was reached at Point C as further crack growth became "spontaneous" by virtue of subsequent nodal force relaxation alone, e.g. the growth between C and D was produced by relaxation of breakable element No. 2, that between D and E by element No. 3 and that between E and F by element No. 4, etc.

The distributions of the normal and shear stresses ahead of the growing crack have been shown in Figures 3.36 & 37 respectively. Although the pattern of these distributions appears to be similar to those shown in Figures 7.23 & 24 for the thermoelastic–plastic cases, an interesting observation is the reversal of the positions for the curves at Point A and Point B. A noticeable relaxation of the normal stress took place during the creep crack growth from Point A to Point B. A more pronounced relaxation phenomenon can be observed in the distribution of the effective stress in Figure 7.38 in which the level of this stress at all stages of the crack growth appears to be less than the initial value at Point A.

Although relaxation of stresses took place during the creep crack growth, the strain continued to climb as depicted in Figure 7.39 for the effective strain.

The shift of the inelastic zones ahead of the growing crack is illustrated in Figure 7.40. The shape of these zones produced by creep deformation appears to be somewhat different from those due to elastic–plastic loadings. Again, a substantial shrinking of the inelastic zones from the initial state at

213

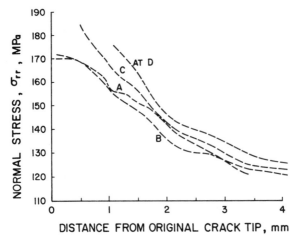

Figure 7.36 Distribution of normal stress ahead of a growing crack due to creep.

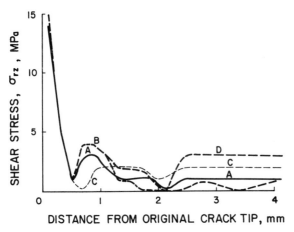

Figure 7.37 Distribution of shear stress ahead of a growing crack due to creep.

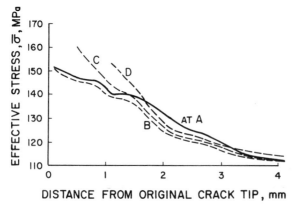

Figure 7.38 Variation of effective stress ahead of a growing crack due to creep.

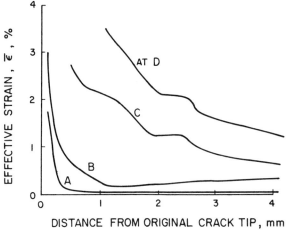

Figure 7.39 Variation of effective strain ahead of a growing crack due to creep.

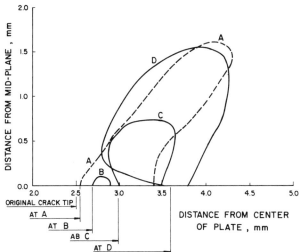

Figure 7.40 Shift of inelastic zones with a growing crack due to creep.

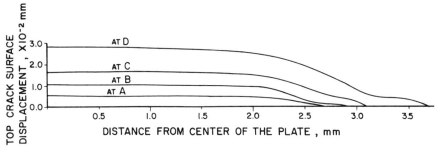

Figure 7.41 Profile of a crack at various stages of growth.

Point A to Point B due to stress relaxation is observed, which is unique in the creep-fracture process.

The profiles of the growing crack have been shown in Figure 7.41. Similar blunting of the crack tips has been observed.

References

1 Knott, J. F. 1973. *Fundamentals of fracture mechanics.* New York: John Wiley.
2 Brock, D. 1974. *Elementary engineering fracture mechanics.* Leyden: Noordhoff.
3 Griffith, A. A. 1925. The theory of rupture, *Proc. Int. Conf. Appl. Mech. Delft,* 1924, J. Waltman, Jr (ed.), 55–63.
4 Irwin, G. R. 1957. Analysis of stresses and strains near the end of a crack traversing a plate. *J. Appl. Mech., ASME Trans.* **24,** 361–4.
5 Irwin, G. R., J. A. Kies and H. L. Smith 1958. Fracture strengths relative to onset and arrest of crack propagation. *Proc. Am. Soc. Testing and Mat.* **58,** 640–60.
6 Dugdale, D. S. 1960. Yielding of steel sheets containing slits, *J. Mech. & Phys. Solids* **8,** 100–4.
7 Wells, A. A. 1963. *Application of fracture mechanics at and beyond general yield.* Report M13/63, British Welding Research Association.
8 Rice, J. R. 1968. Mathematical analysis in the mechanics of fracture. In *Fracture: an advanced treatise.* H. Liebowitz (ed.). New York: Academic Press, Vol. 2, pp. 191–311.
9 Hutchinson, J. W. 1968. Singular behavior at the end of a tensile crack in a hardening material. *J. Mech. & Phys. Solids* **16,** 13–31.
10 Rice, J. R. and G. F. Rosengren 1968. Plane strain deformation near a crack tip in a power-law hardening material. *J. Mech. & Phys. Solids* **16,** 1–12.
11 Paris, P. C. 1977. Fracture mechanics in the elastic–plastic regime. *Flaw growth and fracture.* ASTM STP 631, American Society for Testing and Materials, 3–27.
12 Rice, J. R. 1968. A path independent integral and the approximate analysis of strain concentration by notches and cracks. *J. Appl. Mech. ASME Trans.* **35,** 379–86.
13 Budiansky, B. and J. R. Rice 1973. Conservation laws and energy-release rates. *J. Appl. Mech.* March, 201–3.
14 Landes, J. D., J. A. Begley and G. A. Clarke (eds.), 1979. *Elastic–plastic fracture.* STP 668, American Society for Testing and Materials.
15 Landes, J. D. and J. A. Begley 1976. *A fracture mechanics approach to creep crack growth.* ASTM STP 590, American Society for Testing and Materials, 128–48.
16 Harper, M. P. and E. G. Ellison 1977. The use of the C^*-parameter in predicting creep crack propagation rates. *J. Strain Anal.* **12**(3), 167–79.
17 Rice, J. R. 1978. Some computational problems in elastic–plastic crack mechanics. In *Numerical methods in fracture mechanics* A. R. Luxmoore and D. R. J. Owen (eds.), 434–5.
18 Lin, T. H. 1968. Theory of inelastic structures. New York: John Wiley. p. 112.
19 McMeeking, R. M. 1977. Path dependence of the J-integral and the role of J as a parameter characterizing the near tip field. *Flaw growth and fracture.* ASTM STP 631, American Society for Testing and Materials.
20 Paris, P. C., H. Tada, A. Zahoor and H. Ernst 1979. The theory of instability of tearing mode of elastic–plastic crack growth. In J. D. Landes, J. A. Begley and G. A. Clarke (eds.). *Elastic–plastic fracture.* STP 668, American Society for Testing and Materials.

21 Heyer, R. H. 1973. Crack growth resistance curves (*R*-curves). *Literature review, fracture toughness evaluation by R-curve methods.* ASTM STP 527, American Society for Testing and Materials, 1-16.

22 Carman, C. M. 1973. Plane stress fracture testing using center cracked panels. *Literature review, fracture toughness evaluation by R-curve methods.* ASTM STP 527, American Society for Testing and Materials, 62-84.

23 Sullivan, A. M., C. N. Freed and J. Stoop 1973. Comparison of *R*-curves determined from different specimen types. *Literature review, fracture toughness evaluation by R-curve methods.* ASTM STP 527, American Society for Testing and Materials, 85-104.

24 McClintock, F. A. and G. R. Irwin 1965. Plasticity aspects of fracture mechanics. *Fracture toughness testing and its applications.* ASTM STP 381, American Society for Testing and Materials, 84-113.

25 Rice, J. R. 1975. Elastic–plastic models for stable crack growth in mechanics and mechanisms of crack growth. *Proc. Conf. at Cambridge, England*, April 1973, M. J. May (ed.), British Steel Corporation Physical Metallurgy Centre Publication, 14-39.

26 Kobayashi, A. S., D. E. Maiden, B. J. Simon and S. Ilda 1969. *Application of finite element analysis method to two-dimensional problems in fracture mechanics*, ASME Paper No. 69-WA/PVP-12, American Society of Mechanical Engineers.

27 Chan, S. K., I. S. Tuba and W. K. Wilson 1970. On the finite element method in linear fracture mechanics. *Eng. Fracture Mech.* **2**, 1-17.

28 Watwood, V. B. 1969. The finite element method for prediction of crack behavior. *Nucl. Engng. & Des.* **11**, 323-32.

29 Wilson, W. K. 1972. Some crack tip finite elements for plane elasticity. *Stress analysis and growth of cracks.* ASTM STP 513, American Society for Testing and Materials, 90-105.

30 Byskov, E. 1970. The calculation of stress intensity factors using the finite element method with cracked elements. *Int. J. Fracture Mech.* **6**, 159-67.

31 Hilton, P. D. and G. C. Sih, 1973. Applications of the finite element method to the calculations of stress intensity factors. In *Methods of analysis and solutions of crack problems.* G. C. Sih (ed.), Leyden: Noordhoff. 426-83.

32 Tracey, D. M. 1971. Finite elements for determination of crack tip elastic stress intensity factors. *Engng. Fracture Mech.* **3**, 225-65.

33 Henshell, R. D. 1975. Crack tip finite elements are unnecessary. *Int. J. Num. Meth. Engng.* **9**, 495-507.

34 Barsoum, R. S. 1976. On the use of isoparametric finite elements in linear fracture mechanics. *Int. J. Num. Meth. Engng.* **10**, 25-37.

35 Barsoum, R. S. 1977. Triangular quarter-point elements as elastic and perfectly-plastic crack tip elements. *Int. J. Num. Meth. Engng.* **11**, 85-98

36 Swedlow, J. L., M. L. Williams and W. Yang 1966. Elasto-plastic stresses and strains in cracked plates. *Proc. First Int. Conf. Fracture* at Sendai, Japan, 1965. T. Yokobori, T. Kawasaki and J. L. Swedlow (eds.). Japan Society for Strength and Fracture of Materials, Tokyo, **1**, 259-82.

37 Swedlow, J. L. 1969. Elasto-plastic cracked plates in plane strain. *Int. J. Fracture Mech.* **5**, 33-44.

38 Swedlow, J. L. 1969. Initial comparisons between experiment and theory of the strain fields in a cracked copper plate. *Int. J. Fracture Mech.* **5**, 25-31.

39 Marcal, P. V. and I. P. King 1961. Elastic–plastic analysis of two dimensional stress systems by the finite element method. *Int. J. Mech. Sci.* **9**, 143-55.

40   Miyamoto, H., M. Shiratori and T. Miyoshi 1972. Elastic–plastic response at the tip of a crack. In *Mechanical behavior of materials, Proc. Int. Conf. Fracture.* Kyoto, Japan, **1**, 433–45.

41   Larsson, S. G. and A. J. Carlsson 1973. Influence of non-singular stress terms and specimen geometry on small-scale yielding at crack tips in elastic–plastic materials. *J. Mech. & Phys. Solids* **21**, 263–77.

42   Levy, N., P. V. Marcal, W. J. Ostergren and J. R. Rice 1971. Small scale yielding near a crack in plane strain: a finite element analysis. *Int. J. Fracture Mech.* **7**, 143–56.

43   Rice, J. R. and D. M. Tracey 1973. Computational fracture mechanics. In *Numerical and computer methods in structural mechanics.* S. J. Fenves, N. Perrone, A. Robinson and W. C. Schnobrich (eds.). New York: Academic Press, 585–623.

44   Tracey, D. M. 1976. Finite element solutions for crack tip behavior in small scale yielding. *J. Engng. Mat. Technol., ASME Trans.* **98**, 146–51.

45   McMeeking, R. M. 1977. Finite deformation analysis of crack tip opening in elastic–plastic materials and implications for fracture initiation. *J. Mech. Phys. Solids* **25**, 357–81.

46   Rice, J. R., R. M. McMeeking, D. M. Parks and E. P. Sorenson 1979. Recent finite element studies in plasticity and fracture mechanics. *Computer Methods in Applied Mechanics and Engineering.* **17/18**, 411–42.

47   Wells, A. A. 1969. Crack opening displacements from elastic–plastic analysis of externally notched tension bars. *Engng. Fracture Mech.* **1**, 399–410.

48   Turner, C. E. and J. S. T. Cheung 1972. Computation of post-yield behavior in notch bend and tension test pieces. *J. Strain Anal.* **7**, 303–12.

49   Sumpter, J. D. G. and C. E. Turner 1976. Use of the *J* contour integral in elastic–plastic fracture studies by finite-element methods. *J. Mech. Engng. Sci.* **18**, 97–112.

50   Parks, D. M. 1978. Virtual crack extension in a general finite element technique for *J*-integral evaluation. In *Numerical methods in fracture mechanics.* A. R. Luxmoore and M. J. Owen (eds.), University College of Swansea, 464–78.

51   Miyamoto, H. and K. Kageyama 1978. Extension of *J*-integral to the general elasto-plastic problem and suggestion of a new method for its evaluation. In *Numerical methods in fracture mechanics.* A. R. Luxmoore and M. J. Owen (eds.), University College of Swansea, 479–86.

52   Kobayashi, A. S., S. T. Chiu and R. Beenorkes 1973. A numerical and experimental investigation on the use of the *J*-integral. *Engng. Fracture Mech.* **5**, 293–305.

53   Anderson, H. 1973. A finite-element representation of stable crack-growth. *J. Mech. & Phys. Solids.* **21**, 337–56.

54   Hsu, T.-R. and A. W. M. Bertels 1976. Propagation and opening of a through crack in a pipe subjected to combined cyclic thermomechanical loading. *J. Press. Vess. Technol., ASME Trans.* **98**, 17–25.

55   De Koning, A. U. 1977. *Fracture 1977, Proc. 4th Int. Conf. Fracture,* University of Waterloo, June 1977, University of Waterloo Press, **3**, 25–31.

56   Varanasi, S. R. 1977. Analysis of stable and catastrophic crack growth under rising load. *Flaw growth and fracture.* ASTM STP 631, American Society for Testing and Materials, 507–19.

57   De Koning, A. U., D. P. Rooke and C. Wheeler 1978. Energy dissipation during stable crack growth in aluminum alloy 2024-T3. *Proc. Int. Conf. Num. Meth. Fracture Mech.* A. R. Luxmoore and D. R. J. Owen (eds.). Swansea, 525–36.

REFERENCES

58 Light, M. F. and A. R. Luxmoore 1977. A numerical investigation of post-yield fracture. *J. Strain Anal.* **12**, 293-304.
59 Light, M. F. and A. R. Luxmoore 1977. Crack extension forces in elasto-plastic stress fields. *J. Strain Anal.* **12**, 305-9.
60 Belie, R. G. and J. N. Reddy 1980. Direct prediction of fracture for two-dimensional plane stress structures. *Computers & Structures* **11**, 49-53.
61 Lee, J. D. and H. Liebowitz 1978. Considerations of crack growth and plasticity in finite element analysis. *Computers & Structures* **8**, 403-10.
62 Kfouri, A. F. and K. J. Miller 1976. Crack separation energy rates in elastic-plastic fracture mechanics. *Proc. Inst. Mech. Engrs.* **190**(48), 571-84.
63 Rice, J. R. and E. P. Sorensen 1978. Continuing crack-tip deformation and fracture for plane-strain crack growth in elastic–plastic solids. *J. Mech. Phys. Solids* **26**, 163-86.
64 Shih, C. F., H. G. deLorenzi and W. R. Andrews 1979. Studies on crack initiation and stable crack growth. In *Elastic–plastic fracture.* J. D. Landes, J. A. Begley and G. A. Clarke (eds.). STP 668, American Society for Testing and Materials, 65-120.
65 Kanninen, M. F., E. E. Rybicki, R. B. Stonesifer, D. Broek, A. R. Rosenfield, C. W. Marshall and G. T. Hahn 1979. Elastic–plastic fracture mechanics for two-dimensional stable crack growth and instability problems. In *Elastic–plastic fracture.* J. D. Landes, J. A. Begley and G. A. Clarke (eds.). STP 668, American Society for Testing and Materials, 121-50.
66 Sorensen, E. P. 1979. A numerical investigation of plane strain stable crack growth under small-scale yielding conditions. In *Elastic–plastic fracture.* J. D. Landes, J. A. Begley and G. A. Clarke (eds.). STP 668, American Society for Testing and Materials, 151-74.
67 D'Escatha, Y. and J. C. Devaux 1979. Numerical study of initiation, stable crack growth, and maximum load, with a ductile fracture criterion based on the growth of holes. In *Elastic–plastic fracture.* J. D. Landes, J. A. Begley and G. A. Clarke (eds.). STP 668, American Society for Testing and Materials, 229-48.
68 Newman, J. C., Jr 1977. Finite element analysis of crack growth under monotonic and cyclic loading. *Cyclic stress–strain and plastic deformation aspects of fatigue crack growth.* ASTM STP 637, American Society for Testing and Materials, 56-80.
69 Ohji, K., K. Ogura and Y. Ohkubo 1975. Cyclic analysis of a propagating crack and its correlation with fatigue crack growth. *Engng. Fracture Mech.* **7**, 457-64.
70 Socie, D. F. 1977. Prediction of fatigue crack growth in notched members under variable amplitude loading histories. *Engng. Fracture Mech.* **9**, 849-65.
71 Rolfe, S. T. and J. M. Barsom 1977. *Fracture and fatigue control in structures.* Englewood-Cliffs, NJ: Prentice-Hall. Ch. 4.
72 Schaeffer, B. J., H. W. Liu and J. S. Ke 1971. Deformation and the strip necking zone in a cracked steel sheet. *Experim. Mech.* **11**, 172-5.
73 Gavigan, W. J., J. S. Ke and H. W. Liu 1973. Local and gross deformations in cracked metallic plates. *Int. J. Fracture.* **9**, 255-66.
74 Evans, W. T., M. F. Light and A. R. Luxmoore 1980. An experimental and finite element investigation of fracture in aluminum thin plates. *J. Mech. & Phys. Solids.* **28**, 167-89.
75 Dieter, G. E. 1976. *Mechanical metallurgy.* New York: McGraw-Hill.
76 Blackburn, W. S., A. D. Jackson and T. K. Hellen 1977. An integral associated with the state of a crack tip in a non-elastic material. *Int. J. Fracture.* **13**(2), 183-99.

77 Ainsworth, R. A., B. K. Neale and R. H. Price 1978. Fracture behaviour in the presence of thermal strains. *Proc. Conf. Tolerance of Flaws in Pressurized Components*, Institute of Mechanical Engineering, May 16-18, London, 171-8.

78 Kishimoto, K., S. Aoki and M. Sakata 1980. On the path independent integral. *J. Engng. Fracture Mech.* **13**, 841-50.

79 Wilson, W. K. and I. W. Yu 1979. The use of the *J*-integral in thermal stress crack problems. *Int. J. Fracture* **15**(4), 377-87.

80 Chen, W. H. and K. T. Chen 1981. On the study of mixed mode thermal fracture using modified J_k-integrals. *Int. J. Fracture* **17**, 99-103.

81 Dugdale, D. S. 1960. Yielding of steel sheets containing slits. *J. Mech. Phys. Solids* **8**, 100-4.

82 Bazant, Z. P. and L. Cedolin 1979. Blunt crack band propagation in finite element analysis. *J. Engng. Mech. Division, ASCE Trans.* **105**, 297-315.

83 Nair, P. and K. L. Reifsnider 1974. Unimod: an application oriented finite element scheme for the analysis of fracture mechanics problems. *Fracture Analysis*, ASTM STP 560, American Society for Testing and Materials, 211-25.

84 Hsu, T.-R. and Y. J. Kim 1979. *On slow crack growth in fuel cladding by finite element analysis*. Paper C3/12, 5th International Conference on Structural Mechanics in Reactor Technology.

85 Kim, Y. J. and T.-R. Hsu 1982. A numerical analysis on stable crack growth under increasing load. *Int. J. Fracture* **20**, 17-32.

86 Dorn, W. S. and D. D. McCracken 1972. Numerical methods with Fortran IV case study. New York: John Wiley.

87 Kim, Y. J. 1981. *Stable crack growth in ductile materials—a finite element approach*, PhD thesis, University of Manitoba, Winnipeg, Canada.

88 Erdogan, F. and G. C. Sih 1963. On the crack extension in plates under plane loading and transverse shear. *J. Basic Engng., ASME Trans.* **85**, 519-27.

89 Sih, G. C. 1973. Some basic problems in fracture mechanics and new concepts. *Engng. Fracture Mech.* **5**, 365-77.

90 Palaniswamy, K. and W. G. Knauss 1972. Propagation of a crack under general, in-plane tension. *Int. J. Fracture* **8**, 114-17.

91 Hellen, T. K. 1975. On the method of virtual crack extensions. *Int. J. Num. Meth. Engng.* **9**, 187-207.

92 Sih, G. C. 1974. Fracture mechanics applied to engineering problems—strain energy density fracture criterion. *Engng. Fracture Mech.* **6**, 361-86.

93 Argyris, J. H. and S. Kelsey 1960. Energy theorems and structural analysis. London: Butterworth. 5.

94 Liu, Y. J. and T.-R. Hsu 1985. A general treatment of creep crack growth. *Engng. Fracture Mech.* **21**(3), 437-52.

95 Siverns, M. J. and A. T. Price 1973. Crack propagation under creep conditions in a quenched $2\frac{1}{4}$ chromium-1 molybdenum steel. *Int. J. Fracture* **9**, 199.

96 Neate, G. J. and M. J. Siverns 1974. The application of fracture mechanics to creep crack growth. *Int. Conf. Creep & Fatigue in Elevated Temperature Applications* **1**, 234.

97 Floreen, S. 1975. The creep fracture of wrought nickel-base alloys by a fracture mechanics approach. *Metall. Trans.* **6A**, 1741.

98 Kenyon, J. L., G. A. Webster, J. C. Radon and C. E. Turner 1974. An investigation of the application of fracture mechanics to creep cracking. *Int. Conf. Creep & Fatigue in Elevated Temperature Applications* **1**, 156.

220

99 Yokobori, T., T. Kawasaki and M. Horiguchi 1976. Creep crack propagation in austenitic stainless steel at elevated temperature. *Proc. 3rd Nat. Congress on Fracture.* Law Tatry Slovakia.

100 Taira, S., R. Ohtani and A. Nitta 1974. Creep crack initiation and propagation in an 18 Cr-8 Ni stainless steel. *Proc. 1973 Symp. Mech. Behavior Mat.* 211.

101 Taira, S. and R. Ohtani 1974. Creep crack propagation and creep rupture of notched specimens. *Int. Conf. Creep & Fatigue in Elevated Temperature Applications* 1, 213.

102 Taira, S. and R. Ohtani 1978. Crack propagation in creep. *Proc. 2nd Int. Conf. Mech. Behavior Mat.* Boston. 409-65; special volume, American Society of Metals, 155-82.

103 Haigh, J. R. 1975. The growth of fatigue cracks at high temperature under predominately elastic loading. *Engng. Fracture Mech.* 271-84.

104 Vitek, V. 1977. A theory of the initiation of creep crack growth. *Int. J. Fracture* 13, 39-50.

105 Pilkington, R., D. Hutchinson and C. L. Jones 1974. High-temperature crack opening displacement measurements in a ferrite steel. *Metall. Sci. J.* 8, 237-41.

106 Nikbin, K. M., G. A. Webster and C. E. Turner 1976. Relevance of non-linear fracture mechanics to creep cracking. *ASTM STP 601*, American Society for Testing and Materials, 47-62.

107 Kachanov, L. 1981. Crack growth under conditions of creep and damage. In *Creep in structures.* A. R. S. Ponter and D. R. Hayhurst (eds.). Proc. 3rd Symp. Int. Union Theoret. and Appl. Mech., 520-4, Springer-Verlag.

108 Kubo, S., K. Ohji and K. Ogura 1979. An analysis of creep crack propagation on the basis of the plastic singular stress field. *Engng. Fracture Mech.* 11, 315-29.

109 McCartney, L. N. 1980. Derivation of crack growth laws for linear viscoelastic solids based upon the concept of a fracture process zone. *Int. J. Fracture* 16, 4.

110 To, K. C. 1975. A phenomenological theory of subcritical creep crack growth under constant loading in an inert environment. *Int. J. Fracture* 11, 641-8.

111 Vitek, V. 1977. A theory of the initiation of creep crack growth. *Int. J. Fracture* 13, 39-50.

112 Fu, L. S. 1980. Creep crack growth in technical alloys at elevated temperature—a review. *Engng. Fracture Mech.* 13, 307-30.

113 Riedel, H. and J. R. Rice 1980. Tensile cracks in creeping solids. *Fracture mechanics: twelfth conference.* ASTM STP 700, American Society for Testing and Materials, 112-30.

114 Broberg, K. B. 1971. Crack-growth criteria and non-linear fracture mechanics. *J. Mech. & Phys. Solids* 19, 407-18.

115 Broberg, K. B. 1975. Energy methods in statics and dynamics of fracture. *J. Japan Soc. Strength and Fracture Materials* 10(2), 33-45.

116 Penny, R. K. and D. L. Marriott 1971. *Design for creep.* New York: McGraw-Hill. Ch. 4, 47.

117 Taira, S., R. Ohtani and T. Kitamura 1979. Application of *J*-integral to high-temperature crack propagation: Part 1—creep crack propagation. *ASME Trans.* 101, 154-61.

118 Liu, Y. J., T.-R. Hsu and Z. H. Zhai 1983. On the numerical evaluation of C^*-integrals for creep fracture analysis. *Proc. Int. Symp. Fracture Mechanics,* Beijing, China, November 22-25.

119 Hsu, T.-R. and Z. H. Zhai 1984. A finite element algorithm for creep crack growth. *Engng. Fracture Mech.* 20, 521-33.

8

THERMOELASTIC–PLASTIC STRESS ANALYSIS BY FINITE STRAIN THEORY

8.1 Introduction

The problem of solids undergoing large deformation is a common one in engineering practice. Various forms of metal forming such as deep drawing and extrusion are just a few obvious examples. Others involve the processes of solidification of metal casting, forging and welding of structures. In the latter processes the excessive distortion of shapes and the associated residual stresses are the primary concern of the engineer. Large deformations of the solid structure can also occur due to drastic change of the environment alone. Figure 8.1 shows a set of typical stress–strain curves for an aluminum bar at various temperatures. It can be readily demonstrated that a substantial

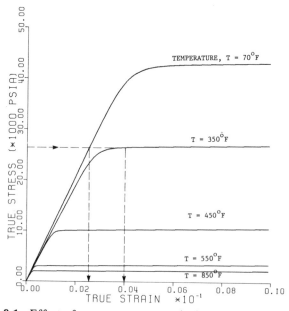

Figure 8.1 Effect of temperature on strain for 6061 aluminum alloy.

222

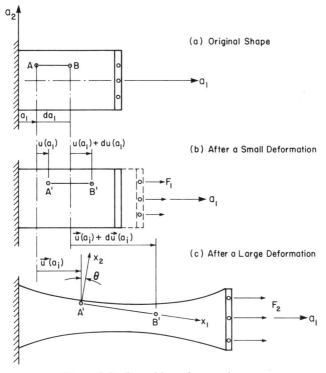

Figure 8.2 Stretching of a membrane.

increase of deformation in the bar can result from a significant temperature increase. Large changes of environmental conditions are becoming increasingly common for structures in modern engineering applications. The associated large deformation must be considered in the engineering analysis.

The major complication in dealing with the finite strain of solids is the substantial change in the structure's geometry. This change is so large that extreme difficulty arises in keeping consistency in the reference frames. Figure 8.2 illustrates the stretching of a thin membrane. The original load-free state of the membrane is shown in Figure 8.2a. Let us take a close look at an arbitrarily chosen pair of two closely located particles A and B in the membrane and choose a coordinate system (a_1, a_2) as the reference frame. Figure 8.2b shows the shape of the membrane after it is slightly stretched by a set of forces F_1 along the a_1 direction. One may observe that particle A is slightly displaced from its original position A to A' by an amount $u(a_1)$, whereas particle B is displaced to B'. The small elongation between A and B can be assumed to be $du(a_1)$ and the average strain between A and B is $du(a_1)/da_1$ following the definition given in Section 3.1. This simple expression, however, is no longer valid when the membrane is excessively stretched by a set of large forces F_2 as illustrated in Figure 8.2c. Two obvious

223

phenomena can be observed in this situation: (1) The displacements of particle A to A' and B to B' are substantial and cannot be ignored; (2) the elongation of AB to A'B' is not only excessive but also exhibits a noticeable rotation about the original coordinates a_1 and a_2. The usual simple definition of strain as stated in Section 3.1 can no longer be applied because (1) the rotational component of the shape (dimensional) change of AB must be considered, and (2) the substantial displacement of particle A itself has to be taken into account in the evaluation of the strain. Accurate computation of stresses has also become extremely difficult for the same reasons. A substantially different theory therefore has to be developed to handle this situation.

Attempts to solve this type of problem have been made by several prominent researchers. Contemporary classical treatments have been developed mainly by Hill[1], Truesdell and Toupin[2] and Prager[3]. Comprehensive elaboration of finite strain theory is available, e.g. in Fung[4] and Malvern[5]. Solutions by the finite element method have been developed in the recent years, e.g. by Larsen and Popov[6], and McMeeking and Rice[7].

Two types of formulation are commonly used in conjunction with finite strain theory; (1) Lagrangian formulation, and (2) Eulerian formulation. For the case illustrated in Figure 8.2, for example, one may either base all mathematical formulations on stresses, strains and equilibrium on the Lagrangian (original) coordinate system, i.e. (a_1, a_2), or choose the Eulerian (current) coordinate system, i.e. (x_1, x_2) to be the reference frame. The cumbersome derivation, e.g. the rotational components, may be avoided in the latter formulation, which represents considerable reduction in computational effort, although all computed quantities eventually have to be transformed back to the reference frame describing the original configuration.

The approach adopted in this chapter is based on the updated Lagrangian formulation with the reference position fixed at the previous step solution. This method was referred to as Eulerian or the modified updated Lagrangian by McMeeking and Rice[7]. The main advantage of this method is its simplicity in implementation using an existing small strain finite element formulation like the one given in Chapter 3. However, a major disadvantage is that it requires extremely small time (load) steps to achieve accurate results.

Since the objective of this chapter is to demonstrate how the existing small strain theory described in Chapter 3 can be extended to accommodate the finite strain formulations, much of the detailed derivation of the key equations, as well as fundamental concepts, have been left out. The reader can find these missing links in Refs. 4 and 5. Furthermore, for the sake of clarity, the usual rules of tensorial transformation of physical quantities have not been strictly followed. One such noticeable violation is that no distinction is made between the covariant and contravariant transformations.

8.2 Lagrangian and Eulerian coordinate systems

As illustrated in Figures 8.2a & c, points A and B have displaced substantially from their respective original positions after an excessive stretching. The geometry of the membrane bears little resemblance between the two states. It is therefore convenient to introduce another set of local coordinates, i.e. (x_1, x_2, x_3) in Figure 8.2c to describe the deformed (or current) state of the structure (membrane). These new current coordinates are commonly referred to as the "Eulerian" coordinate system. The coordinates (a_1, a_2, a_3) which describe the original geometry of the structure, and to be used as references, are referred to as the "Lagrangian" coordinate system.

For a general case in a curvilinear coordinate system, the following terminologies are used with reference to Figure 8.3.

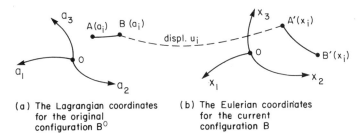

(a) The Lagrangian coordinates
    for the original
    configuration B^0

(b) The Eulerian coordinates
    for the current
    configuration B

Figure 8.3 Coordinate systems used in finite strain analysis.

Lagrangian (original) coordinates: a_i: (a_1, a_2, a_3)
Eulerian (current) coordinates: x_i: (x_1, x_2, x_3)
    From Figure 8.3, it is obvious that:

$$x_i = a_i + u_i \tag{8.1}$$

with

$$a_i = a_i(x_1, x_2, x_3) \tag{8.2a}$$

and

$$x_i = x_i(a_1, a_2, a_3) \tag{8.2b}$$

It is apparent from (8.1) that $x_i \approx a_i$ for small strain cases in which the deformation vector, u_i, is indeed infinitesimal.

8.3 Green and Almansi strain tensors

Two types of strain tensors are frequently used in the finite strain theory: the Green strain tensor and the Almansi strain tensor. The former is usually referred to the Lagrangian coordinate system and the latter to the Eulerian coordinate system.

The following expressions are presented in a summarized form. Detailed derivations are available in Refs. 2–5.

In a general coordinate system

Consider a case similar to that of a stretched membrane illustrated in Figure 8.2: the relationship of the Lagrangian coordinates a_i and the Eulerian coordinates x_i given in (8.1) can be described graphically as in Figure 8.4.

The distance between points A and B in a Cartesian coordinate system a_i is

$$dS_0^2 = da_1^2 + da_2^2 + da_3^2$$

and in a general curvilinear coordinate system, a_i:

$$dS_0^2 = a_{ij}\, da_i\, da_j \tag{8.3}$$

Likewise, the distance between points A' and B' in the "deformed" state in a general coordinate system x_i is

$$dS^2 = X_{ij}\, dx_i\, dx_j \tag{8.4}$$

in both (8.3) and (8.4), a_{ij} and X_{ij} are the respective Euclidean metric tensors for a_i and x_i. It can be shown that

$$a_{ij} = X_{ij} = \delta_{ij}$$

for Cartesian coordinates, where δ_{ij} is the Kronecker delta. Using the relationship given in (8.2a) and (8.2b), one can obtain

$$dS_0^2 = a_{ij}\frac{\partial a_i}{\partial x_l}\frac{\partial a_j}{\partial x_m}\, dx_l\, dx_m \tag{8.5}$$

and

$$dS^2 = X_{ij}\frac{\partial x_i}{\partial a_l}\frac{\partial x_j}{\partial a_m}\, da_l\, da_m \tag{8.6}$$

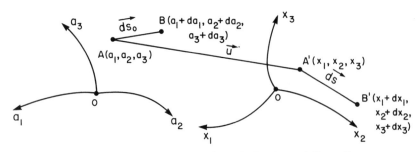

Figure 8.4 Change of dimensions during large deformation.

Green's strain tensor[†] is defined as[1,3,4]

$$2E_{ij}\,da_i\,da_j = dS^2 - dS_0^2 \tag{8.7}$$

or

$$E_{ij} = \tfrac{1}{2}\left(X_{pq}\frac{\partial x_p}{\partial a_i}\frac{\partial x_q}{\partial a_j} - a_{ij}\right) \tag{8.8}$$

with reference to the Lagrangian coordinates, a_i.

However, if the strain tensor is referred to Eulerian coordinates instead, then the following expression can be derived for the Almansi strain tensor:

$$\varepsilon_{ij} = \tfrac{1}{2}\left(X_{ij} - a_{pq}\frac{\partial a_p}{\partial x_i}\frac{\partial a_q}{\partial x_j}\right) \tag{8.9}$$

The Almansi strain tensor, as shown above, can be used to give the Cauchy strain tensor for small strains.

It is worth noting that both the Green and Almansi strain tensors are symmetric, i.e.

$$E_{ij} = E_{ji} \quad \text{and} \quad \varepsilon_{ij} = \varepsilon_{ji}$$

In a Cartesian coordinate system

Since the Euclidean metric tensors a_{ij} and X_{ij} in (8.8) and (8.9) become identical and equal to the Kronecker delta δ_{ij} in a Cartesian coordinate system, the Green and Almansi strain tensors can easily be condensed as

$$\begin{aligned}
E_{ij} &= \frac{1}{2}\left(\delta_{pq}\frac{\partial x_p}{\partial a_i}\frac{\partial x_q}{\partial a_j} - \delta_{ij}\right) \\
&= \frac{1}{2}\left(\frac{\partial u_j}{\partial a_i} + \frac{\partial u_i}{\partial a_j} + \frac{\partial u_p}{\partial a_i}\frac{\partial u_p}{\partial a_j}\right)
\end{aligned} \tag{8.10}$$

and

$$\begin{aligned}
\varepsilon_{ij} &= \frac{1}{2}\left(\delta_{ij} - \delta_{pq}\frac{\partial a_p}{\partial x_i}\frac{\partial a_q}{\partial x_j}\right) \\
&= \frac{1}{2}\left(\frac{\partial u_j}{\partial x_i} + \frac{\partial u_i}{\partial x_j} - \frac{\partial u_p}{\partial x_i}\frac{\partial u_p}{\partial x_j}\right)
\end{aligned} \tag{8.11}$$

If we let u, v and w be the corresponding displacement components along a_1, a_2 and a_3 in the Lagrangian coordinate system, or x_1, x_2 and x_3 in the

---

† Physical interpretation of E_{ij} and ε_{ij} for the finite strain cases as shown in (8.8) and (8.9) is not as simple as that in the small strain theory. Attempts at such physical interpretation may be found on pp. 97–99 of Fung[4].

Eulerian coordinate system, the following familiar expressions have been derived:

$$E_{11} = \frac{\partial u}{\partial a_1} + \frac{1}{2}\left[\left(\frac{\partial u}{\partial a_1}\right)^2 + \left(\frac{\partial v}{\partial a_1}\right)^2 + \left(\frac{\partial w}{\partial a_1}\right)^2\right]$$

$$E_{12} = \frac{1}{2}\left(\frac{\partial u}{\partial a_2} + \frac{\partial v}{\partial a_1}\right) + \frac{1}{2}\left(\frac{\partial u}{\partial a_1}\frac{\partial u}{\partial a_2} + \frac{\partial v}{\partial a_1}\frac{\partial v}{\partial a_2} + \frac{\partial w}{\partial a_1}\frac{\partial w}{\partial a_2}\right), \text{ etc.}$$

Likewise, for the Almansi strain tensor:

$$\varepsilon_{11} = \frac{\partial u}{\partial x_1} - \frac{1}{2}\left[\left(\frac{\partial u}{\partial x_1}\right)^2 + \left(\frac{\partial v}{\partial x_1}\right)^2 + \left(\frac{\partial w}{\partial x_1}\right)^2\right]$$

$$\varepsilon_{12} = \frac{1}{2}\left(\frac{\partial u}{\partial x_2} + \frac{\partial v}{\partial x_1}\right) - \frac{1}{2}\left(\frac{\partial u}{\partial x_1}\frac{\partial u}{\partial x_2} + \frac{\partial v}{\partial x_1}\frac{\partial v}{\partial x_2} + \frac{\partial w}{\partial x_1}\frac{\partial w}{\partial x_2}\right), \text{ etc.}$$

The reader may easily recognize the fact that the higher order terms of the derivatives of the displacement components in the second portion of the above formulation can be neglected for small deformation of solids, which leads to the following familiar expressions similar to those given in (3.6):

$$E_{11} = \frac{\partial u}{\partial a_1}, \quad E_{12} = \frac{1}{2}\left(\frac{\partial u}{\partial a_2} + \frac{\partial v}{\partial a_1}\right), \ldots$$

and

$$\varepsilon_{11} = \frac{\partial u}{\partial x_i}, \quad \varepsilon_{12} = \frac{1}{2}\left(\frac{\partial u}{\partial x_2} + \frac{\partial v}{\partial x_1}\right), \ldots$$

One may further conclude that for the case of small strain, i.e. $u_i \to 0$ in (8.1) and $x_i \approx a_i$, no significant distinction exists between the two types of strain tensor. Both strain tensors are then equivalent to the Cauchy strain tensor as defined in (3.6).

8.4 Lagrangian and Kirchhoff stress tensors

We have learned from the previous sections that, because of the excessive linear and rotational displacements between pairs of particles in the solid after a finite deformation process, the usual definition of Cauchy strain can no longer describe the geometric changes of the solid properly and special care must be taken in specifying the reference frame. Consequently, the Green strain tensor, E_{ij} has been derived with reference to the solid's original configuration B^0 and the Almansi strain tensor ε_{ij} on the basis of the current configuration B. It should not be too difficult for the reader to envisage the necessity of modifying the usual definition of stresses in the solid as stresses are generally geometrically dependent. This argument can also be supported by the graphical illustration in Figure 8.2. It is apparent that the stress acting on the cross-sectional area of the membrane after a finite deformation represented by A'B' in Figure 8.2c does not have the same magnitude in comparison to the same stress acting on the same cross-sectional area in Figure 8.2b, which represents the configuration of the membrane after a small stretching from its original configuration. Proper description of stresses in the solid after a finite deformation is thus necessary.

Following the detailed derivation provided in several prominent references, e.g. pp. 436–8 of Fung[4], three commonly used stress tensors are available when dealing with the finite deformation of a solid:

Lagrangian stresses:
$$T_{ji} = \frac{\rho_0}{\rho} \frac{\partial a_j}{\partial x_m} \sigma_{mi} \tag{8.12}$$

Kirchhoff stresses;

$$S_{ji} = \frac{\rho_0}{\rho} \left[\sigma_{ji} - \left(\delta_{jq} \frac{\partial u_i}{\partial x_p} + \delta_{ip} \frac{\partial u_j}{\partial x_q} - \frac{\partial u_i}{\partial x_p} \frac{\partial u_j}{\partial x_q} \right) \sigma_{qp} \right] \tag{8.13}$$

in which σ_{ji} are the Eulerian stress components. Physically, these stress tensors have the following significance:

(1) Eulerian stresses σ_{ji} represent the current configuration B in coordinates x_i;
(2) Kirchhoff stresses S_{ji} represent the current configuration B in coordinates a_i; and
(3) Lagrangian stresses T_{ji} represents the original configuration B^0 in coordinates a_i.

Special attention should be given to the facts that:

(1) T_{ji} is not symmetric, i.e. $T_{ji} \neq T_{ij}$; hence this stress tensor has little practical use in analysis;
(2) It can be readily proved that $\sigma_{ji} = S_{ji}$ in a convective coordinate system, i.e. $a_i \equiv x_i$.

8.5 Equilibrium in the large

There is a certain class of problems for which the equilibrium conditions given in (3.10) in a solid are satisfied under all circumstances, including the current deformed state. Such equilibrium conditions are usually referred to as "equilibrium in the large". An example of this class of problem is a solid deformed following a linear elastic path. The extremely small displacements of the solid make the change of the solid's geometry insignificant and hence the equilibrium state of the stresses is maintained at all time.

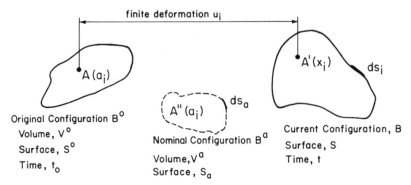

Figure 8.5 Equilibrium in the large.

Consider now a more general case as illustrated in Figure 8.5. Let the original and current configurations of the solid be represented respectively by B^0 and B. The equilibrium condition at B without the presence of a body force takes the form:

$$\sigma_{ij,j} = 0 \quad \text{in } V \quad (8.14)$$

and the boundary conditions:

$$F_j = \sigma_{ij} n_j \quad \text{on } S \quad (8.15)$$

where σ_{ij} are the true stress components in the Eulerian (current) coordinates x_i and n_j are the outward normal vectors.

However, if one wants to express the equilibrium conditions with respect to B^0 in the Lagrangian coordinates a_i, an uncertainty arises, because according to (8.1), $a_i = x_i - u_i$ with undefined u_i in the configuration $B^\dagger$. It is thus not possible to transform the equilibrium conditions in (8.14) into the coordinate (a_i) system. A nominal state which assumes an intermediate configuration between B^0, and B which is assumed to be infinitesimally distinct from B^0 is adopted to solve this problem. All quantities in this state are assumed to be known.

Thus by referring to Figure 8.5, the relationship between the nominal stresses S_{ij} and the current stress σ_{ij} can be established through the principle of force equilibrium:

$$\sigma_{ij}\, dS_i = S_{ij}\, dS_a$$

The nominal stresses S_{ij} are also referred to as the first Piola–Kirchhoff stress tensor.

---

† The displacement vector in the current state, u_i, after all, are the unknown quantities to be determined, but not the given conditions.

It can be shown[2] that

$$S_{ij} = \frac{\rho_0}{\rho} \frac{\partial a_i}{\partial x_m} \sigma_{mj} \qquad (8.16)$$

and

$$S_{ij,j} = 0 \qquad \text{in } V^a \qquad (8.17)$$

with

$$F_j = S_{ij} n_j^a \qquad (8.18)$$

8.6 Equilibrium in the small

Physically, "equilibrium in the small" means that the solid is still in equilibrium even after an "infinitesimal" incremental deformation from the current equilibrium state B. Such equilibrium conditions are normally imposed upon the incremental plastic deformation theory described in Chapter 3. This type of problem is sometimes referred to as "the rate boundary value problem".

Referring to the case illustrated in Figure 8.6, a solid in its original configuration B^0 is deformed into a new configuration B, satisfying the equilibrium conditions given in (8.14). Such conditions are sought now in yet another configuration B' resulting from application of an infinitesimal incremental load (i.e. $\Delta x_i \to 0$, $\Delta t \to 0$ in Figure 8.6).

Since the nature of the problem is clearly history-dependent, as in all plastically deforming solids, time increments have become an important element in the analysis and the "rates", rather than the absolute quantities, are therefore used in all the following formulations.

By referring to the current state B with the coordinate system x_i, the rates of the deformation tensors can be defined as follows:

$$\text{deformation rate:} \qquad \dot{\varepsilon}_{ij} = \tfrac{1}{2}(v_{i,j} + v_{j,i}) \qquad (8.19)$$

$$\text{spin rate:} \qquad \omega_{ij} = \tfrac{1}{2}(v_{i,j} - v_{j,i}) \qquad (8.20)$$

The reader should be reminded at this stage that because of the rate-sensitive nature of the problem involved, the precise description of particle

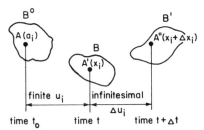

Figure 8.6 Equilibrium in the small.

231

positions and also the relative distance between particles in a deforming solid become time-dependent. The coordinates which describe these positions and distances have to be clearly defined in order to provide a true representation of the deformation behavior of the solid. The following terminologies have been introduced by prominent researchers, e.g. Truesdell and Toupin[2]:

$$\text{spatial coordinates:} \quad x_i = x_i(a_i, t) \qquad (8.21)$$

$$\text{material coordinates:} \quad a_i = a_i(x_i(t)). \qquad (8.22)$$

Physical interpretations of these two types of coordinates are illustrated in Figure 8.2. It is apparent that the coordinates x_i may vary from time to time whereas the coordinates a_i are fixed in space and can be used for the evaluation of true deformation of the solid (material).

As time t is an independent variable for all physical quantities such as σ_{ij}, ε_{ij} and u_i involved in the analysis, it is important to properly describe the rate of change of these quantities in the above two coordinate systems. The material derivatives which relate these changes in the two coordinate systems will satisfy this particular requirement.

The material derivative is defined to be

$$(\dot{\phi}) = \frac{\partial \phi}{\partial t} \qquad (8.23)$$

when ϕ is associated with the material coordinates a_i, and

$$(\dot{\phi}) = \frac{\partial \phi}{\partial t} + V_k \phi_{,k} \qquad \text{sum over } k \qquad (8.24)$$

when ϕ is associated with the spatial coordinates x_i.

The expressions in (8.23) and (8.24) ensure the rate of ϕ to be invariant to the two coordinate systems used in the finite deformation analysis.

The equilibrium conditions for the solid in the state B' in Figure 8.6 can be derived by introducing the material derivative for the stress components as given in Truesdell and Toupin[2]:

$$(\dot{\sigma}_{ij,k}) = \dot{\sigma}_{ij,k} - \sigma_{ij,m} V_{m,k} \qquad \text{sum over } m$$

Upon substitution of (8.14) into the above expression we obtain

$$(\dot{\sigma}_{ij,j}) = \dot{\sigma}_{ij,j} - \sigma_{ij,k} V_{k,j} = 0$$

the conditions for the "equilibrium in the small" can thus be established as:

$$\dot{\sigma}_{ij,j} - \sigma_{ij,k} V_{k,j} = 0 \qquad (8.25)$$

The rate-dependent equilibrium in terms of the nominal stresses given in (8.17) still takes the simple form:

$$\dot{S}_{ij,j} = 0 \qquad (8.26)$$

as S_{ij} are referred to the material coordinates a_i.

For formulations that are based on convective coordinates, i.e. $a_i = x_i$ and $B^a = B$ in Figure 8.5, the equilibrium conditions in terms of S_{ij} become[1]

$$\dot{S}_{ij} = \dot{\sigma}_{ij} + \sigma_{ij} V_{k,k} - \sigma_{jk} V_{i,k} \tag{8.27}$$

All derivations in the remainder of this chapter will be made on the basis of the convective coordinate system.

8.7 The boundary conditions

Proper adjustment of boundary conditions is also necessary for solids subject to large deformation. An obvious example is the case of a solid with its surface subject to hydrostatic pressure loading. Excessive rotations of the boundary can violate the normality of the load application, leading to significant error in the computation.

While correct adjustment of the boundary condition is not possible, due to lack of information of the boundary deformation at the current load step, an approximate course of action, however, can be taken on the nominal state instead[8]. Thus,

$$\dot{F}_j = n_i \dot{S}_{ij} \qquad \text{on } S_f$$

for the traction on the boundary S_f, and

$$\dot{U}_i = V_i \qquad \text{on } S_u$$

for the prescribed rate of deformation on the boundary S_u.

Of the surface traction boundary conditions commonly used in analysis, two appear to be of special importance. These are:

(1) Surface load:

$$\dot{F}_j = \dot{Q}_j$$

(2) Hydrostatic pressure[9]:

$$\dot{F}_j = -\dot{q}n_j - q(v_{k,k}n_j - g_{jk}n_i v_{i,k})$$

where q is the hydrostatic pressure and g_{jk} is the metric tensor.

The above two types of boundary loading conditions can be lumped into the single expression

$$\dot{F}_j = \dot{\theta}_j + m_i \lambda_{ijkl} v_{l,k} \tag{8.28}$$

in which $m_i =$ directional outward normal to the surface and θ, λ are the respective prescribed forces and hydrostatic pressures.

8.8 The constitutive equation

The physical meaning of the constitutive equation for a material can be concisely described as follows, according to Truesdell and Toupin[2]:

"the conditions to define ideal materials"

A typical example is that "the stress may be determined from the strain alone in a perfectly elastic body".

Despite the simplicity of the definition of a constitutive equation, the single most important factor in such a description of an ideal material is that the response of the material must be independent of the observer. The reader needs little persuasion to conclude that the elongation of a spring induced by a mass as observed on the Earth is substantially more than would be observed on the Moon.

Similar ambiguous observations on the material's response can also be demonstrated by the case of a stretching membrane as illustrated in Figure 8.2c. If an observer observes the deformation of the membrane in the reference frame (x_1, x_2), an elongation of the membrane equals $\overline{A'B'} - \overline{AB}$ is observed. However, when the observer "jumps" out of the reference frame (x_1, x_2) and takes an overall view on the deformed membrane, he or she not only observes the elongation between particles A and B but also the substantial translation of the particle A, as well as noticeable rotation of the particle pair AB. The observer thus draws completely different conclusions for the same material response depending on the observation point.

It is therefore clear that extreme care must be taken when deriving a constitutive equation for a solid subject to a finite deformation process. Whatever the constitutive equation derived, it must be independent of the observer, or to be more specific, independent of reference frames.

The Jaumann derivative defined as follows (p. 155 in Prager[3]) is one such quantity that is invariant to the observation position:

$$D\sigma_{ij}/Dt = \dot{\sigma}_{ij} - \sigma_{ik}\omega_{jk} - \sigma_{kj}\omega_{ik} \qquad (8.29)$$

The reader is urged to prove that $D\sigma_{ij}/Dt \equiv 0$ in the case of pure rotational deformation.

The constitutive equation proposed by Hill[10] has been used for the solid subject to finite deformation:

$$\frac{D\sigma_{ij}}{Dt} + \sigma_{ij}V_{k,k} = \lambda V_{k,k}g_{ij} + 2\mu\dot{\varepsilon}_{ij} - \frac{4\mu^2}{2\mu + h}(m_{kl}\dot{\varepsilon}_{kl})m_{ij} \qquad (8.30)$$

where λ, μ = Lamé's constants, and h = a positive scalar and a measure of the current rate of work hardening.

234

The unit outward normal tensor to the local yield surface m_{ij} for the case of isotropic hardening material is equivalent to[8]

$$m_{ij} = \frac{\partial F}{\partial \sigma'_{ij}} \bigg/ \left| \frac{\partial F}{\partial \sigma'_{ij}} \right| \qquad (8.31)$$

in which F is the plastic potential function and σ'_{ij} are the deviatoric stress components, both defined in Chapter 3.

For material obeying von Mises yield function, i.e. $\partial F/\partial \sigma'_{ij} = \sigma'_{ij}$ as given in (3.40) and also using the relationship $|\partial F/\partial \sigma_{ij}| = \sqrt{2}\bar{\sigma}/3$, the constitutive equation (8.30) in Cartesian coordinates can be reduced to

$$d\sigma_{ij} = 2G\left(\frac{v}{1 - 2v}\delta_{ij}\delta_{kl} + \delta_{ik}\delta_{jl} - \frac{\sigma'_{ij}\sigma'_{kl}}{S}\right) d\varepsilon_{kl} \qquad (8.32)$$

where

$$S = \frac{2}{3}\bar{\sigma}^2\left(1 + \frac{H'}{3G}\right)$$

$$H' = d\bar{\sigma}/d\bar{\varepsilon}_p$$

with the substitution of $\mu = G$, $\lambda = 2Gv/(1 - 2v)$ and $H' = 3h/2$ into (8.30). An important condition used in deriving (8.32) from (8.30) is that

$$\sigma_{ij} V_{k,k} = 0 \qquad (8.33)$$

which leads to

$$\dot{\sigma}_{ij} = d\sigma_{ij}/dt = D\sigma_{ij}/Dt$$

for the left-hand side of (8.32).

The reader may readily recognize the fact that the constitutive equation for finite strain analysis given in (8.30) is indeed identical to the similar equation for small strain theory in (3.50) with no thermal or strain rate effects, provided that the condition in (8.33) is satisfied. The expression in (8.32) is in fact identical to that derived by Yamada et al.[11] for small strain theory.

8.9 Equations of equilibrium by the principle of virtual work

The complexity of transforming the volume and surfaces of the solid from one configuration to another, i.e. B^0, B' and B^a etc. during finite deformation makes accurate variational process such as described in Chapter 3 for small strain theory virtually impossible. An alternative approach is to ensure equilibrium by the principle of virtual work as proposed by Larsen and Popov[6].

As the finite deformation of a solid is rate sensitive, the primary unknown quantity should be the incremental velocity, δv_j, rather than the incremental displacement δu_j as used in small strain analysis. Thus, by expressing the rate of work induced by δv_j to be[8]

$$\int_v \dot{S}_{ij}\delta(v_{j,i})\,dv = \int_s \dot{F}_i\,\delta v_i\,dS_f \tag{8.34}$$

in which S_{ij} is the nominal stress tensor.

On using (8.27) for $\dot{S}_{ij}$ and (8.29) for the $\dot{\sigma}_{ij}$, the following equilibrium equations can be derived:

$$\int_v\left[\left(\frac{D\sigma_{ij}}{Dt} + \sigma_{ij}v_{k,k}\right)\delta\dot{\varepsilon}_{ij} - \tfrac{1}{2}\sigma_{ij}\delta(2\dot{\varepsilon}_{ik}\dot{\varepsilon}_{kj} - v_{k,i}v_{k,j})\right]dv = \int_s \dot{F}_i\,\delta v_i\,dS_f \tag{8.35}$$

This equation will form the basis for the subsequent finite element formulation. The reader will find it to be identical to that shown in Eqn. (5) in McMeeking and Rice[7].

8.10   Finite element formulation

The derivation of element equations in a discretized solid follows that described in the previous chapters. The primary quantity in this case is the velocity vector, which becomes the quantity to be discretized. Much of the following derivations were described in a technical report by Cheng and Hsu[12]. Thus, letting:

$$V_i = \{V\} = \partial\{U\}/\partial t = [N]\{\dot{u}\} \tag{8.36}$$

where $\{V\}$ = element rate of deformation, or velocity vector, $\{U\}$ = element displacement vector, $\{\dot{u}\}$ = nodal rate of deformation, and $\{N\}$ = interpolation function. By referring to (3.95), one may establish the strain rates in an element to be

$$\{\dot{\varepsilon}\} = [B]\{\dot{u}\} \tag{8.37}$$

in which $[B]$ is the matrix relating the element strains and the nodal displacements as shown in (3.6) in the Cartesian coordinates. As well, according to (3.6), the entries of the $[B]$ matrix can be expressed as

$$[B_{ij}] = \tfrac{1}{2}[N_i]_{,j} + \tfrac{1}{2}[N_j]_{,i}$$

from which the following relationships exist:

$$V_i = [N_i]\{\dot{u}\} \tag{8.38a}$$

and

$$\dot{\varepsilon}_{ij} = [B_{ij}]\{\dot{u}\} \tag{8.38b}$$

It is apparent from the description in Section 8.8 of the constitutive equation that (8.30) in fact can be expressed in the form

$$\frac{D\sigma_{ij}}{Dt} + \sigma_{ij}v_{k,k} = C^{ep}_{ijkl}\dot{\varepsilon}_{kl} \tag{8.39}$$

where C^{ep}_{ijkl} are the elastoplasticity coefficients and equal to $[C_{ep}]$ in (3.49).

The substitution of (8.39) for the first term on the left-hand side of (8.35) yields:

$$\int_v \left(\frac{D\sigma_{ij}}{Dt} + \sigma_{ij}v_{k,k}\right)\delta\dot{\varepsilon}_{ij}\,\mathrm{d}v = \{\delta\dot{u}\}^T\left(\int_v [B]^T[C_{ep}][B]\,\mathrm{d}v\right)\{\dot{u}\} \tag{8.40}$$

By using (8.38), the second term in (8.35) can be expressed as

$$-\tfrac{1}{2}\int_v \sigma_{ij}\delta(2\dot{\varepsilon}_{ik}\dot{\varepsilon}_{kj} - v_{k,i}v_{k,j})\,\mathrm{d}v = -2\{\delta\dot{u}\}^T\left(\int_v [B_{ki}]^T\{\sigma\}[B_{kj}]\,\mathrm{d}v\right)\{\dot{u}\}$$

$$+ \{\delta\dot{u}\}^T\left(\int_v [N_k]^T_{,i}\{\sigma\}[N_k]_{,j}\,\mathrm{d}v\right)\{\dot{u}\} \tag{8.41}$$

By following a similar procedure and with the substitution of (8.28), the right-hand side of (8.35) can be shown to be

$$\int_s \dot{F}_i\,\delta v_i\,\mathrm{d}S_f = \{\delta\dot{u}\}^T\int_s [N_i]^T\{\dot{\theta}\}\,\mathrm{d}S_f$$

$$+ \{\delta\dot{u}\}^T\left(\int_s m_i\lambda_{ijkl}[N_i]^T[N_l]_{,k}\,\mathrm{d}S_f\right)\{\dot{u}\} \tag{8.42}$$

Substituting (8.40), (8.41) and (8.42) into (8.35) gives

$$([K_1] + [K_2] + [K_3])\{\dot{u}\} = \{\dot{P}\} \tag{8.43}$$

with the three stiffness matrices

$$[K_1] = \int_v [B]^T[C_{ep}][B]\,\mathrm{d}v \tag{8.44}$$

$$[K_2] = \int_v ([N_k]^T_{,i}\{\sigma\}[N_k]_{,j} - 2[B_{ki}]^T\{\sigma\}[B_{kj}])\,\mathrm{d}v \tag{8.45}$$

$$[K_3] = -\int_s m_i\lambda_{ijkl}[N_i]^T_{,k}[N_j]\,\mathrm{d}s \tag{8.46}$$

and the load matrix:

$$\{\dot{P}\} = \int_s [N]^T\{\dot{\theta}\}\,\mathrm{d}S_f \tag{8.47}$$

The element equation shown above is identical to that given by McMeeking and Rice[7] except for an additional stiffness matrix $[K_3]$ in the present derivation. This matrix becomes significant for boundaries that are subject to hydrostatic pressure loadings.

The reader will immediately recognize the fact that the stiffness matrix $[K_1]$ is indeed identical to that for the small strain elastic-plasticity theory as shown in (3.10a). Detailed formulations of the other two stiffness matrices, however, have yet to be derived.

8.11 Stiffness matrix $[K_2]$

Physically the stiffness matrix $[K_2]$ given in (8.45) characterizes the rate dependence of the finite strain theory. As the primary structural geometry for the base TEPSAC code is axisymmetric, a fixed cylindrical coordinate system $x_i(x_1, x_2, x_3)$ with axes coinciding respectively with the (r, z, θ) coordinates has been adopted for the following expressions.

For constant strain triangular torus elements such as illustrated in Figure 3.27, the same expression for matrix $[B]$ given in (3.111) can be used, i.e.

$$[B] = [G][h]$$

with

$$[G] = \begin{bmatrix} 0 & 1 & 0 & 0 & 0 & 0 \\ 0 & 0 & 0 & 0 & 0 & 1 \\ 1/r & 1 & z/r & 0 & 0 & 0 \\ 0 & 0 & 1 & 0 & 1 & 0 \end{bmatrix} \tag{8.48}$$

and $[h]$ is given in (3.106).

Furthermore, the interpolation function $[N]$ is already shown in (3.10a) to be

$$[N] = [R][h]$$

in which the matrix $[R]$ can be deduced from (3.97):

$$[R] = \begin{bmatrix} 1 & r & z & 0 & 0 & 0 \\ 0 & 0 & 0 & 1 & r & z \end{bmatrix} \tag{8.49}$$

with

$$[N_3]_{,3} = [R_3]_{,3}[h]$$

where $[N_3]_{,3}$ are related to $\varepsilon_{\theta\theta}$ as

$$\varepsilon_{\theta\theta} = \frac{1}{r}\frac{\partial u_\theta}{\partial \theta} + \frac{u_r}{r} = \frac{u_r}{r} = [N_3]_{,3}\{u\}$$

The $[K_2]$ matrix in (8.45) can thus be reduced to

$$[K_2] = [h]^{\mathrm{T}}\left(\int_v \{\sigma\}([R_k]^{\mathrm{T}}_{,i}[R_k]_{,j} - 2[G_{ki}]^{\mathrm{T}}[G_{kj}])\,\mathrm{d}v\right)[h]$$

$$= [h]^{\mathrm{T}}\left(\int_v [D]\,\mathrm{d}v\right)[h] \tag{8.50}$$

Summing up the terms associated with the non-vanishing components of $\{\sigma\}$, the matrix $[D]$ takes the final form

$$[D] = -\begin{bmatrix} 0 & 0 & 0 & 0 & 0 & 0 \\ 0 & \sigma_{rr} & 0 & 0 & \sigma_{rz} & 0 \\ 0 & 0 & \frac{1}{2}(\sigma_{rr} - \sigma_{zz}) & 0 & 0 & \sigma_{rz} \\ 0 & 0 & 0 & 0 & 0 & 0 \\ 0 & \sigma_{rz} & 0 & 0 & -\frac{1}{2}(\sigma_{rr} - \sigma_{zz}) & 0 \\ 0 & 0 & \sigma_{rz} & 0 & 0 & \sigma_{zz} \end{bmatrix}$$

$$- \sigma_{\theta\theta}\begin{bmatrix} 1/r^2 & 1/r & z/r^2 & 0 & 0 & 0 \\ 1/r & 1 & z/r & 0 & 0 & 0 \\ z/r^2 & z/r & z^2/r^2 & 0 & 0 & 0 \\ 0 & 0 & 0 & 0 & 0 & 0 \\ 0 & 0 & 0 & 0 & 0 & 0 \\ 0 & 0 & 0 & 0 & 0 & 0 \end{bmatrix} \tag{8.51}$$

8.12 Stiffness matrix $[K_3]$

The hydrostatic boundary conditions described in (8.28) can be elaborated to give

$$\lambda_{ijkl} = -q(g_{ij}g_{kl} - g_{il}g_{jk})$$

$$= -qe_{ik\gamma}e_{\alpha\beta\gamma}g_{\alpha j}g_{\beta l} \tag{8.52}$$

in which $e_{ik\gamma}$ and $e_{\alpha\beta\gamma}$ are the alternating tensors defined in Jeffreys[13] (p. 12) and g_{ij} are the metric tensors. Equation (8.46) can then be expressed as

$$[K_3] = -q\int ((n_1[N_2]^{\mathrm{T}}_{,2} + n_1[N_3]^{\mathrm{T}}_{,3} - n_2[N_2]^{\mathrm{T}}_{,1})[N_1]$$

$$+ (n_2[N_1]^{\mathrm{T}}_{,1} + n_2[N_3]^{\mathrm{T}}_{,3} - n_1[N_1]^{\mathrm{T}}_{,2})[N_2])\,\mathrm{d}S \tag{8.53}$$

in the $x_i(i = 1, 2, 3)$ coordinates.

239

Since the displacement functions are only functions of the nodal points which lie on the boundary surface, $[N]$ and $[N_i]_{,j}$ are modified as follows:

$$[\bar{N}] = \begin{bmatrix} \alpha & 0 & \beta & 0 \\ 0 & \alpha & 0 & \beta \end{bmatrix} \tag{8.54}$$

where
$$\alpha = \frac{1}{r_i z_j - r_j z_i}(z_j r - r_j z)$$

$$= \frac{1}{\zeta}(z_j r - r_j z)$$

$$\beta = -\frac{1}{r_i z_j - r_j z_i}(z_i r - r_i z)$$

$$= -\frac{1}{\zeta}(z_i r - r_i z)$$

$$\zeta = r_i z_j - r_j z_i$$

r_i, z_i, r_j, z_j are the nodal coordinates of the boundary nodes I and J respectively, as shown in Figure 8.7.

The modified interpolation function $[\bar{N}]$ in (8.54) relates the displacement vector of the boundary element in Figure 8.7 to the corresponding nodal displacements:

$$U_r^s = \alpha u_r^I + \beta u_r^J \tag{8.55a}$$

$$U_z^s = \alpha u_z^I + \beta u_z^J \tag{8.55b}$$

and

$$[\bar{N}_1]_{,1} = \frac{1}{\zeta}[z_j \ 0 \ -z_i \ 0]$$

$$[\bar{N}_1]_{,2} = \frac{1}{\zeta}[-r_j \ 0 \ r_i \ 0]$$

$$[\bar{N}_2]_{,1} = \frac{1}{\zeta}[0 \ z_j \ 0 \ -z_i]$$

$$[\bar{N}_2]_{,2} = \frac{1}{\zeta}[0 \ -r_j \ 0 \ r_i] \tag{8.56}$$

$$[\bar{N}_3]_{,3} = \frac{1}{\zeta}\left[\left(z_j - \frac{r_j z}{r}\right) \ 0 \ \left(-z_i + \frac{r_i z}{r}\right) \ 0\right]$$

Substituting (8.56) into (8.53) and then taking integrating along the boundary surface, we finally arrive at:

240

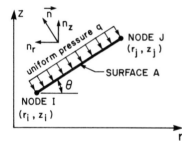

Figure 8.7 Boundary of an element subject to hydrostatic pressure.

$$[K_3] = \frac{q}{\xi^2} \begin{bmatrix} S_1 & S_2 & S_4 & S_5 \\ S_3 & 0 & S_6 & 0 \\ S_7 & S_8 & S_{10} & S_{11} \\ S_9 & 0 & S_{12} & 0 \end{bmatrix} \tag{8.57}$$

in which

$$S_1 = n_r I_1$$
$$S_2 = (n_z z_j + n_r r_j)I_2 + n_z I_1$$
$$S_3 = -(n_z z_j + n_r r_j)I_2$$
$$S_4 = n_r I_3$$
$$S_5 = -(n_z z_j + n_r r_j)I_4 + n_z I_3$$
$$S_6 = (n_z z_j + n_r r_j)I_4$$
$$S_7 = S_4$$
$$S_8 = -S_9 + n_z I_3$$
$$S_9 = (n_z z_i + n_r r_i)I_2$$
$$S_{10} = n_r I_5$$
$$S_{11} = -S_{12} + n_z I_5$$
$$S_{12} = -(n_z z_i + n_r r_i)I_4$$

with $I_1 = \frac{1}{3}(C_r z_j - C_z r_j)^2 A^3$
$$\qquad + (r_i z_j - r_j z_i)(C_r z_j - C_z r_j)A^2 + (r_i z_j - r_j z_i)^2 A$$
$$I_2 = \frac{1}{3}C_r(C_r^2 z_j - C_r C_z r_j)A^3$$
$$\qquad + \frac{1}{2}[r_i(C_r z_j - C_z r_j) + C_r(r_i z_j - r_j z_i)]A^2 + r_i(r_i z_j - r_j z_i)A$$
$$I_3 = -\frac{1}{3}(C_r z_i - C_z r_i)(C_r z_j - C_z r_j)A^3$$
$$\qquad - \frac{1}{2}(C_r z_i - C_z r_i)(r_i z_j - r_j z_i)A^2$$
$$I_4 = \frac{1}{3}C_r(C_r z_i - C_z r_i)A^3 + \frac{1}{2}r_i(C_r z_i - C_z r_i)A^2$$
$$I_5 = \frac{1}{3}(C_r z_i - C_z r_i)^2$$

241

It is apparent from Figure 8.7 that the boundary surface area with unit width is

$$A = (r_j - r_i)^2 + (z_j - z_i)^2$$

and $C_r = \sin \theta$, $C_z = \cos \theta$.

Unlike the matrices $[K_1]$ and $[K_2]$, the matrix $[K_3]$ as shown in (8.57) is unsymmetric. The physical meaning of this matrix is the change of nodal forces due to the boundary pressure q while one particular node undergoes a unit displacement.

8.13 Constitutive equations for thermoelastic–plastic stress analysis

The reader may have noticed that in spite of the complexity of the element equation and the expressions for the stiffness matrices as shown in (8.43) to (8.46), the constitutive equations for the finite strain cases in (8.30) have been shown to resemble those for the small strain theory given in (3.50) for isothermal conditions. The constitutive equations for the case of finite strain of a solid with thermal effect can thus be derived following a similar procedure to that for the small strain case described in Chapter 3.

The stress and strain conjugate used in the small strain cases was the Cauchy stress and strain. The conjugate for the finite strain case however becomes the Kirchhoff stress and Green strain as described in Sections 8.3 and 8.4. We have also learned from Section 8.5 that the first Piola–Kirchhoff stress is a more suitable stress tensor for the case of finite strain elastic problems. For the situation of finite plasticity such as described in Section 8.6, the second Piola–Kirchhoff stress tensor defined as:

$$\sigma_{ij} = \frac{\rho}{\rho_0} \frac{\partial x_i}{\partial a_k} \frac{\partial x_j}{\partial a_l} S_{kl} \tag{8.58}$$

in which σ_{ij} and S_{kl} are the respective true Cauchy and 2nd Piola Kirchhoff stress tensors, is a more suitable substitute for the stresses (see for instance Section 5.3 of Malvern[5]). The strain conjugate for this case, of course, is still the Green strain tensor defined in (8.8). The reader will find that the same procedure described in Section 3.10 can then be followed for the derivation of the constitutive equations for the finite strain theory.

Referring to Figure 8.6 once again, let the incremental Green strain components be expressed as

$$\Delta E_{ij} = \tfrac{1}{2}(\Delta u_{i,j} + \Delta u_{j,i} + \Delta u_{k,i} \Delta u_{k,j}) \qquad \text{(sum over } k) \tag{8.59}$$

where Δu_i are the incremental displacement components (sum over k).

The total thermoelastic–plastic strain increment given in (8.59) can be expressed as the summation of

$$\Delta E_{kl} = \Delta E_{kl}^e + \Delta E_{kl}^p + \Delta E_{kl}^T + \Delta E_{kl}^{Te} \qquad (8.60)$$

in which ΔE_{kl}^e, ΔE_{kl}^p are the respective incremental elastic and plastic strain components, ΔE_{kl}^T is the incremental thermal strain, and ΔE_{kl}^{Te} are the incremental strain due to temperature-dependent material properties. Equation (8.60) is similar to (3.38) for the case of small strain without the effect of strain-rate-dependent material properties.

The plastic potential function for the finite strain case takes a similar form to that for small strain:

$$F = F(S_{ij}, K, T) \qquad (8.61)$$

where

$$K = \text{work hardening parameter}$$

$$= \int S_{ij}' \, dE_{ij}^p \qquad (8.62)$$

and

$$S_{ij}' = \text{deviatoric 2nd Piola–Kirchhoff stress}$$

$$= S_{ij} - \tfrac{1}{3}\delta_{ij} S_{kk} \qquad \text{(sum over } k\text{)}$$

All other basic formulations used for the small strain theory described in Chapter 3 can be adopted here with the substitution of S_{ij} and ΔS_{ij} for the respective σ_{ij} and $\Delta\sigma_{ij}$, and E_{ij}, ΔE_{ij} for ε_{ij} and $\Delta\varepsilon_{ij}$ respectively. Thus, the constitutive equation for the finite strain thermoelastic–plastic cases can be expressed as

$$\{dS\} = [C_{ep}]\{dE\} - [C_{ep}](\{a\}\,dT + \frac{[C_e]^{-1}}{\partial T}\{S\}\,dT)$$

$$- \frac{[C_e]\{S\}}{S}\frac{\partial F}{\partial T}\,dT \qquad (8.63)$$

which is similar to (3.50) without the effect of strain-rate-dependent material properties.

The matrix $[C_{ep}]$ has the usual meaning of being the elastoplastic matrix and equals $[C_e] - [C_p]$, where $[C_e]$ is the elasticity matrix which is identical to those shown in (3.7) and (3.8). The general form of the plastic matrix $[C_p]$ is the same as that shown in Section 3.11 with only the substitution of σ_{ij}' by S_{ij}'. The equivalent stiffness H' here takes a slightly different form:

$$H' = \partial\bar{S}/\partial\bar{E}_p \qquad (8.64)$$

which requires the definition and availability of the equivalent $\bar{S}$–$\bar{E}$ curves.

8.14 The finite element formulation

Although it was demonstrated in the last section that the constitutive equation for the finite strain thermoelastic–plastic cases takes an identical form to that for the small strain cases, except for the stress–strain conjugates. The determination of H' in (8.64) requires the $\bar{S}$-$\bar{E}$ relation of the material, which is not available for most cases. One possible approximation is to take extremely small load increments and update the nodal coordinates at the end of each load step. This exercise will effectively treat, in approximation, the current state value at B' configuration to be equal to that at the original state B^0. In this case, the following equalities exist:

$$\frac{\text{2nd Piola–Kirchhoff}}{\text{stress } S_{ij}} = \frac{\text{1st Piola–Kirchhoff}}{\text{stresses } S_{ij}^{a}} = \frac{\text{Cauchy}}{\text{stresses } \sigma_{ij}}$$

and

$$\text{Green strains } E_{ij} = \text{Cauchy strains, } \varepsilon_{ij}$$

The finite element formulations presented in Chapter 3 can be adopted with the additional stiffness $[K_2]$ and $[K_3]$ given in (8.50) and (8.57).

8.15 The computer program

As mentioned early in this chapter, there are several well-established methods of handling finite strain elastic–plastic analyses. The present approach was chosen for its simplicity and foremostly for its ease of incorporation into the already existing TEPSAC code for small strains. Indeed, the solution procedure for the finite strain case is again very similar to that used for the small strain case described in Chapter 3. Additional subroutines, however, had to be constructed to compute the matrix $[D]$ in (8.51), for the $[K_2]$ matrix given in (8.50), and the stiffness matrix $[K_3]$ in (8.57). These two subroutines can be attached to the base TEPSAC code for the finite strain option.

8.16 Numerical examples

The fundamental difference between the two strain theories in the thermo-elastic–plastic analysis has been described in detail in Section 8.1. As will be demonstrated by the first numerical example of the stretching of a long cylinder, there are certain physical features of a deforming solid, which just cannot be predicted by the use of the small strain theory. Improper use of the small strain theory for a solid undergoing large deformation will produce a drastically wrong result. The second example demonstrates the applicability of the present theory in a metal extrusion problem.

8.16.1 Stretching of a long cylinder[12]

The geometry, dimensions and the finite element idealization of the cylinder are illustrated in Figure 8.8. Due to the symmetry of the geometry and loading, only a quarter of the cylinder needed to be included in the computation. A total of 133 nodes and 108 quadrilateral torus elements were used in the model. In order to control the initiation of the necking of the cylinder, the central one-third of the length was slightly thinned. The following material properties were used in the computation:

Young's modulus, $E = 71\,724$ MPa (10 400 ksi)
Poisson ratio, $v = 0.33$
initial yield strength, $\sigma_y = 276$ MPa (40 ksi)
kink stress, $\sigma_k = 379.30$ MPa (55 ksi)—Eqn. (3.24)
plastic modulus, $E' = 454.50$ MPa (65.9 ksi)—Eqn. (3.24)
stress power, $n = 10$—Eqn. (3.24)

Tensile forces P were applied at the ends of the cylinder, on which the incremental pressure boundary conditions were prescribed. Although the geometry, boundary conditions and stress–strain laws used in the present

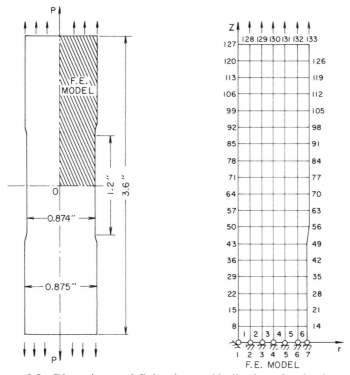

Figure 8.8 Dimensions and finite element idealization of a circular rod.

example differ from those used in some published literatures, e.g. McMeeking and Rice[7], some quantitative as well as qualitative comparison of results can still be made, as will be summarized below:

(1) The tensile stress at P_{max} is in agreement with the well-known formula, i.e.

$$\sigma_{max} = E_t \quad \text{for incompressible material}$$

and

$$\sigma_{max} > E_t \quad \text{for compressible material}$$

in which σ_{max} is the tensile stress at P_{max} and E_t is the tangent modulus shown in (3.63).

It has been shown from the present computation that $\sigma_{max} = 466\,MPa$ (67 600 psi) in element No. 1, whereas $E_t \approx E' = 454.5\,MPa$. The above criteria were satisfied.

(2) Figures 8.9 & 10 present the change of radial displacement at node No. 7 and the longitudinal strain of the cylinder respectively. It is readily seen that the cylinder starts elastic unloading at the ends and then spreads from there toward the central portion. This behavior is in agreement with the results shown in McMeeking and Rice[7], Needleman[14] and Chen[15].

(3) The shape of stress distribution at the necking section such as shown in Figure 8.11 is rather similar to those shown in Needleman[14] and Bridgman[16].

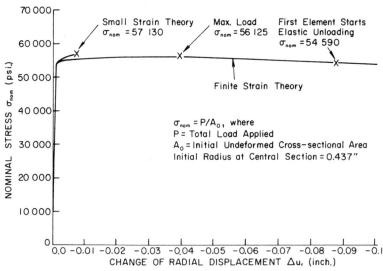

Figure 8.9 Contraction of diameter in the rod.

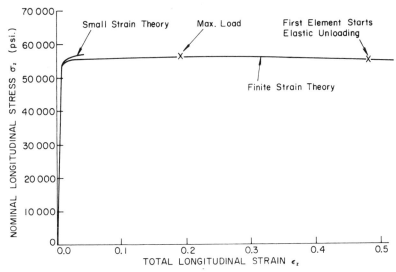

Figure 8.10 Axial strain in the rod.

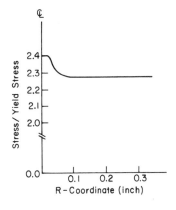

Figure 8.11 Stress distribution in the necking section.

The same problem was also computed by the small strain theory by TEPSA code. The results using this theory are included in Figures 8.9 & 10. It is apparent that the small strain theory has severe limitations when dealing with a solid undergoing large deformation. Another interesting comparison of results by both theories was made on the stress concentration factor F defined and shown in Figure 8.12. It can be seen from this figure that not only the small strain theory failed to predict the elastic unloading of some portion of the material after reaching maximum loading, but also was totally incapable of predicting the instability of the material after a large amount of deformation, which is a common phenomenon in the large elastic–plastic deformation of solids.

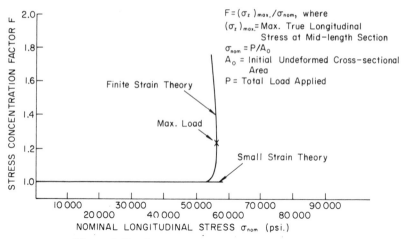

$F = (\sigma_z)_{max.}/\sigma_{nom},$ where
$(\sigma_z)_{max.} =$ Max. True Longitudinal
Stress at Mid-length Section
$\sigma_{nom} = P/A_0$
$A_0 =$ Initial Undeformed Cross-sectional Area
$P =$ Total Load Applied

Figure 8.12 Stress concentration factor in a rod.

8.16.2 Extrusion of metal

The second numerical illustration presented here deals with the determination of the driving pressure required to extrude a metal billet made of aluminum alloy 6061-T6 through a rigid die. Physical representation of the problem is shown in Figure 8.13. A relatively soft metal billet is confined in a rigid container and is to be "pushed" through the die by a rigid ram placed behind the billet. Because of the symmetry of the geometry of the billet, only half of it is included in the finite element idealization. A total of 368 triangular torus elements with 216 nodes were used as shown in Figure 8.14.

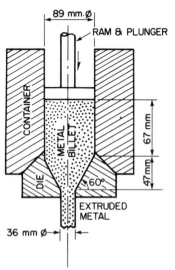

Figure 8.13 Typical set-up for metal extrusion.

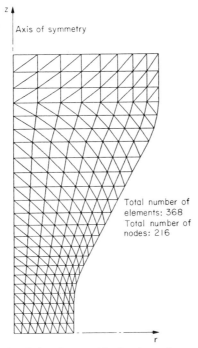

Figure 8.14 Finite element idealization of a metal billet.

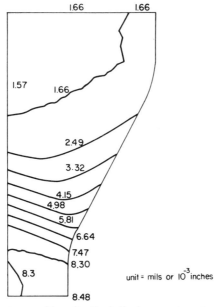

Figure 8.15 Contours of longitudinal displacement vector in a metal billet.

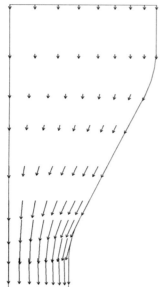

Figure 8.16 Trace of velocity vectors.

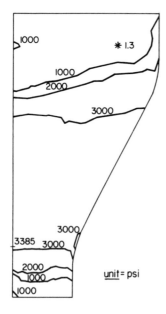

Figure 8.17 Contours of effective stress.

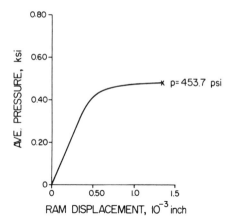

Figure 8.18 Variation of average driving pressure under the ram.

The extrusion process was carried out by the downward movement of the rigid ram as illustrated in Figure 8.13. As a first attempt, no friction was assumed to exist between the billet and containing walls. This frictionless boundary condition may be achieved by intensive lubrication of the con-tacting surfaces[17]. Figure 8.15 shows the longitudinal displacement contours of the billet with a downward 1.66×10^{-3} inches (0.042 mm) movement of the ram. The stress–strain curves of the billet material are identical to those shown in Figure 8.1, with the following key parameters at 550°F (287.8°C):

Young's modulus, E = 8970 ksi (61 862 MPa)
Poisson ratio, v = 0.355
initial yield strength, σ_y = 3000 psi (20.7 MPa)
plastic modulus, E' = 46.94 ksi (323.7 MPa)
kink stress, σ_k = 3000 psi (20.7 MPa)
stress power, n = 10

Figure 8.16 shows the velocity vectors in the billet, and Figure 8.17 shows the effective stress contours. The minimum pressure required to cause the metal to flow was found to be 453.7 psi (3.13 MPa) as illustrated in Figure 8.18. Exact quantitative comparison of results of metal extrusion by the present analysis and experimental measured values cannot be made easily. However, the displacement contours shown in Figure 8.15 were quite similar to those measured by Medrano et al.[18]

251

References

1   Hill, R. 1957. On uniqueness and stability in the theory of finite elastic strain. *J. Mech. & Phys. Solids* **5**, 229–41.
2   Truesdell, C. and R. Toupin 1960. The classical field theories. In *Encyclopedia of physics*. S. Flügge (ed.). Vol. III/1; *Principles of classical mechanics and field theory*. Springer-Verlag, 226–858.
3   Prager, W. 1961. *Introduction to mechanics of continua*. Boston: Ginn.
4   Fung, Y. C. 1965. *Foundation of solid mechanics*. Englewood Cliffs, NJ: Prentice-Hall. Chs. 4, 16.
5   Malvern, L. E. 1969. *Introduction to mechanics of a continuum medium*. Englewood Cliffs, NJ: Prentice-Hall.
6   Larsen, P. K. and E. P. Popov 1974. A note on incremental equilibrium equations and approximate constitutive relations in large inelastic deformations. *ACTA Mech.* **19**, 1–14.
7   McMeeking, R. M. and J. R. Rice 1975. Finite element formulations for problems of large elastic–plastic deformation. *Int. J. Solids & Structures* **11**, 601–16.
8   Cheng, S. Y. 1970. *Bifurcation in elastic–plastic solids under an axial load and a lateral pressure*. PhD thesis, University of Waterloo, Waterloo, Canada.
9   Hill, R. 1962. Uniqueness criteria and extremum principles in self-adjoint problems of continuum mechanics. *J. Mech. & Phys. Solids* **10**, 185–94.
10 Hill, R. 1958. A general theory of uniqueness and stability in elastic–plastic solids. *J. Mech. & Phys. Solids* **6**, 236–49.
11 Yamada, Y., N. Yoshimura and T. Sakurai 1968. Plastic stress–strain matrix and its application for the solution of elastic–plastic problems by the finite element method. *Int. J. Mech. Sci.* Pergamon Press, **10**, 343–54.
12 Cheng, S. Y. and T.-R. Hsu 1977. *Finite element elasto-plastic stress analysis of solids by finite strain theory*. Thermomechanics Lab. report No. 77-10-48, University of Manitoba, Winnipeg, Canada.
13 Jeffreys, H. 1963. *Cartesian tensors*. Cambridge: Cambridge University Press.
14 Needleman, A. 1972. A numerical study of necking in circular cylindrical bars. *J. Mech. & Phys. Solids* **20**, 111–27.
15 Chen, W. H. 1971. Necking of a bar. *Int. J. Solids Structures* **7**, 685–717.
16 Bridgman, P. W. 1964. *Studies in large plastic flow and fracture*. Cambridge, Mass.: Harvard University Press.
17 Bikales, N. M. 1971. *Extrusion and other plastic operations*. New York: John Wiley.
18 Medrano, R., P. Gilles, C. Hinesley and H. Conrad 1971. Application of visco-plasticity techniques to axisymmetric extrusions. In *Metal forming*. A. L. Hoffmanner (ed.). Plenum Press, 85–113.

9

COUPLED THERMOELASTIC-PLASTIC STRESS ANALYSIS

9.1 Introduction

One of a few natural phenomena that have been known to us as long as our civilization is that the shapes of matters change when they are heated up. Simple methods for computing the thermal dilatation of solids induced by temperature changes are taught in most elementary schools. Indeed, we have translated this shape change of material into thermal strains which have been accounted for in the overall stress analysis of solids as described in Chapter 3. In addition to the change of solid geometries, thermal effects also affect material properties, including plastic yielding and creep behavior as described respectively in Chapters 3 and 4.

A question arising from the observation that temperature always induces strains (or deformations) is whether the converse is true. In other words, does the deformation of a solid induced by mechanical means produce heat in the solid? Take, for an example, an aluminum rod of one inch diameter and 6 inches long. Suppose a temperature rise from 70°F to 200°F takes place in a medium with the following average material properties: density = 0.0975 lb/in^3, specific heat = 0.225 Btu/lb °F and linear thermal expansion coefficient = 13.22×10^{-6}/°F. This temperature rise should produce an elongation of 0.013 inch in the rod. A simple calculation also shows that an amount of heat of 13.44 Btu is required to raise the temperature of the rod by 130°F. If the reverse trend described above be true, then a 0.013 inch elongation on the aluminum rod by a tensile tester should, in theory, produce an equivalent heat of 13.44 Btu, which in turn should raise the temperature of the rod from 70°F to 200°F. The reality, of course, is not so. Elongation of metal rods never produces a temperature rise of as much as 130°F, which can be easily felt by one's hand. Can we conclude then that mechanical deformation is incapable of producing a temperature change in material? The answer to this is not positive either, as there is strong experimental evidence provided by Farren and Taylor[1] and Dillon[2,3] that mechanical deformation of solids did indeed produce heat and measurable temperature rises in the solid. The truth of the fact is that although a thermal input can produce a sizeable mechanical deformation (e.g. a 130°F temperature rise to produce a 0.013 inch elongation in an aluminum rod), the converse effect is significantly smaller (a 0.013 inch elongation of a rod by mechanical force produces only a fractional degree of temperature rise, but nowhere near 130°F).

Realizing the fact that mechanical deformation can indeed produce a heat source (or sink) in the solid, the exact solutions to the problems which involve simultaneous applications of mechanical and thermal loads such as those described in Chapter 3 become extremely complicated as the thermal field would affect the mechanical strain field and vice versa. The two field analyses must therefore be solved simultaneously in a "coupled" manner. The equivalent heat source (or sink) terms resulting from the mechanical loads and appearing in the thermal energy equation are regarded as "thermo-mechanical coupling factors".

Although the intricacy of the thermomechanical coupling effect has been realized by researchers for many years, little effort has been made in searching for effective solution methods. It is the author's view, as well as that of some others[4], that a unified approach is lacking, especially in the theoretical treatment for coupled thermomechanical problems, although the topic of coupled thermoelasticity is relatively well documented[4–6]. Little consistency, however, appears to exist among the various published theoretical formulations of the subject of coupled thermoplasticity[7–12]. This inconsistency was recognized by researchers as early as 1970[13].

The rather slow progress and inconsistent approaches to the coupled thermomechanics of solids was partly due to the fact that such a coupling effect did not appear to be significant earlier when engineers seldom stretched their design effort beyond the elastic limit of materials. Thus, many important applications, including nuclear reactor components and high performance engines, can be reasonably analyzed without having to consider the coupling factor, as deformations in these structures were kept extremely small. However, with the availability of sophisticated analytical tools such as the finite element method, demands for utilizing a material's strength beyond its elastic limit have been increasing steadily. The coupling effect in some practical applications in metal forming, fatigue and ratcheting due to excessive local yieldings becomes a significant design consideration. A comprehensive treatment, preferably in the form of the finite element expression, thus appears to be highly desirable.

9.2 The energy balance concept

In the numerical example presented in Section 9.1, we realize that a temperature rise of 130°F will produce an elongation of 0.013 inch in an aluminum rod, but an equal amount of elongation by means of a mechanical load cannot produce nearly such a temperature rise in the rod. The process is obviously an irreversible one. The energy introduced to the solid by heat input is not equal to that which can be produced by mechanical loads

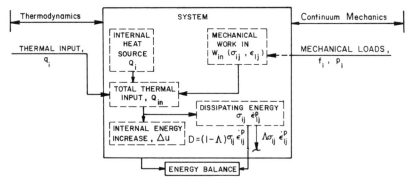

Figure 9.1 Schematic diagram of energy balance in a coupled thermomechanical system.

producing the same amount of deformation in the solid. The irreversibility of the thermal–mechanical energy conversion has prompted most published thermomechanical coupling formulations to be based on the theory of irreversible thermodynamics.

The rather diversified approaches that have been proposed by various researchers[13] in the subject of irreversible thermodynamics have made it one of the least adaptable classical theories for practical engineering applications. The second law of thermodynamics, which is meant to describe this subject, is often mentioned but not taught in the undergraduate classes nearly as thoroughly as its counterpart, the first law. The reader will find that most thermomechanical coupling formulations (see for instance Refs. 7–12) are beyond the usual level of comprehension of practicing engineers. The intention of this chapter is therefore to present this subject as simply and clearly as possible.

The relationship between thermodynamics and continuum mechanics can be better described in a block diagram as illustrated in Figure 9.1.

Before we take a closer look at the system described in Figure 9.1, let us first review the first law of thermodynamics, which expresses the energy balance in a simple closed system as illustrated in Figure 9.2. The heat introduced to the system is converted so as to: (1) increase the internal energy

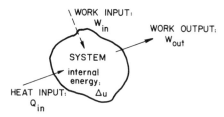

Figure 9.2 Energy balance in a simple closed system.

255

Δu, and (2) produce work, W_{out}. The physical phenomenon can be expressed mathematically as

$$Q_{in} = \Delta u + W_{out}$$

or

$$\Delta u = Q_{in} - W_{out}$$

For the case where mechanical work is applied to the system, (W_{in}), rather than is produced by the system (W_{out}), then the energy equation may be revised and expressed in terms of rates to give:

$$\Delta \dot{u} = \dot{Q}_{in} + \dot{W}_{in} \tag{9.1}$$

The reader is reminded that other possible forms of energy, e.g. flow, kinetic and potential energies have been neglected in the above formation.

Now if we look back at Figure 9.1, we may identify the following relations for a small element (i) in the solid of volume dv and surface ds:

$$\dot{Q}_{in} = \int_v Q_i \, dv + \int_s q_i n_i \, ds$$

$$\dot{W}_{in} = \int_v f_i v_i \, dv + \int_s p_i v_i \, ds$$

where v_i = rate of displacement vectors; f_i, p_i = respective components of body force and surface traction.

On substituting the above expressions into (9.1), we may show that the rate of the increment of internal energy in the system is

$$\int_v \dot{u} \, dv = \int_v Q_i \, dv + \int_s q_i n_i \, ds + \int_v f_i v_i \, dv + \int_s p_i v_i \, ds \tag{9.2}$$

By equating the $\dot{W}_{in}$ to the strain energy rate as illustrated in Figure 9.1, we obtain

$$\int_v f_i v_i \, dv + \int_s p_i v_i \, ds = \int_v \sigma_{ij} \dot{\varepsilon}_{ij} \, dv$$

a special local form of the energy balance can be derived either by substituting the above relation into (9.2) or, as shown in Dillon[8],

$$\dot{u} = \sigma_{ij} \dot{\varepsilon}_{ij} + q_{i,i} + Q_i \tag{9.3}$$

The energy balance equation (9.3) is derived on the basis of a strictly reversible energy balance concept. In reality, however, it is not quite true. Not all the mechanical energy that goes into the solid can be converted into thermal energy as illustrated in the foregoing section. A great portion of this energy is dissipated elsewhere, as shown in Figure 9.1. This dissipation of energy must be accounted for in the overall energy balance in a coupled thermomechanical system.

Since the process is an irreversible one, the second law of thermodynamics has to be used, which indicates that[14]:

$$T \, dS > dQ \qquad (9.4a)$$

where T = absolute temperature, S = entropy, and Q = heat flow in the solid, or

$$dS = d_e S + d_i S \qquad (9.4b)$$

in which $d_e S$ is the increment of entropy derived from the transfer of heat from external sources across the boundary of the system, and the term $d_i S$ is due to changes inside the system which is always greater than zero.

The inequality in (9.4a) leads to the definition of an important term in the thermomechanical coupling relation, namely the "internal dissipation variable", D, which can be expressed mathematically as

$$T \, dS = dQ + D$$

or, in another form,

$$T\dot{S} = q_{i,i} + Q_i + D \qquad (9.5)$$

The energy balance for the overall thermomechanical system illustrated in Figure 9.1 can thus be established with $\dot{u}$ in (9.3) and D in (9.5).

9.3 Derivation of the coupled heat conduction equation

Although the conceptual formulations of the energy balance in a coupled thermomechanical system have been established, expressions which allow the use of common engineering quantities remain to be derived. An effective approach to this problem has been the use of the Helmholtz free energy as will be used in the subsequent derivations.

Now, the Helmholtz free energy F is usually defined as[5, 15]

$$F = F(\varepsilon_{ij}^e, T) = u - ST \qquad (9.6)$$

where ε_{ij}^e denotes the elastic strain components.

In some published works, e.g. by Lehmann[16], another state variable related to the work-hardening parameter K, as described in Chapter 3 for the plastic deformation, has been included in the free energy, F. We will, however, follow the expression in (9.6), as the irreversible nature of the plastic deformation of solids has been lumped into the internal dissipation variable D defined in (9.5). The physical meaning of D will be elaborated later in this chapter.

By differentiating (9.6) with respect to time t, we get

$$\dot{F} = \dot{u} - S\dot{T} - \dot{S}T \tag{9.7}$$

With the substitution of $\dot{u}$ in (9.3) into the above expression, the following relationship is obtained:

$$\dot{F} = \sigma_{ij}\dot{\varepsilon}_{ij} + q_{i,i} + Q_i - S\dot{T} - \dot{S}T$$

Further, by replacing the terms of heat input with D given in (9.5), we may obtain the following expression for the free energy increase:

$$\dot{F} = \sigma_{ij}\dot{\varepsilon}_{ij} - S\dot{T} - D \tag{9.8}$$

Since the free energy F is assumed to be a function of two independent state variables, ε_{ij}^{e} and T, as shown in (9.6), the following expression is valid:

$$\dot{F} = \frac{\partial F}{\partial \varepsilon_{ij}^{e}}\dot{\varepsilon}_{ij}^{e} + \frac{\partial F}{\partial T}\dot{T} \tag{9.9}$$

From the concept of the elastic energy and the definition of entropy, the following two expressions exist:

$$\partial F/\partial \varepsilon_{ij}^{e} = \sigma_{ij} \tag{9.10a}$$

$$\partial F/\partial T = -S \tag{9.10b}$$

Equation (9.10b) yields, with the aid of (9.9),

$$T\dot{S} = -T\frac{d}{dt}\left(\frac{\partial F}{\partial T}\right) = -T\frac{\partial^2 F}{\partial T \partial \varepsilon_{ij}^{e}}\dot{\varepsilon}_{ij}^{e} - T\frac{\partial^2 F}{\partial T^2}\dot{T} \tag{9.11}$$

Again, by following the usual definitions given in thermodynamics,

$$\text{specific heat,} \quad \rho_0 C_{\mathrm{p}} = T\frac{\partial S}{\partial T} = -T\frac{\partial^2 F}{\partial T^2} \tag{9.12a}$$

and the thermal modulus tensor,

$$\beta_{ij} = \frac{\partial \sigma_{ij}}{\partial T} = \frac{\partial^2 F}{\partial T \partial \varepsilon_{ij}^{e}} \tag{9.12b}$$

which leads to

$$T\dot{S} = \rho_0 C_{\mathrm{p}}\dot{T} - \beta_{ij}T\dot{\varepsilon}_{ij}^{e}$$

with the aid of the relationship given in (9.11).

The above expression, when combined with (9.5), yields

$$\rho_0 C_{\mathrm{p}}\dot{T} - \beta_{ij}T\dot{\varepsilon}_{ij}^{e} - D = q_{i,i} + Q_i \tag{9.13}$$

258

If one substitutes the Fourier law of heat conduction defined in Section 2.2.1 into the above equation, i.e. $q_i = -kT_{,i}$ or $q_{i,i} = -(kT_{,i})_{,i}$, in which k is the thermal conductivity of the solid, a more familiar form of equation results:

$$(kT_{,i})_{,i} + Q_i + \beta_{ij}T\dot{\varepsilon}_{ij}^e + D = \rho_0 C_p \dot{T} \qquad (9.14)$$

Equation (9.14) is virtually identical to the heat conduction equation given in Chapter 2 except for the following two extra heat source terms:

$$\beta_{ij}T\dot{\varepsilon}_{ij}^e = \text{thermoelastic coupling factor}$$

and

$$D = \text{thermoplastic coupling factor}$$

The heat conduction equation (9.14) involving these coupling terms is therefore called the "coupled heat conduction equation".

9.4 Coupled thermoelastic–plastic stress analysis

The effect of thermal input to the stress fields in a solid can be considered to contribute in the following three aspects: (1) dilatation-induced thermal stress; (2) properties change; and (3) change in the yield surface when the solid is loaded beyond its elastic limit. The mathematical formulation of all these three aspects has been described in Chapter 3. The coupling aspect with the thermal analysis derived in the foregoing section, however, has not been demonstrated.

Consider a differential quantity in a small element as follows:

$$(Tq_i)_{,i} = Tq_{i,i} + q_i T_{,i}$$

or

$$Tq_{i,i} = (Tq_i)_{,i} - q_i T_{,i} \qquad (9.15)$$

Integrating the above expression for the entire solid, we obtain

$$\int_v Tq_{i,i}\,dv = \int_v (Tq_i)_{,i}\,dv - \int_v q_i T_{,i}\,dv$$

which, on using the Gauss–Green theorem

$$\int_v (Tq_i)_{,i}\,dv = \int_s Tq_i n_i\,ds$$

becomes

$$\int_v Tq_{i,i}\,dv = -\int q_i T_{,i}\,dv + \int_s Tq_i n_i\,ds \qquad (9.16)$$

where n_i is the outward normal to the surface s.

By substituting $q_{i,i}$ given in (9.13) into (9.16), we obtain the following coupled thermal and mechanical relationship:

$$\int_v T[\rho_0 C_p \dot{T} - \beta_{ij} T \dot{\varepsilon}_{ij}^e - D - Q_i]\, dv = -\int_v q_i T_{,i}\, dv + \int_s T q_i n_i\, ds$$

(9.17)

Now, if we consider the universal equilibrium conditions for the solid subject to mechanical loads, we obtain

$$\int_v \dot{\sigma}_{ij} \dot{\varepsilon}_{ij}\, dv = \int_v \dot{f}_i v_i\, dv + \int_s \dot{p}_i v_i\, ds$$

(9.18)

where v_i = rate of deformation vectors.

The constitutive equation for a solid subject to combined thermal and mechanical loads, as has been shown in Chapter 3, can be expressed in tensorial form as

$$\dot{\sigma}_{ij} = C_{ijkl}^{ep} \dot{\varepsilon}_{kl} + \gamma_{ij} \dot{T}$$

(9.19)

where

$$C_{ijkl}^{ep} = C_{ijkl}^{e} - C_{ijkl}^{p}$$

with

$$C_{ijkl}^{e} = C_{ijkl}^{e}(T), \qquad \text{the elasticity matrix}$$

$$C_{ijkl}^{p} = C_{ijkl}^{p}(\sigma_{ij}, T), \qquad \text{the plasticity matrix}$$

$$\gamma_{ij} = \gamma_{ij}(\sigma_{ij}, T, C_{ijkl}^{e}, C_{ijkl}^{p}), \qquad \text{generalized thermal moduli}$$

The term $\beta_{ij} T \dot{\varepsilon}_{ij}^e$ in (9.14) can be expressed in term of total strain rates $\dot{\varepsilon}_{ij}$ as

$$\beta_{ij} T \dot{\varepsilon}_{ij}^e = \bar{\beta}_{ij} \dot{\varepsilon}_{ij} - \bar{\gamma}_{ij} \dot{T}$$

(9.20)

with

$$\bar{\beta}_{ij} = \bar{\beta}_{ij}(\sigma_{ij}, T, C_{ijkl}^{e}, C_{ijkl}^{p})$$

$$\bar{\gamma}_{ij} = \bar{\gamma}_{ij}(\sigma_{ij}, T, C_{ijkl}^{e}, C_{ijkl}^{p})$$

Substitution of (9.19) into (9.18) and (9.20) into (9.17) together with the Fourier law of heat conduction yields the following two important equations:

$$\int_v (C_{ijkl}^{ep} \dot{\varepsilon}_{kl}) \dot{\varepsilon}_{ij}\, dv + \int_v \gamma_{ij} \dot{T} \dot{\varepsilon}_{ij}\, dv = \int_v \dot{f}_i v_i\, dv + \int_s \dot{p}_i v_i\, ds$$

(9.21)

$$\int_v (\rho_0 C_p + \gamma) T \dot{T}\, dv - \int_v T \bar{\beta}_{ij} \dot{\varepsilon}_{ij}\, dv - \int_v D T\, dv - \int_v Q_i T\, dv$$

$$= \int_v k_{ij} T_{,i} T_{,j}\, dv + \int_s T q_i n_i\, ds$$

(9.22)

260

The above equations can be expressed in matrix form:

$$\int_v \{\dot{\varepsilon}\}^T [C_{ep}]\{\dot{\varepsilon}\}\, dv + \int_v \{\dot{\varepsilon}\}^T \{\gamma\}\dot{T}\, dv = \int_v \{V\}^T \{\dot{f}\}\, dv + \int_s \{V\}^T \{\dot{p}\}\, dv \tag{9.23}$$

$$\int_v (\rho_0 C_p + \bar{\gamma})\{T\}^T\{\dot{T}\}\, dv - \int_v \{\bar{\beta}\}^T \{\dot{\varepsilon}\}\{T\}\, dv$$

$$= \int_v D\{T\}^T\, dv + \int_v Q(\mathbf{r})\{T\}\, dv$$

$$+ \int_v \{T_{,i}\}^T [k]\{T_{,j}\}\, dv + \int_s \{q\}^T \{n\}\{T\}\, ds \tag{9.24}$$

where the elastic–plastic matrix $[C_{ep}]$ and the thermal conductivity matrix $[k]$ have been shown in Section 3.11 and (2.39), (2.43) respectively. As the reader will find (9.23) and (9.24) are inter-related, and they cannot be solved separately. These equations are thus referred to as the "coupled thermo-elastic–plastic equations".

9.5 Finite element formulation

The volume integrals in (9.23) and (9.24) form the principle of discretization which is expressed mathematically as follows:

$$\int_v \cdots dv = \sum_m \int_{v_m} \cdots dv_m$$

and

$$\int_s \cdots ds = \sum_n \int_{s_n} \cdots ds_n$$

are directly applicable for the finite element formulation with dv_m, ds_n, m and n denoting the respective element volumes and surfaces and the total number of elements and element surfaces.

Thus, by following the element discretization scheme given in (3.107) for the mechanical displacement components, we have

$$\{U(\mathbf{r}, t)\} = \{R(\mathbf{r})\}^T [h]\{u(t)\} \tag{9.25}$$

which leads to

$$\{V(\mathbf{r}, t)\} = \{R(\mathbf{r})\}^T [h]\{\dot{u}(t)\} \tag{9.26}$$

and

$$T_m(\mathbf{r}, t) = \{b(\mathbf{r})\}^T \{T(t)\} = \{T(t)\}^T \{b(\mathbf{r})\} \tag{9.27}$$

for the temperature in the elements, in which $\{b(\mathbf{r})\}$ is the interpolation function for the temperature field. The matrices $\{u(t)\}$ and $\{T(t)\}$, as usual, are the respective nodal displacements and temperature in the discretized solid.

The strain matrix in (9.23) can be expressed following (3.111) as

$$\{\varepsilon(\mathbf{r}, t)\} = [G(\mathbf{r})][h]\{u(t)\} \tag{9.28}$$

and the temperature gradient matrix can be expressed as

$$\{\nabla T(\mathbf{r}, t)\} = \{a(\mathbf{r})\}^{\mathrm{T}}\{T(t)\} \tag{9.29}$$

Upon substituting the relationships given in (9.26) to (9.29) into (9.23) and (9.24), the following expressions can be obtained:

$$\sum_m \{\dot{u}\}^{\mathrm{T}}([K_u]\{\dot{u}(t)\} + [M_T]\{\dot{T}(t)\} - \{\dot{L}(t)\}) = 0 \tag{9.30}$$

$$\sum_m \{T\}^{\mathrm{T}}([C]\{\dot{T}(t)\} + [M_u]\{\dot{u}(t)\} - \{D\} - \{Q\} - [K_T]\{T(t)\}) = 0 \tag{9.31}$$

where $\quad [K_u]$ = element mechanical stiffness matrix

$$= [h]^{\mathrm{T}}\left(\int_{v_m} [G(\mathbf{r})]^{\mathrm{T}}[C_{\mathrm{ep}}][G(\mathbf{r})]\,\mathrm{d}v_m\right)[h] \tag{9.32}$$

$[M_T]$ = element thermal stiffness matrix

$$= [h]^{\mathrm{T}}\int_{v_m} [G(\mathbf{r})]^{\mathrm{T}}\{\gamma\}\{b(\mathbf{r})\}\,\mathrm{d}v_m \tag{9.33}$$

$\{\dot{L}\}$ = mechanical load matrix

$$= [h]^{\mathrm{T}}\left(\int_{v_m} \{R(\mathbf{r})\}\{\dot{f}\}\,\mathrm{d}v_m + \int_{s_n} \{R(\mathbf{r})\}\{\dot{p}\}\,\mathrm{d}s_n\right) \tag{9.34}$$

$[C]$ = heat capacity matrix

$$= \int_{v_m} \{b(\mathbf{r})\}(\rho_0 C_{\mathrm{p}} + \bar{\gamma})\{b(\mathbf{r})\}^{\mathrm{T}}\,\mathrm{d}v_m \tag{9.35}$$

$[K_T]$ = conductivity matrix

$$= \int_{v_m} \{a(\mathbf{r})\}[k]\{a(\mathbf{r})\}^{\mathrm{T}}\,\mathrm{d}v_m \tag{9.36}$$

$[M_u]$ = thermomechanical coupling matrix

$$= -\left(\int_{v_m} \{b(\mathbf{r})\}\{\bar{\beta}\}^{\mathrm{T}}[G(\mathbf{r})]\,\mathrm{d}v_m\right)[h] \tag{9.37}$$

262

$\{Q\}$ = thermal load matrix

$$= \int_{v_m} \{b(\mathbf{r})\}^T Q_m(\mathbf{r})\, dv_m + \int_{s_n} \{b(\mathbf{r})\}\{q\}^T\{n\}\, ds_n \qquad (9.38)$$

$\{D\}$ = dissipation matrix

$$= \int_{v_m} D\{b(\mathbf{r})\}^T\, dv_m \qquad (9.39)$$

The finite element formulation for the entire solid can be achieved by expressing (9.30) and (9.31) in global form to give

$$[K_u]\{\dot{u}(t)\} + [M_T]\{\dot{T}(t)\} = \{\dot{L}(t)\} \qquad (9.40)$$

$$[M_u]\{\ddot{u}(t)\} + [C]\{\dot{T}(t)\} = [K_T]\{T(t)\} + \{Q(t)\} + \{D\} \qquad (9.41)$$

9.6 The $\{\gamma\}$ matrix

The solution of simultaneous equations (9.40) and (9.41) requires the formulations for the $\{\gamma\}$ matrix in (9.33) and $\bar{\gamma}$ value in (9.35). These matrices can in fact be formulated by referring to the constitutive equation given in (3.50) for a solid undergoing thermoelastic–plastic deformation.

Thus, by expressing (3.50) with the absence of strain-rate effect, we obtain

$$\{\dot{\sigma}\} = [C_{ep}]\{\dot{\varepsilon}\} - [C_{ep}]\{a\}\dot{T} - [C_{ep}]\frac{\partial[C_e]^{-1}}{\partial T}\{\sigma\}\dot{T} - \frac{[C_e]\{\partial F/\partial\} \,\partial F}{S}\,\dot{T}$$
$$(9.42)$$

where

$$[C_{ep}] = [C_e] - [C_p]$$

$$S = \tfrac{4}{9}(3G + H')\bar{\sigma}^2 \qquad (3.56)$$

$$[C_p] = \frac{2G}{S_0}\,[\text{SYM}] \qquad (3.57)$$

with the symmetric matrix [SYM] given in (3.51 a),

$$S_0 = \tfrac{2}{3}\bar{\sigma}^2\left(1 + \frac{H'}{3G}\right) \qquad (3.58)$$

The matrix $\{\gamma\}$ can be expressed immediately by comparing (9.42) with (9.19) to give

$$\{\gamma\} = -[C_{ep}]\left(\{a\} + \frac{\partial[C_e]^{-1}}{\partial T}\{\sigma\}\right) - \frac{[C_e]\{\partial F/\partial\} \,\partial F}{S}\,\frac{\partial F}{\partial T} \qquad (9.43)$$

263

The determination of the matrix $\bar{\gamma}$ requires first that the constitutive relation for the nonlinear thermoelastic deformation given in (3.15), without the strain-rate effect, be recalled:

$$\{\dot{\sigma}\} = [C_e]\{\dot{\varepsilon}^e\} - [C_e]\left(\{a\} + \frac{\partial[C_e]^{-1}}{\partial T}\{\sigma\}\right)\dot{T} \qquad (3.15)$$

Subtracting (3.15) from (9.42) yields

$$\{\dot{\varepsilon}^e\} = [C_e]^{-1}[C_{ep}]\{\dot{\varepsilon}\} - [C_e]^{-1}[C_p]\left(\{a\} + \frac{\partial[C_e]^{-1}}{\partial T}\{\sigma\}\right)\dot{T} - \left\{\frac{\partial F}{\partial \sigma}\right\}\frac{\partial F}{\partial T}\dot{T}/S \qquad (3.15a)$$

Again, the comparison of (3.15a) and (9.20) leads to the following expressions:

$$\bar{\gamma} = -(\{\beta\}^T T)\left([C_e]^{-1}[C_p]\left(\{a\} + \frac{\partial[C_e]^{-1}}{\partial T}\{\sigma\}\right) - \left\{\frac{\partial F}{\partial \sigma}\right\}\frac{1}{S}\frac{\partial F}{\partial T}\right) \qquad (9.44)$$

and also

$$\{\bar{\beta}\}^T = (\{\beta\}^T T)[C_e]^{-1}[C_{ep}] \qquad (9.45)$$

for the evaluation of the thermomechanical coupling matrix in (9.37).

9.7 The thermal moduli matrix, $\{\beta\}$

It is seen from (9.44) and (9.45) that the determination of $\{\bar{\gamma}\}$ and $\{\bar{\beta}\}$ requires the evaluation of the thermal moduli matrix $\{\beta\}$.

The definition of $\{\beta\}$ (i.e. of β_{ij}) was first given in (9.12b) as

$$\beta_{ij} = \partial\sigma_{ij}/\partial T$$

and hence

$$\beta_{ij} = \partial\dot{\sigma}_{ij}/\partial\dot{T}$$

Differentiating the constitutive equation in (9.42) provides the following expression for $\{\beta\}$:

$$\{\beta\} = \partial\{\dot{\sigma}\}/\partial\dot{T} = -[C_{ep}]\left(\{a\} + \frac{\partial[C_e]^{-1}}{\partial T}\{\sigma\}\right) - \frac{[C_e]\{\partial F/\partial\sigma\}}{S}\frac{\partial F}{\partial T} \qquad (9.46)$$

which is identical to (9.43).

It is thus interesting to note that:

$$\{\beta\} = \{\gamma\} \qquad (9.47)$$

9.8 The internal dissipation factor

Early in this chapter, we examined the irreversible nature of the thermo-mechanical process. As an example, a temperature rise of 130°F can produce an elongation of 0.013 inch in a 6 inch long aluminum rod, whereas the same elongation in the rod by mechanical forces raises the temperature of the rod by only a small fraction of the expected level of 130°F. A net loss of mechanical energy is the obvious reason for this imbalance. This loss of mechanical energy has been attributed to the internal dissipating energy D as illustrated in Figure 9.1 and later was included in the formulations.

Although internal energy dissipation was recognized by scientists as early as 1925[1], mathematical formulations have not been made available until fairly recently.

In 1962, Dillon reported some interesting observations made from experiments on the cyclic twisting of circular bars[2]. He discovered that cyclic dilation of the specimen alone induced no temperature rise whereas twisting the specimen in alternating clockwise and counterclockwise directions produced heat in both actions. From these experiments, he concluded that deviatoric deformations were the main reason for generating heat in the solid, even in an elastically deforming solid. In a subsequent paper[8], he identified work done to the system by plastic deformation as the major contribution to the heat effect and proposed a mathematical expression for the internal dissipation rate:

$$D = \sigma_{ij}\dot{\varepsilon}_{ij}^{p} = \sigma'_{ij}\dot{\varepsilon}'^{p}_{ij} \tag{9.48}$$

in which σ'_{ij} and ε'_{ij} are the respective deviatoric stress and strain components defined in (3.17a) and (3.18a).

Since there is a general acceptance that not all the mechanical work produced by the plastic deformation can be converted to the thermal energy in the solid, the quantity D which appears to be an additional heat source in the coupled heat conduction equation, (9.14) would, in a real sense, be a fraction of what is given in (9.48), A larger portion of that work is believed to have been spent in the change of material's microscopic structure, e.g. to produce dislocation, work hardening, etc.[17, 18]. Consequently, various factors have been proposed and applied to the quantity given in (9.48) for the numerical evaluation of D. Lee[19] proposed a variable factor η, so that:

$$D = \eta\sigma_{ij}\dot{\varepsilon}_{ij}^{p} \tag{9.49}$$

with η varying slowly from 0.9 to unity with increasing plastic deformation.

A similar expression was proposed by Lehmann[11] with $\eta < 1.0$ and is a function of both temperature, T, and the work hardening parameter, K, of the material.

Raniecki and Sawczuk[10] and Mroz and Raniecki[20] proposed a slightly different form for the energy dissipation:

$$D = \sigma_{ij}\dot{\varepsilon}_{ij}^{p} - \pi\dot{K} \qquad (9.50)$$

where π is called the conjugate to the internal state variable of hardening parameter K in Raniecki and Sawczuk's model, whereas in Mroz and Raniecki's work, the following relation was used:

$$\pi = -\partial D/\partial\dot{K}$$

and is regarded as a constant.

Nied and Batterman[17] proposed yet another form for the D factor:

$$D = (1 - \Lambda)\sigma_{ij}\dot{\varepsilon}_{ij}^{p} \qquad (9.51)$$

in which the parameter $\Lambda = \Lambda(T)$ is a measure of the ratio of energy stored to plastic energy expended under adiabatic conditions. The numerical value of Λ increases with temperature. It can also be regarded as the ratio of the rates of energy stored in the microstructure of the material resulting from the conversion of the kinetic energy to the internal energy during an adiabatic inelastic deformation process.

It is obvious from the above description that an accurate estimate of the magnitude of the internal energy dissipation D is not possible at this time as the factor η, π and Λ in (9.49) to (9.51) cannot be accurately determined without extensive experiments on various materials. In the meantime, engineers are compelled to settle for some intelligent guesses of these values. A range of 0.1 to 0.9 has been used for the parameter Λ in conjunction with (9.51).

9.9 Computation algorithm

It is apparent from (9.40) and (9.41) that the two unknown quantities, the deformation rate vectors $\{\dot{u}\}$ and the rate of temperature increase $\{\dot{T}\}$ are coupled and that they cannot be solved separately. In the finite element computation it is necessary however to lump these two equations into one:

$$[\bar{K}]\left\{\begin{matrix} \dot{u}(t) \\ \dot{T}(t) \end{matrix}\right\} = \{P(t)\} \qquad (9.52)$$

where the thermoelastic–plastic coupled stiffness matrix $[K]$ takes the form, following (9.40) and (9.41),

$$[\bar{K}] = \begin{bmatrix} K_u & M_T \\ M_u & C \end{bmatrix} \qquad (9.53)$$

whereas the overall load matrix is

$$\{P(t)\} = [K_T]\{T(t)\}\dot{L} + \{Q(t)\} + \{D\} \qquad (9.54)$$

The submatrices which constitute the elements of $[\bar{K}]$ and $\{P(t)\}$ are expressed in (9.32) to (9.39).

One unique characteristic of coupled thermoelastic–plastic problems is that the thermal portion can be a transient in which the time, t, is an independent variable. This is not the case, however, for the elastic–plastic deformation portion of the problem. The numerical integration scheme is thus governed by the usual criteria for thermal analyses as described in Chapter 2. One popular scheme for the thermal analysis is the Crank–Nicholson scheme which defines the incremental quantities to be

$$\{\dot{T}(t)\} = \{\dot{T}_{t+(1/2)\Delta t}\} \approx \frac{1}{\Delta t}(\{T_{t+\Delta t}\} - \{T_t\}) = \frac{1}{\Delta t}\{\Delta T\}$$

and

$$\{\ddot{u}(t)\} \approx \frac{1}{\Delta t}\{\Delta u\}, \qquad \{\dot{L}\} \approx \frac{1}{\Delta t}\{\Delta L\}, \qquad \{P(t)\} \approx \frac{1}{\Delta t}\{\Delta P\}$$

which convert (9.52) into the incremental form

$$[\bar{K}]\left\{\begin{matrix} \Delta u \\ \overline{} \\ \Delta T \end{matrix}\right\} = \{\Delta P\} \tag{9.55}$$

The stiffness matrix $[\bar{K}]$ as expressed in (9.53) is nonsymmetric, and special techniques such as Choleski's method[21] can be used effectively for the solution of (9.55).

9.10 Numerical illustration

The following numerical example is presented to illustrate the significance of the thermomechanical coupling effect. A range of numerical values of the parameter Λ varying between 0.1 and 1.0 was selected to determine the dissipation energy D defined in (9.51) as no quantitative assessment of this energy is available at the present time.

The case being considered here concerns the tensile elongation of a rod stretched at both ends by mechanical loads as illustrated in Figure 9.3a. The rod is assumed to have an initial temperature of 20°C and adiabatic thermal boundary conditions over its entire surface. This boundary condition is justified if the loading time is so short that little heat is allowed to dissipate into the surrounding air. Due to symmetry, only one-quarter of the rod needs to be modelled for the finite element analysis as shown in Figure 9.3b. There are 27 quadrilateral four-node isoparametric elements used in this analysis with the interpolation function identical to that shown in (1.18).

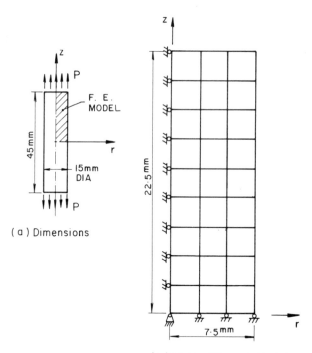

(b) Finite Element Model

Figure 9.3 Dimensions and finite element idealization of a rod subject to uniaxial tension.

Table 9.1 Properties of annealed copper.

Temperature (°C)	E (GPa)	E' (MPa)	σ_k (MPa)	n	k (W/m °C)	α (1/°C)	ρ (kg/m³)	C (kJ/kg °C)	ν
0.0	123	5.36	78.5	9.0	388	2.77×10^{-6}	8.94×10^{3}	0.381	0.34
50.0	123	5.36	78.5	9.0	388	2.77×10^{-6}	8.94×10^{3}	0.371	0.34

The material properties of annealed copper used to establish the constitutive law by (3.24) and other relevant properties are listed in Table 9.1.

Necessary modifications and additional subroutines were made to the base TEPSAC code following the formulations in (9.41), (9.53) and (9.55) with the following special conditions:

$$\{\dot{L}\} = [h]^{\mathrm{T}} \int_{S_n} \{R(\mathbf{r})\}\{\dot{p}\}\, ds_n \tag{9.34}$$

in the absence of the body forces $\{\dot{f}\}$.

$$\{Q\} = 0 \tag{9.38}$$

268

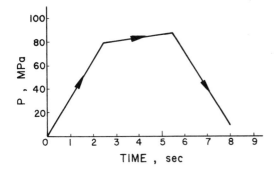

Figure 9.4 The loading history

as $\{q\} = 0$

$$\{\gamma\} = -[C_{ep}]\left(\{a\} + \frac{\partial[C_e]^{-1}}{\partial T}\{\sigma\}\right),\tag{9.43}$$

since $\partial F/\partial T = 0$

$$\bar{\gamma} = -\{\beta\}^T T\left[[C_e]^{-1}[C_p]\left(\{a\} + \frac{\partial[C_e]^{-1}}{\partial T}\{\sigma\}\right)\right]\tag{9.44}$$

as $\partial F/\partial T = 0$

$$\{D\} = \int_{v_m} (1 - \Lambda)\{\sigma\}^T\{\dot{\varepsilon}^p\}\{b(\mathbf{r})\}^T \, dv_m\tag{9.39}$$

with $D = (1 - \Lambda)\sigma_{ij}\dot{\varepsilon}_{ij}^p$.

The applied mechanical loading history is shown in Figure 9.4.

Results obtained from the coupled thermoelastic–plastic stress analysis are presented graphically as follows:

Figure 9.5: the computed effective stress–strain in the elements;

Figure 9.6: the temperature rise in the rod induced by the applied mechanical loads depicted in Figure 9.4 with various Λ values;

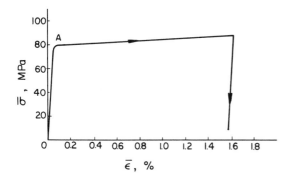

Figure 9.5 Computed effective stress–strain.

Figure 9.7: the temperature rise in the rod vs. the effective strain in the rod with various Λ values;

Figure 9.8: the temperature rise in the rod vs. the degree of plastic deformation measured by the ratio of effective stress to the initial yield strength of the material.

As expected, the temperature rise in the rod induced by the mechanical loading is sensitive to the dissipation energy as shown in Figure 9.6. By referring to Figure 9.1, part of the dissipation energy, $D = (1 - \Lambda)\sigma_{ij}\dot{\varepsilon}_{ij}^P$, contributes to the internal heat source responsible for the temperature rise, whereas the remaining portion, $\Lambda\sigma_{ij}\dot{\varepsilon}_{ij}^P$ is expended to the change of the material's microstructure. Lower Λ values clearly mean less dissipation energy lost to the material and hence more mechanical work could be made available for the temperature rise. It is also interesting to notice from Figure 9.6 to 9.8 that a slight temperature drop occurred during the elastic deformation. A plausible analogy to this phenomenon is the temperature drop of gas during an adiabatic expansion process. Mathematically, one may notice that during the elastic deformation both Q_i and D in (9.14) vanish, and the thermoelastic coupling factor takes a negative sign according to (9.46) for the expression of the thermal moduli matrix $\{\beta\}$. The thermoelastic coupling factor effectively acts like a small heat sink during the tensile deformation. The sign of this term, however, reverses when plastic deformation is initiated, as one may observe from (9.46); again, the numerical values of $[C_{ep}]$ decrease as plastic strain increases. As a result, the thermoelastic coupling factor has now been converted into a small heat source. In the meantime, the thermoplastic coupling factor D which appears in (9.14) and is expressed in (9.51) keeps increasing its numerical values. The net result obviously is the rise of temperature observed in these figures.

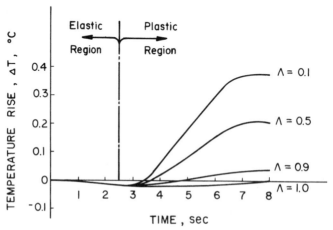

Figure 9.6 Temperature rise in the rod induced by applied mechanical load.

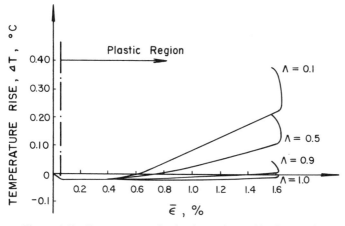

Figure 9.7 Temperature rise in the rod vs. effective strain.

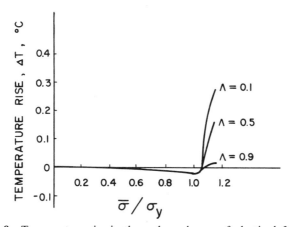

Figure 9.8 Temperature rise in the rod vs. degree of plastic deformation.

9.11 Concluding remarks

Although the fact of deformation-induced thermal effects in solids have been recognized by engineers and scientists for many years, relatively little progress has been made in the quantification of the theory. Research on the subject of coupled thermal–mechanical behavior of solids has been sporadic and no uniform treatment appears to be emerging. Two factors are considered to be responsible for the slow progress: (1) All signs indicate that such a coupling effect is trivial for solids undergoing elastic deformation. Until fairly recently, most structures and machine components were rarely

271

designed to operate beyond the elastic limit of the material, so that the need for intensive study on this subject did not exist; (2) the subject of thermomechanical coupling is closely tied to irreversible thermodynamics. This tie is much stronger when material deforms in the plastic range. Most of the energy produced by both thermal and mechanical sources is converted in the microstructure of the material and cannot be formulated by the established theories for continua.

As quantification of the coupled thermoelastic–plastic stress analysis does not appear to be possible in the present state of the art, it still has a great value to engineers to appreciate this subject in a qualitative sense as summarized below:

(1) While a temperature rise can produce dilatations and shape changes in a solid, only a fraction of the temperature rise can be reproduced by the same amount of mechanical deformation. The thermomechanical deformation process is therefore irreversible.

(2) Cyclic dilatation produces no heat in the material, but cyclic twists always do[2].

(3) The heat source produced by elastic deformation fields is trivial[4], but contributions by deviatoric components of stress and strain during nonlinear elastic or inelastic deformation can be significant[8].

(4) The irreversibility of the thermomechanical deformation process described in (1) is attributed to internal energy dissipation. This dissipative energy is primarily produced by the mechanical work associated with the rate of plastic deformation of the solid ($\sigma_{ij}\dot{\varepsilon}_{ij}^{p}$). Part of this energy is expended in the change of the microstructure of the material, e.g. in the production of dislocations, work hardening, etc.; the remaining part (D factor in the formulation) is converted to heat source in the material[17-20]. Quantitative evaluation of the D factor, however, is not available at this time.

(5) Experiments have revealed that the coupling effect reduces as temperature increases[17]. However, at high temperature, the thermomechanical process becomes more rate sensitive due to more vigorous changes of microstructure of the material[18].

(6) The present work as described in this chapter attempts to present this complex subject in a comprehensive way with finite element formulations given in (9.55). Users are expected to make an intelligent estimate of the numerical values of the D factor in order to obtain numerical solutions. It is hoped that better understanding, and therefore more accurate methods for determining this factor, will soon become available through intensive research efforts.

References

1 Farren, W. S. and G. I. Taylor 1925. The heat developed during plastic extension of metals. *Proc. R. Soc. (Lond.)* **107**, 422.
2 Dillon, O. W. Jr. 1962. A nonlinear thermoelasticity theory. *J. Mech. & Phys. Solids* **10**, 123–31.
3 Dillon, O. W. Jr. 1962. An experimental study of the heat generated during torsional oscillations. *J. Mech. & Phys. Solids* **10**, 235–44.
4 Boley, B. A. and J. H. Weiner 1960. *Theory of thermal stresses.* New York: John Wiley. Ch. 1.
5 Fung, Y. C. 1965. *Foundation of solid mechanics.* Englewood Cliffs, NJ: Prentice-Hall. Chs. 12, 14.
6. Nowacki, W. 1962. *Thermoelasticity.* Reading, Mass.: Addison-Wesley. Chs. 1, 5.
7 Oden, J. T. 1972. *Finite elements on nonlinear continua.* New York: McGraw-Hill. Chs. 12–14, 19, 20.
8 Dillon, O. W. Jr. 1963. Coupled thermoplasticity. *J. Mech. & Phys. Solids* **11**, 21–33.
9 Perzyna, P. and A. Sawczuk 1973. Problems of thermoplasticity. *Nucl. Engng. & Des.* **24**, 1–55.
10 Raniecki, B. and A. Sawczuk 1975. Thermal effects in plasticity. Part I: coupled theory. *Z. Angew. Math. Mech.* **55**, 333–41.
11 Lehmann, Th. 1977. On the theory of large, non-isothermic, elastic–plastic and elastic–visco-plastic deformation. *Arch. Mech.* **29**(3), 393–409.
12 Lehmann, Th. 1980. Coupling phenomena in thermoplasticity. *Nucl. Engng. & Des.* **57**, 323–32.
13 Kestin, J. and J. R. Rice 1970. Paradoxes in the application of thermodynamics to strained solids. In *A critical review of thermodynamics.* Baltimore: Mono Book, 275–98.
14 Yourgrau, W., A. van der Merwe and G. Raw 1966. Treatise on irreversible and statistical thermophysics. New York: Macmillan, 10–18.
15 Sears, F. W. 1953. *An introduction to thermodynamics, the kinetic theory of gases and statistical mechanics*, 2nd edn. Reading, Mass.: Addison-Wesley, 159.
16 Lehmann, Th. 1972. Some thermodynamic considerations of phenomenological theory of non-isothermal elastic–plastic deformations. *Arch. Mech.* **24**, 975–89.
17 Nied, H. A. and S. C. Batterman 1972. On the thermal feedback reduction of latent energy in the heat conduction equation. *Materials Sci. Engng.* **9**, 243–5.
18 Lehmann, Th. 1979. Coupling phenomena in thermoplasticity. *Trans. 5th Int. Conf. Structure Mech. Reactor Technol.* Berlin. Paper L1/1.
19 Lee, E. H. 1969. Elastic-plastic deformations at finite strains. *J. Appl. Mech.* **36**, 1–6.
20 Mroz, Z. and B. Raniecki 1976. On the uniqueness problem in coupled thermoplasticity. *Int. J. Engng. Sci.* **14**, 211–21.
21 Ralston, A. 1965. *A first course in numerical analysis.* New York: McGraw-Hill. Ch. 9.

10

APPLICATION OF THERMOMECHANICAL ANALYSES IN INDUSTRY

10.1 Introduction

A computer code under the name of TEPSA was developed in the early 1970s to handle the elastic–plastic stress analysis on two-dimensional planar or three-dimensional axisymmetric structures subject to combined thermal and mechanical loads. The thermomechanical coupling effect as described in Chapter 9 was neglected in that analysis and the temperature and stress or strain fields were solved separately by the respective formulations presented in Chapters 2 and 3. The fact that both thermal and mechanical analyses were amalgamated into a single code with automatic data transfer has made it an effective tool for the simulation of the thermomechanical behavior of thermal nuclear reactor fuel elements[1] and other advanced engineering analyses involving simultaneous thermal and mechanical loadings.

Several other versions of the TEPSA-based codes have since been developed to extend its application into other areas, as outlined in Table 10.1. The user's guide and listing of the TEPSAC code have been attached in Appendices 4 and 5 respectively.

Table 10.1   Versions of TEPSA-based codes.

Code name	Main functions	Chapters for reference
TEPSAC	Thermoelastic–plastic creep stress analysis	2, 3 & 4
TEPSA-A	Elastic–plastic stress analysis for axisymmetric structure subject to nonaxisymmetric loads	3 & 5
TEPSA-D	Thermoelastodynamic stress analysis	3 & 6
TEPSA-P	Heat conduction analysis with phase change	2 and modifications described in Section 10.2
TEPSA-F	Elastic–plastic fracture analysis with thermal effects	2, 3 & 7
TEPSA-L	Elastic–plastic stress analysis by finite strain theory	3 & 8
CTEPSA	Coupled thermoelastic–plastic stress analysis	2, 3 & 9

The various versions of the TEPSA code have been used to solve a variety of engineering problems, as will be outlined in the following sections.

10.2 Thermal analysis involving phase change

Many practical engineering problems involve a change of phase of materials from a liquid state to solid or vice versa. One obvious example is the welding of metals in which the union of two pieces of material is accomplished through local melting and fusion of material. One natural consequence of this process is the undesirable residual stresses induced in the structure. Another practical example is the precision casting of structures. The solidification of the molten metal often results in excessive distortion in the final shape of the structure. In recent years, the renewed interest and increasing exploration activities for oil and gas in Northern regions, such as the North Sea, Alaska, the Beaufort Sea and East Siberia have prompted many engineers and scientists to focus their efforts on the search for accurate analytical methods to study the melting and fusion of ice structures.

Two major problems make the thermomechanical stress analysis of solids involving phase changes difficult to handle. These are: (1) the method of accounting the latent heat absorption or release during the phase change; (2) the kinematics of the phase boundaries. The abrupt change of properties during phase change such as illustrated in Figure 10.1 for most materials will undoubtedly cause numerical instability in the analysis.

Attempts have been made to solve this type of problem by the classical method using the Fourier heat conduction equation in (2.4) with the latent heat and moving solid–liquid interface included in the boundary conditions (p. 191 in Boley and Weiner[2]). Only a limited number of problems involving simple geometries have been solved this way. Numerical solutions have also been developed in recent years. The use of the finite element method has been reported in two papers[3,4]. Both approaches require the use of special elements and algorithms.

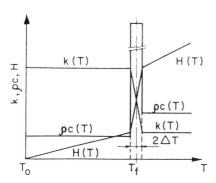

Figure 10.1 Thermophysical properties for ice–water.

A different concept which could be readily adopted by the finite element formulation in Chapter 2 and thus the TEPSA code has been proposed[5]. This concept was formulated on the basis of integrating the heat capacity of the material over a small region of phase temperature T_f, as illustrated in Figure 10.1. Mathematically, such integration will result in the change of the enthalpy of the material, which is a much more smooth transition during the phase change than that of the heat capacity as shown in the figure. The enthalpy can be expressed as

$$H(T) = \int_{T_0}^{T} \rho C(T) \, dT \tag{10.1}$$

where ρ, $C(T)$ are the respective density and specific heat of the material and T_0 is the initial temperature.

The enthalpy-temperature relation (H-T curve) for many engineering materials can be established by reference to appropriate handbooks.

The enthalpy is discretized by relating each element to the correponding nodes following a similar scheme to that used for the element temperature described in (2.30), i.e.

$$H_m(r, z, t) = [N(r, z)]\{H(t)\} \tag{10.2}$$

in which the interpolation function has the same form as in (2.36) and the elements of $\{H(t)\}$ are the enthalpies at the nodes.

Differentiating (10.1) leads to the relation

$$\rho C = dH/dT$$

from which the element heat capacity can be expressed as

$$(\rho C)_{r,z} \simeq \frac{1}{3}\left(\frac{\partial\{H_m\}}{\partial r}\middle/\frac{\partial\{T_m\}}{\partial r} + \frac{\partial\{H_m\}}{\partial z}\middle/\frac{\partial\{T_m\}}{\partial z}\right) \tag{10.3}$$

for a triangular toroidal element as illustrated in Figure 2.6.

It is thus possible to handle the elements undergoing phase changes by determining the nodal values of $\{H(t)\}$ corresponding to the nodal temperature $\{T(t)\}$ computed from the last time instance from the H-T curve, and then evaluating $(\rho C)_{r,z}$ from (10.3). The value of $(\rho C)_{r,z}$ so found can be substituted into the heat capacitance matrix given in (2.28) and (2.44) or in the form

$$[C_e] = \int_v (\rho C)_{r,z}[N(r, z)]^T[N(r, z)] \, dv \tag{10.4}$$

The above algorithm was incorporated into the TEPSA-P version and was used successfully to predict the fusion rate of ice and ice cover with layers of oil with experimental verifications[6].

10.3 Thermoelastic–plastic stress analysis

The following three examples will illustrate the use of the TEPSA code for the solution of problems involving complex thermomechanical loading conditions.

10.3.1 Thermochemical stresses in a gear tooth[7]

The TEPSA code was first used to check the stress concentrations in the fillets of a gear tooth established by empirical formula and photoelastic investigation as described by Dolan and Broghamer[8]. The test set-up for their photoelastic model is illustrated in Figure 10.2 with the finite element idealization shown in Figure 10.3. The gear tooth was assumed to be made of mild carbon steel.

The stress–strain curves of the gear tooth material at five selected temperatures are given in Figure 10.4. The parameters listed in the insert were used to determine the material constitutive laws approximated by (3.24). The stress concentrations at the fillet computed by the TEPSA code were correlated with those proposed by Dolan and Broghamer[8] as demonstrated in Figure 10.5. The isoclinic contours computed by the TEPSA code are shown in Figure 10.6. These contours show excellent correlation to those obtained by photoelastic observation as described in the same reference.

The analysis was later extended to plastic loading with more complex thermal conditions imposed on the two sides of the gear tooth. As illustrated in Figure 10.7, the front face of the gear was subjected to an increasing mechanical load P and the temperature of the surrounding gas T_g began to rise upon reaching the maximum pressure. Both faces were in contact with the hot gas with heat transfer coefficients H_1 and H_2 given in the same figure. The following thermophysical properties were used in the computation: $k = 30$ Btu/ft-hr-°F; $c = 0.112$ Btu/lb-°F and $\rho = 0.285$ lb/in^3. Typical results produced by the TEPSA code are shown in Figures 10.8 & 9 for the respective isothermal contours and the spreading of the plastic zones at various instants.

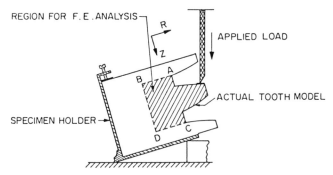

Figure 10.2 Test rig for a photoelastic investigation on a gear tooth.

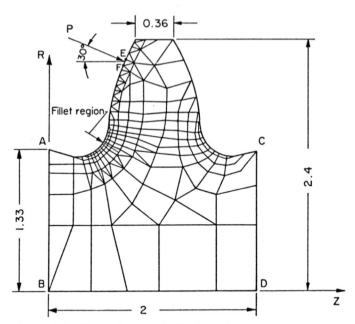

Figure 10.3 Finite element idealization of a gear tooth structure (dimensions in inches).

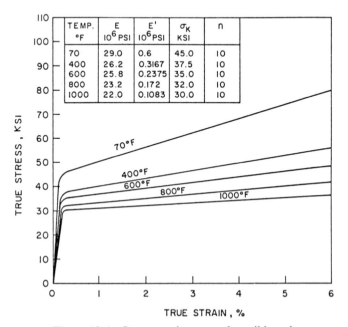

TEMP. °F	E 10⁶ PSI	E' 10⁶ PSI	σ_K KSI	n
70	29.0	0.6	45.0	10
400	26.2	0.3167	37.5	10
600	25.8	0.2375	35.0	10
800	23.2	0.172	32.0	10
1000	22.0	0.1083	30.0	10

Figure 10.4 Stress–strain curves for mild steel.

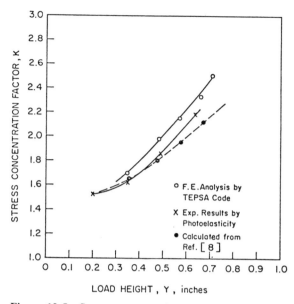

Figure 10.5 Stress concentration in gear tooth fillets.

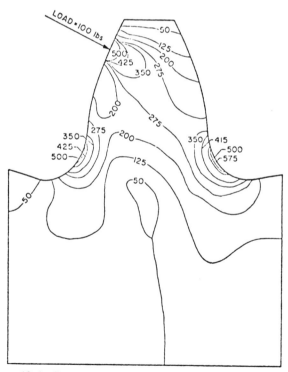

Figure 10.6 Stress contours (isoclinics) by the TEPSA output.

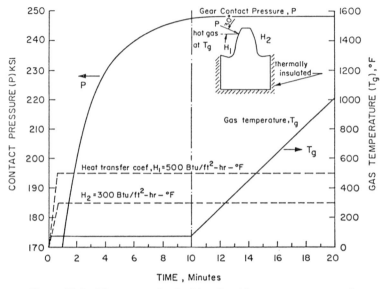

Figure 10.7   Thermomechanical loading history on a gear tooth.

10.3.2   Residual stresses/strains in a pressure vessel[9]

Theoretical formulations presented in Chapter 3 have demonstrated that permanent damage can be induced in a structure in the form of residual stresses or strains after a complete elastic–plastic load cycle. For most materials these residual strains acccumulate with each load cycle due to the Bauschinger effect. The following numerical example will demonstrate that such accumulation of residual strains is possible in a thick-walled cylindrical vessel as a result of cyclic thermomechanical loadings. It will also show that the well-known ratcheting effect often observed in pressure vessels can indeed be predicted by the TEPSA code analysis.

The dimensions of the pressure vessel in this particular case study has an inside diameter of 20 inches and a wall thickness of 1 inch. The vessel is made of mild steel with the stress–strain curves shown in Figure 10.4. Repeated thermomechanical load cycles are applied to the vessel with two complete cycles illustrated in Figure 10.10. This figure presents the variations of the internal pressure P; the bulk fluid temperatures T_i and T_o; the heat transfer coefficients H_i and H_o with subscripts i and o denoting the respective inside and outside surfaces. The thermophysical properties of the vessel material for the computation were identical to those used in the previous example on the gear tooth. It was obvious that the kinematic hardening rule was a suitable choice for this analysis. Interesting results have been given in Hsu and Too[9]. Particularly worth noting is the accumulation of the axial residual strain in both the inner and outer surfaces of the vessel as shown in Figure 10.11. It is interesting to note that not only the magnitudes of residual strains at both surfaces of the vessel accumulate after each cycle of loading, but the difference of these strains between these two surfaces increases as well. This difference of longitudinal residual strains is attributed to a well-known phenomenon called the ratcheting effect. TEPSA code has demonstrated, by this numerical example, its ability to predict such an effect.

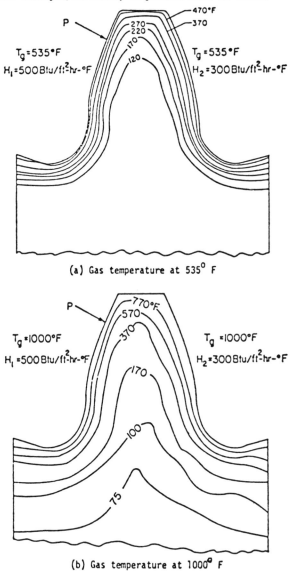

(a) Gas temperature at 535° F

(b) Gas temperature at 1000° F

Figure 10.8 Isothermal contours in a gear tooth.

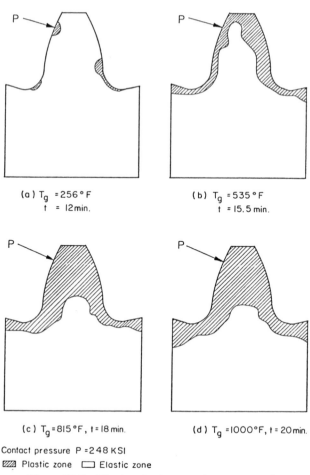

(a) $T_g = 256°F$
    $t = 12\,min.$

(b) $T_g = 535°F$
    $t = 15.5\,min.$

(c) $T_g = 815°F$, $t = 18\,min.$

(d) $T_g = 1000°F$, $t = 20\,min.$

Contact pressure $P = 248\,KSI$

▨ Plastic zone   ▢ Elastic zone

Figure 10.9 Elastic–plastic zones in a gear tooth.

10.3.3 Strengthening of sheet metal by thermal shock

A new concept for strengthening metal sheet was first developed by the author in the early seventies[10]. This idea was intended to reinforce the rim of small holes in metal sheets by introducing opposite residual stress with respect to the applied stresses in that region. The desirable forms of residual stresses could be introduced and "locked" into the predetermined locations by appropriate thermal shocks produced by a powerful laser. The feasibility of this idea was investigated both experimentally and analytically on a metal strip containing a small hole of 3.18 mm diameter. The thickness of the strip ranged from 1.6 to 3.2 mm. A thermal shock by a pulsed laser was applied to one surface of the metal strip in the shape of a ring concentric to the hole. Both types of investigations proved that compressive tangential residual

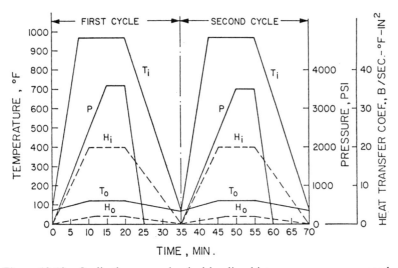

Figure 10.10 Cyclic thermomechanical loading history on a pressure vessel.

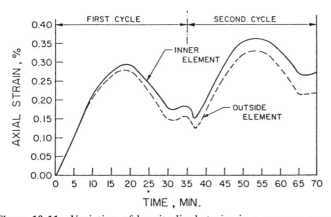

Figure 10.11 Variation of longitudinal strains in a pressure vessel.

stress had been generated in the rim and the nearby region of the hole and significant improvement of fatigue life of the perforated sheet due to tensile load cycles was also observed[10–12]. A quantitative study on the effectiveness of this process was made and reported in Hsu *et al.*[13]. The specimen used in this study was the bare 2024-T3 aluminum sheet of 0.41 mm thickness. It was 25.4 mm wide and 127 mm long with a small hole of the same size as described before located at the geometric center of the strip. Three K-type thermocouples of 50 µm in diameter were spot welded to strategic locations at each face of the specimen. A thermal shock by a ruby laser beam was

283

applied to one face of the strip in the shape of a 6.35 mm inner diameter ×
9.53 mm outer diameter annulus concentric to the hole. The peak energy
output from the laser was measured to be 72 joules with a pulse duration of
1.5 ms. A 40 kV and 20 mA X-ray beam was used to measure the residual
stresses in the thermally shocked specimen, along with the temperature
measurements given by the six thermocouples.

The TEPSA code was used to assess the temperature variations and the
induced residual stresses. Axisymmetric geometry was used with the z-axis
perpendicular to the flat faces. The upper face was designated to be the side
where thermal shock was applied. This model was chosen so that the amount
of penetration of the thermal shock by the laser beam could be assessed from
the variation of the induced residual stresses across the thickness of the strip.
A total of 321 elements and 374 nodes were used in the finite element model
as described in Hsu et al.[13]. Temperature-dependent material properties used
in the computation are given in Figure 10.12. Although the peak energy

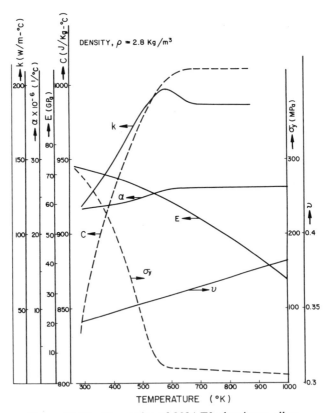

Figure 10.12 Properties of 2024-T3 aluminum alloy.

output from the laser was measured to be 72 joules, only 10% of that energy could actually be absorbed by the material due to back reflection from the polished metal surface. The thermal energy absorbed by the aluminum strip was dissipated to the surrounding air through convective heat transfer from all exposed faces. Heat transfer coefficients covering a range of 0.309 to 11.4 W/m^2-°C were used in the natural convective boundary conditions in the form described in Section 2.9.2(b). These coefficients were estimated using empirical formulae and the measured metal surface temperatures were used to determine the thermophysical properties of the surrounding air.

Correlation of key results from the TEPSA code prediction and those measured can be summarized as in Table 10.2. The permanent residual stress at the rim was computed to be −28 MPa at the shocked face and was reduced to −16 MPa at the back face for the 0.41 mm thick aluminum strip.

Table 10.2 TEPSA code predictions.

	TEPSA	Measured	% error
peak temperature at 4 mm from center	428°K	417°K	2.6
average residual stress at the shocked zone	−225 MPa	−288 MPa	21

10.4 Thermoelastic–plastic stress analysis by TEPSAC code

The use of the TEPSAC code for the thermo-creep-fracture analysis has been described in Chapter 7. Two examples will be presented here. The first example deals with the creep of ice due to constant mechanical load with experimental verification on the results. The second will illustrate an important role that creep plays in the low cycle fatigue analysis.

10.4.1 Creep of ice at constant stress[14]

Creep deformation of ice covers under constant load presents a serious problem for the use of these covers for transportation and shelter purposes. A numerical example which deals with the "sinking" of a pipe into an ice cover will illustrate the application of the TEPSAC code for the solution of this type of problem. Detailed description of this example is available in Liu and Hsu[14]. Only a brief outline and some of the results will be reiterated here.

The geometries and dimensions of the ice foundation and the tube are illustrated in Figure 10.13. A 34 kg dead weight was applied to the steel tube resting on the ice surface. A finite element idealization was applied to the ice foundation with refined meshes near the loading area, which was under the tube wall. The following creep law for ice was used in the analysis with numerical values of parameters given in[15]:

$$\dot{\varepsilon}^c = 1.5 \times 10^{-12} \, N_t (\sigma/\sigma_0)^n \exp\left(Q_c T / 273 R T_0\right) \tag{10.5}$$

285

where

N_t = total number of mobile dislocations
  = 24 000 and 46 000/cm^2
σ_0 = stress for the initiation of dislocation mobility at
  temperature T_0 (= 0.2 MPa)
R = universal gas constant = 1.986 cal/mol
Q_c = activation energy for creep = 21 660 cal/mol
T_0 = ambient air temperature = 263°K
n = stress power = 3.

The modulus of elasticity of the ice was calculated from the expression[16]

$$E = 52\,423 - 673T$$

Units for E and T in the above expression are kg/cm^2 and °K respectively.

The vertical movement of the tube into the ice surface was measured by a pair of displacement transducers. The amount of movement was correlated with that computed by the TEPSAC code as shown in Figure 10.14. It is interesting to note that almost all measured values fall within the two sets of computed values using the upper and lower bounds of N_t values. The computed deformed shapes of the top surface of the ice foundation at various instants and the relaxation of the effective stress in the loading area have been shown in Figures 10.15 & 16 respectively.

10.4.2 Residual stresses and strains in a pressure vessel induced by low cycle thermomechanical loads[17]

Much power plant equipment is expected to operate under high temperature and pressure conditions for an extended period of time of weeks or months before they are shut down for maintenance or refuelling. During this long

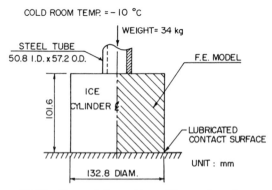

Figure 10.13 Creep deformation of ice under a heavy pipe.

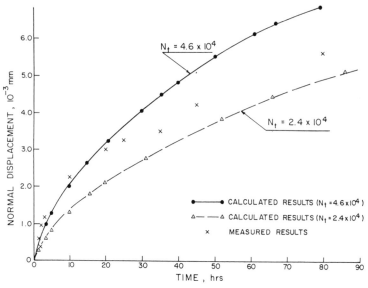

Figure 10.14 Correlation of measured creep deformation of ice with computed results.

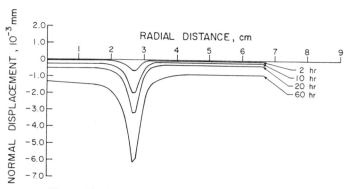

Figure 10.15 Creep deformation of ice surface.

period of operation, little variation of temperature and stress conditions in the structure is expected. It is often referred to as the "hold" time in the study of the low cycle fatigue of materials. The significance of the "hold" time in power cycles has been recognized by engineers for a long time. However, very little is available for the quantitative analysis of its effect on the overall thermomechanical behavior of the structure.

An approach to the low cycle thermomechanical behavior of structure was proposed by Liu and Hsu[17] with the TEPSAC code, using the kinematic

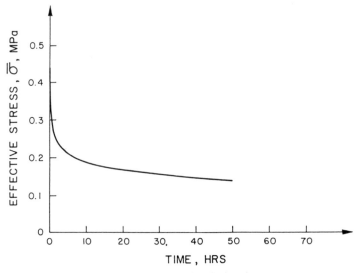

Figure 10.16 Stress relaxation in ice due to creep.

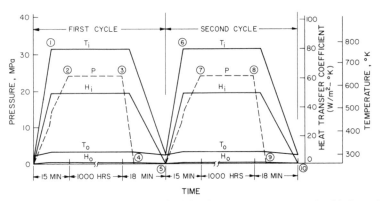

Figure 10.17 Cyclic thermomechanical load to a pressure vessel with long hold periods.

hardening scheme. The rationale was that the thermoelastic–plastic part of the analysis can be used for the start-up and shut-down portions of the load cycles whereas the thermal–creep part of the TEPSAC code can analyze the structure behavior during the long "hold" time. The following numerical example of a pressure vessel will demonstrate the effect of the "hold" time in a quantitative sense.

The geometry and dimensions of the pressure vessel are identical to that used in a previous case illustration of a cyclic thermomechanically loaded pressure vessel as described in Section 10.3.2. The same cyclic loading

288

conditions were used in the present case except that the period of constant pressure loading was extended to 1000 hours as illustrated in Figure 10.17. The same stress–strain relations were adopted for the present example, with the material's creep law identical to that used in (4.10).

Figures 10.18 & 19 show the hoop stresses at the inner and outside surfaces with and without the "hold" time (or creep) effect. The hoop strains and their residual values at the end of each load cycle have been shown in Figure 10.20. Again, the effect of the creep deformation of the material during the hold period is obviously significant, as can be observed from this figure.

10.5 Simulation of thermomechanical behavior of nuclear reactor fuel elements

Like several other major finite element codes[18], the TEPSA code was originally intended for applications in the nuclear industry. The development of this code was initiated in 1972 by the author and his research associate, A. W. Bertels, under sponsorship from Atomic Energy of Canada Ltd (AECL). The primary objective of this development work was to simulate the thermomechanical behavior of the reactor fuel elements under both normal operating and LOCA (loss of coolant accident) conditions. This code was later used as the basis of two separate codes developed by the AECL staff. The FULMOD code which included analytical and empirical modules such as creep, the fission gas release and cracking, was developed to assess the behavior of fuel elements under operating conditions, whereas the other code, under the name of FAXMOD was intended for the purpose of LOCA condition. Empirical modules such as tertiary creep and sheath oxidation have been incorporated into the FAXMOD code.

This section will summarize various aspects of the thermomechanical behavior of fuel elements as computed by the TEPSA code.

10.5.1 General description of a fuel element

Typical nuclear fuel elements (or fuel pins as others may call them) consists of a stack of ceramic fuel pellets made of uranium dioxide or uranium carbide contained in long slender thin-walled cans made of zirconium alloys (or Zircalloy). The primary function of the fuel can (or sheath, cladding) is for the containment of fission products produced by fuel pellets. In the CANDU (CANadian Deuterium Uranium) type of reactor, these fuel elements are clustered together to form a fuel bundle. A few of these bundles are then fastened at the ends to form a fuel string which is placed inside the pressure tube and allows the coolant to flow over the channels provided by the fuel elements. A detailed description of this arrangement may be found in Page[19] with "cut-away" views presented in Figure 10.21.

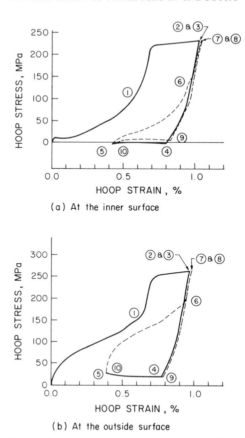

Figure 10.18 Hoop stress–strain in a pressure vessel without creep effect.

A small clearance, e.g. 50 μm, is provided between the pellets and inner surface of the sheath to allow for assembly as well as radial expansion of the pellets during the start-up of the reaction. A small amount of inert gas usually fills in the plenum enclosures.

10.5.2 Thermomechanical loading conditions

As one may well imagine, heat is generated by the fuel pellets at the initiation of the fission process. The heat generated by the fuel pellets is conducted to their periphery across the pellet–cladding interface, then through the sheath and is carried away by the coolant flowing over its surface. As a result of this heat flow pattern, a rather steep temperature gradient is created in the fuel element and serious thermal stress and distortion have become a major concern in the design and operation of the reactor.

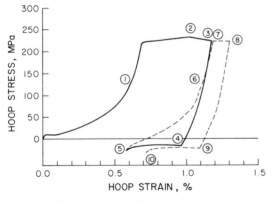

(a) At the inner surface

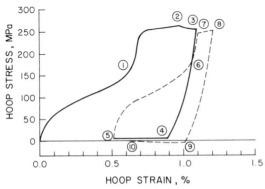

(b) At the outside surface

Figure 10.19 Hoop stress-strain in a pressure vessel with creep effect.

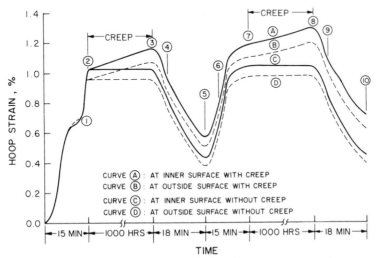

CURVE Ⓐ : AT INNER SURFACE WITH CREEP
CURVE Ⓑ : AT OUTSIDE SURFACE WITH CREEP
CURVE Ⓒ : AT INNER SURFACE WITHOUT CREEP
CURVE Ⓓ : AT OUTSIDE SURFACE WITHOUT CREEP

Figure 10.20 Hoop strain variations in a pressure vessel.

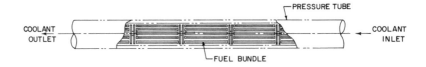

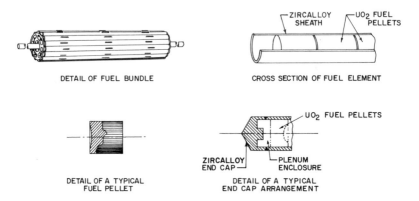

Figure 10.21 Graphic description of CANDU reactor fuel elements.

The stress analysis of fuel element structures is further complicated by the following factors:

(1) A fresh fuel element causes the fuel pellets first to expand radially towards the sheath during the initial period. The pellets and sheath then deform together after the initial contact, resulting in a very complex thermomechanical PCI (pellet–cladding interface) which has to be included in the analytical model.

(2) Fission gas is being released by the fuel pellets during the power generation. This gas, when mixed with the initial filling gas, varies both in quantity and pressure. This pressurized gas mixture has to be contained by the thin fuel sheath.

(3) Because of the anticipated high temperature in the fuel element, all material properties are temperature dependent as described in Section 3.4. The high temperature in the fuel element materials also make the structure vulnerable to creep deformation.

(4) The thermomechanical stresses in certain parts of the fuel element may exceed the yield strengths of the materials and may be followed by unloading.

(5) In addition to the strong PCI, the interactions between the fuel pellets cannot be neglected in the analysis. The well-known fuel ridging is considered to be the result of such interaction.

292

Other loads that cannot be specifically identified in a fuel element have been included in a graphical illustration shown in Figure 10.22.

10.5.3 Thermomechanical behavior of fuel element by TEPSA code

Based on the above assessments of the complex geometry of the fuel elements and the loading conditions, it is not surprising that the finite element method is considered to be the most favorable tool for the modelling and simulation of fuel elements. Computer codes such as ANSYS[18] and BERSAFE[20] have been widely used in the nuclear industry. The following are cases which have been solved by using the TEPSA code or TEPSA-based codes.

(1) PCI and pellet/pellet interactions have been modelled by assigning a thin artificial layer of pseudomaterial functioning as gap elements between the contacting surfaces. As described in detail in Hsu and Bertels[21], this material has unique properties in that it behaves like a

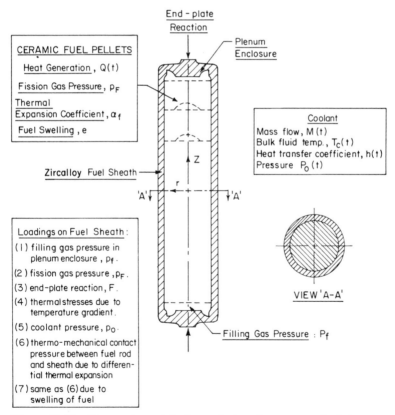

Figure 10.22 Geometric model of a typical nuclear reactor fuel element.

gas (low Young's modulus, E and high yield strength, σ_y) before the contact takes place. The properties are switched to those of incompressible fluid with high E and low σ_y upon contacting and thereafter. This concept of gap elements was also used to simulate the variation of the thermal resistance in the gaps before and after the contact of surfaces as described in Hsu et al.[1].

(2) The potentially devastating fuel sheath ballooning during a LOCA has been modelled by the TEPSA code utilizing the gap element concept[22]. The thermomechanical behavior of a fuel element during a complete power transient including that of PCI at initial contact and separation moments after the initiation of the hypothetical LOCA has been simulated.

(3) The TEPSA-based FULMOD code was also successfully used to predict the fuel sheath deformation[23]. The computed hoop strains have shown excellent correlation with those measured in an in-reactor short power excursion experiment[24]. The well-known sheath ridgings in the same fuel element at various instants were also simulated by the code.

(4) The TEPSA code was used to predict the accumulation of residual stresses in the fuel sheath due to combined cyclic thermomechanical loads[25]. Comparisons were made on the results obtained from both isotropic and kinematic hardening rules as described in Chapter 3.

(5) Attempts were also made, by using the TEPSA code, to predict the deformation behavior of an irradiated fuel sheath[26]. In this paper, modifications were made to the polynomial approximation of the material constitutive law in (3.24) in order to account for the strain softening behavior in a portion of the stress–strain curve for the irradiated Zircalloy. Significant differences in results were obtained when this effect of radiation damage was included in the analysis.

References

1 Hsu, T.-R., A. W. M. Bertels, S. Banerjee and W. C. Harrison 1976. *Theoretical basis for a transient thermal elastic–plastic stress analysis of nuclear reactor fuel elements*. Atomic Energy of Canada Ltd, Rep. AECL-5233.

2 Boley, B. A. and J. H. Weiner 1960. *Theory of thermal stresses*. New York: John Wiley.

3 Bonnerot, R. and P. Jamet 1974. A second order finite element method for the one-dimensional Stefan problem. *Int. J. Num. Meth. Engng.* **8**, 811–20.

4 Wellford, L. C. and R. M. Ayer 1977. A finite element free boundary formulation for the problem of multiphase heat conduction. *Int. J. Num. Meth. Engng.* **11**, 933–43.

5 Comini, G., S. del Guidice, R. W. Lewis and O. C. Zienkiewiez 1974. Finite element solution of non-linear heat conduction problems with special reference to phase change. *Int. J. Num. Meth. Engng.* **8**, 613–24.

6 Hsu, T.-R. and G. Pizey 1981. On the prediction of fusion rate of ice by finite element analysis. *J. Heat Transfer, ASME Trans.* **103**, 727–32.

7 Hsu, T.-R. 1976. *Application of finite element technique to the technology transfer design.* ASME Paper 76-DET-74.

8 Dolan, T. J. and E. L. Broghamer 1942. *A photoelastic study of stresses in gear tooth profiles.* Bulletin No. 335, Engineering Experimental Station, University of Illinois.

9 Hsu, T.-R. and J. J. M. Too 1977. Analysis of residual stresses/strains in pressure vessels due to cyclic thermomechanical loads. *Proc. 3rd Int. Conf. Press. Vess. Technol.* Part I, Tokyo, Japan, American Society of Mechanical Engineers, 83-91.

10 Hsu, T.-R. 1973. Application of the laser beam technique to the improvement of metal strength. *J. Testing & Eval., ASTM* **1**(6), 457-8.

11 Hsu, T.-R. and S. R. Trasi 1976. On the analysis of residual stress induced in sheet metal by thermal shock treatment. *J. Appl. Mech., ASME Trans.*, March, 117-23.

12 Trasi, S. R. and T.-R. Hsu 1978. Improvement of sheet metal strength by localized multiple thermal shocks. *J. Testing & Eval., ASTM* **6**(4), 280-3.

13 Hsu, T.-R., W. J. McAllister and D. A. Scarth 1982. Reinforcement of perforated metal sheets by thermal shock with laser beams. *Exper. Mech.* **22**(8), 302-9.

14 Liu, Y. J. and T.-R. Hsu 1982. On the multidimensional creep deformation of ice by finite element analysis. *J. Energy Resources, ASME Trans.* **104**, 193-8.

15 Michel, B. 1978. *Ice Mechanics*, Les Presses de L'Université Laval, 122-69.

16 Ramseier, R. A. 1971. Mechanical properties of snow ice. *Proc. Inst. Int. Conf. Port & Ocean Engng. under Arctic Conditions* **1**, 192-210.

17 Liu, Y. J. and T.-R. Hsu 1982. *On residual stresses/strains in pressure vessels induced by cyclic thermomechanical loads.* ASME Paper 82-PVP-27.

18 Noor, A. K. 1981. Survey of computer programs for solution of nonlinear structural and solid mechanics problems. *Computers & Structures* **13**, 425-65.

19 Page, R. D. 1977. *Canadian power reactor fuel.* Rep. No. AECL-5609, Atomic Energy of Canada Ltd.

20 BERSAFE 1982. *Berkeley structural analysis by finite elements.* Berkeley Nuclear Laboratories, Berkeley, Glos., England.

21 Hsu, T.-R. and A. W. M. Bertels 1976. Application of elastoplastic finite element analysis to the contact problems of solids. *AIAA J.* **14**(1), 121-2.

22 Hsu, T.-R., A. W. M. Bertels, B. Arya and S. Banerjee 1974. Application of the finite element method to the nonlinear analysis of nuclear reactor fuel behavior. In *Computational methods in nonlinear mechanics.* J. T. Oden *et al.* (eds.). The Texas Institute for Computational Mechanics, University of Texas, Austin, 531-40.

23 Too, J. J. M., T.-R. Hsu and A. W. M. Bertels 1975. FULMOD—an inelastic analysis program to predict the operating behavior of CANDU fuel elements. In *Fuel element analysis.* Y. R. Rashid and F. C. Weiler (eds.). The American Society of Mechanical Engineers.

24 Notley, M. J. F., M. J. Pettigrew and H. Vidal 1972. *Measurements of the circumferential strains of the sheathing of UO2 fuel elements during reactor operation.* Rep. No. AECL-4072, Atomic Energy of Canada Ltd.

25 Hsu, T.-R. 1980. On behavior of fuel elements subject to combined cyclic thermomechanical loads. *Nucl. Engng. & Des.* **56**, 279-87.

26 Hsu, T.-R. and Y. J. Kim 1981. On thermomechanical behavior of irradiated fuel sheath. *Trans. 5th Int. Conf. Structural Mechanics in Reactor Technol.* **C**, Paper No. C2/6, Paris, France.

AREA COORDINATE SYSTEM FOR TRIANGULAR SIMPLEX ELEMENTS

A.1.1 Introduction

Volume and surface integrals over elements with triangular cross sections are frequently involved in finite element formulations. Expressions for the element thermal conductivity, heat capacitance and thermal forces matrices in (2.27), (2.28) and (2.29) respectively all require such integrations. Exact integration with the usual global nodal coordinates involves extremely tedious mathematical manipulations. A special set of local coordinates, known as "area coordinate system" developed by Felippa (see Ref. 12, Chapter 2), has proved to be the most effective method for this purpose.

A.1.2 Definition of area coordinates

Instead of expressing the position of the nodes of a triangular element in the global coordinates as shown in Figure 2.13, these nodes can be specified by the area coordinates, L_i ($i = 1, 2, 3$) defined as follows:

$$L_i = \frac{\text{perpendicular distance from one side, } S_i}{\text{altitude, } h_i} \qquad (A.1.1)$$

These are graphically represented in Figure A.1.1.

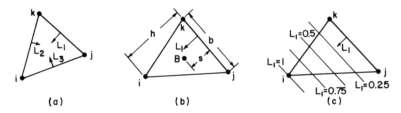

Figure A.1.1 Definition of area coordinates.

A.1.3 Special characteristics

Following the above definition, it can be readily seen that

$$L_1 = \begin{cases} 1 & \text{at node } i \text{ (opposite node)} \\ 0 & \text{at node } j \text{ and } k \end{cases}$$

Likewise,

$$L_2 = \begin{cases} 1 & \text{at node } j \text{ (opposite node)} \\ 0 & \text{at node } i \text{ and } k \end{cases}$$

and

$$L_3 = \begin{cases} 1 & \text{at node } k \\ 0 & \text{at node } i \text{ and } j \end{cases}$$

It is not difficult to prove that

$$L_1 + L_2 + L_3 = 1 \qquad (A.1.2)$$

These local coordinates can be related to the global coordinates by the expressions

$$r = r_i L_1 + r_j L_2 + r_k L_3 \qquad (A.1.3)$$

$$z = z_i L_1 + z_j L_2 + z_k L_3 \qquad (A.1.4)$$

For a triangular element with interpolation functions N_1, N_2 and N_3 defined after (2.37) to be:

$$N_1 = N_i = L_1$$

$$N_2 = N_j = L_2$$

$$N_3 = N_k = L_3 \qquad (A.1.5)$$

then the following relations are useful in evaluating the volume and surface integrals:

$$\int_l N_i^a N_j^b \, dl = \int_l L_1^a L_2^b \, dl = \frac{a! \, b!}{(a + b + 1)!} l \qquad (A.1.6)$$

$$\int_A N_i^a N_j^b N_k^c \, dA = \int_A L_1^a L_2^b L_3^c \, dA = \frac{a! \, b! \, c!}{(a + b + c + 2)!} 2A \qquad (A.1.7)$$

where l and A are the respective length and cross-sectional area of the element.

NUMERICAL ILLUSTRATION ON THE IMPLEMENTATION OF THERMAL BOUNDARY CONDITIONS

A.2.1 Description of the problem

The following numerical example is presented to illustrate the procedures of implementing various boundary conditions by the two distinct methods described in Sections 2.9.1 and 2.9.2. Also, the resultant thermal equilibrium equation in the forms given in (2.47) will be solved by both the two-level explicit method and the mid-interval scheme as outlined in Section 2.8.

The reader will find that results of the nodal temperature increments are virtually identical regardless of the approach used.

The physical conditions describing the example were designed to accommodate all three types of boundary conditions specified in Section 2.2.4, yet the problem was made simple enough to be handled by a pocket calculator. The graphical representation of the example is given in Figure A.2.1 about a section of a pressure vessel made of AISI 1010 steel. It was initially at a uniform temperature of 20°C before it was exposed to a set of thermal boundary conditions as illustrated in Figure A.2.1a. Due to symmetry, only the top half of the section needed to be considered in the finite element model, which included four triangular toroidal elements as depicted in Figure A.2.1b. Element no. 1 was clearly an interior element whereas the boundary conditions for the other three elements can be specified as follows:

element 2: heat flux $q = 6\,\text{kW/m}^2$ across side 2-3
element 3: convective boundary on side 3-4 with $h = 6\,\text{W/m}^2\text{-°K}$ and $T_0 = 20°\text{C}$
element 4: prescribed surface temperature $T_s = 20°\text{C}$ on side 4-1.

A.2.2 Computation of $[K]$ and $[C]$ matrices

The following thermophysical properties from Reference 13 of Chapter 2 were used in the computation:

density $= 7832\,\text{kg/m}^3$
specific heat, $C = 434\,\text{J/kg-°K}$
thermal conductivity, $k = 64\,\text{W/m-°K}$

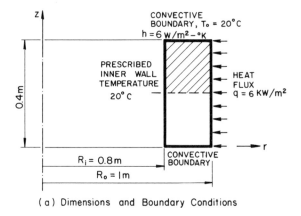

(a) Dimensions and Boundary Conditions

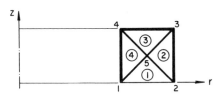

(b) Finite Element Idealization

Figure A.2.1 Thermal boundary conditions for a pipe section.

Table A.2.1 Element descriptions.

Element no.	Nodal designation			Nodal coordinates, m					
	i	j	k	r_i	r_j	r_k	z_i	z_j	z_k
1	5	1	2	0.9	0.8	1.0	0.1	0	0
2	5	2	3	0.9	1.0	1.0	0.1	0	0.2
3	5	3	4	0.9	1.0	0.8	0.1	0.2	0.2
4	5	4	1	0.9	0.8	0.8	0.1	0.2	0

The element descriptions and nodal designation and coordinates are given in Table A.2.1.

The matrix elements of the interpolation functions N_i, N_j and N_k can be calculated by using (2.35) with the cross-sectional area of all elements taken to be 0.01 m² (Table A.2.2), from which the interpolation function $[N]$ was formulated and the matrix $[B]$ then established following (2.37).

Table A.2.2 Matrix elements of the interpolation functions.

Element no.	h_{2i}	h_{2j}	h_{2k}	h_{3i}	h_{3j}	h_{3k}
1	0	−5.0	5.0	10.0	−5.0	−5.0
2	−10.0	5.0	5.0	0	−5.0	5.0
3	0	5.0	−5.0	−10.0	5.0	5.0
4	10.0	−5.0	−5.0	0	5.0	−5.0

The thermal conductivity matrices given in (2.43) for each element were formulated as shown below:

$$[K_e]_1 = 2\pi \times \frac{2.7}{3} \times 0.01 \times 64 \begin{bmatrix} 100 & -50 & -50 \\ & 50 & 0 \\ \text{SYM} & & 50 \end{bmatrix} = 2\pi \begin{bmatrix} K_{11,1} & K_{12,1} & K_{13,1} \\ & K_{22,1} & K_{23,1} \\ \text{SYM} & & K_{33,1} \end{bmatrix}$$

(nodes: 5, 1, 2)

$$\text{(A.2.1a)}$$

$$[K_e]_2 = 2\pi \times \frac{2.9}{3} \times 0.01 \times 64 \begin{bmatrix} 100 & -50 & -50 \\ & 50 & 0 \\ \text{SYM} & & 50 \end{bmatrix} = 2\pi \begin{bmatrix} K_{11,2} & K_{12,2} & K_{13,2} \\ & K_{22,2} & K_{23,2} \\ \text{SYM} & & K_{33,2} \end{bmatrix}$$

(nodes: 5, 2, 3)

$$\text{(A.2.1b)}$$

$$[K_e]_3 = 2\pi \times \frac{2.7}{3} \times 0.01 \times 64 \begin{bmatrix} 100 & -50 & -50 \\ & 50 & 0 \\ \text{SYM} & & 50 \end{bmatrix} = 2\pi \begin{bmatrix} K_{11,3} & K_{12,3} & K_{13,3} \\ & K_{22,3} & K_{23,3} \\ \text{SYM} & & K_{33,3} \end{bmatrix}$$

(nodes: 5, 3, 4)

$$\text{(A.2.1c)}$$

$$[K_e]_4 = 2\pi \times \frac{2.5}{3} \times 0.01 \times 64 \begin{bmatrix} 100 & -50 & -50 \\ & 50 & 0 \\ \text{SYM} & & 50 \end{bmatrix} = 2\pi \begin{bmatrix} K_{11,4} & K_{12,4} & K_{13,4} \\ & K_{22,4} & K_{23,4} \\ \text{SYM} & & K_{33,4} \end{bmatrix}$$

(nodes: 5, 4, 1)

$$\text{(A.2.1d)}$$

The overall thermal conductivity matrix for the entire structure was formulated by summing all element conductivity matrices given in (A.2.1a–d) following a standard procedure:

$K_{11} = K_{22,1} + K_{33,4} = 55.46$ $K_{12} = K_{23,1} = 0$

$K_{13} = 0$ $K_{14} = K_{23,4} = 0$

$K_{15} = K_{12,1} + K_{13,4} = -55.46$ $K_{22} = K_{33,1} + K_{22,2} = 59.73$

$K_{23} = K_{23,2} = 0$ $K_{24} = 0$

$K_{25} = K_{13,1} + K_{12,2} = -59.7$ $K_{31} = 0$

$K_{32} = K_{23,2} = 0$ $K_{33} = K_{33,2} + K_{22,3} = 59.7$

$K_{34} = K_{23,3} = 0$ $K_{35} = K_{13,2} + K_{12,3} = -59.7$

$K_{44} = K_{33,3} + K_{22,4} = 55.46$ $K_{45} = K_{13,3} + K_{12,4} = -55.46$

$K_{55} = K_{11,1} + K_{11,2} + K_{11,3} + K_{11,4} = 230.4$

The overall thermal conductivity matrix therefore takes the form

$$[K] = \begin{bmatrix} 55.47 & 0 & 0 & 0 & -55.46 \\ & 59.73 & 0 & 0 & -59.73 \\ & & 59.73 & 0 & -59.73 \\ & & & 55.46 & -55.47 \\ \text{SYM} & & & & 230.4 \end{bmatrix} \quad (A.2.2)$$

The heat capacitance matrix $[C_e]$ in each element was determined by (2.44) as shown below:

$$[C_e] = \frac{2\pi \times 0.01 \times 7832 \times 434}{60} \begin{bmatrix} C_{11} & C_{12} & C_{13} \\ & C_{22} & C_{23} \\ \text{SYM} & & C_{33} \end{bmatrix}$$

with the elements $C_{11}, C_{12}, \ldots, C_{33}$ as given in Table A.2.3.

Table A.2.3 Calculating the heat capacitance matrix.

Element no.	$C_{11}=6r_i+$ $2r_j+2r_k$	$C_{12}=2r_i+$ $2r_j+r_k$	$C_{13}=2r_i+$ r_j+2r_k	$C_{22}=2r_i+$ $6r_j+2r_k$	$C_{23}=r_i+$ $2r_j+2r_k$	$C_{33}=2r_i+$ $2r_j+6r_k$
1	9.0	4.4	4.6	8.6	4.5	9.4
2	9.4	4.8	4.8	9.8	4.9	9.8
3	9.0	4.6	4.4	9.4	4.5	8.6
4	8.6	4.2	4.2	8.2	4.1	8.2

The overall capacitance matrix was obtained by assembling the element capacitance matrices listed above following an identical procedure used for the assembly of element thermal conductivity matrices:

$$[C] = 2\pi \times 566.51 \begin{bmatrix} 16.8 & 4.5 & 0 & 4.1 & 8.6 \\ & 19.2 & 4.9 & 0 & 9.4 \\ & & 19.2 & 4.5 & 9.4 \\ & & & 16.5 & 8.6 \\ \text{SYM} & & & & 36.0 \end{bmatrix} \quad (A.2.3)$$

We can now proceed to solve the problem.

A.2.3 Solution with the boundary conditions included in $[K_e]$ and $\{F_e\}$

As already outlined in Section 2.9.1, the additional thermal conductivity matrix $[K_e']$ for the element with convective boundary (side 3–4 of Element

301

no. 3) was evaluated by the last term in (2.52) as

$$[K_e']_3 = 2\pi \begin{bmatrix} 0 & 0 & 0 \\ & 0.38 & 0.18 \\ \text{SYM} & & 0.34 \end{bmatrix}$$

The overall thermal conductivity matrix could be modified by summing up the above matrix to the $[K_e]$ given in (A.2.2):

$$[K_e] = \begin{bmatrix} 55.4 & 0 & 0 & 0 & -55.46 \\ & 59.7 & 0 & 0 & -59.7 \\ & & 60.08 & 0.18 & -59.7 \\ & & & 55.8 & -55.46 \\ \text{SYM} & & & & 230.4 \end{bmatrix} \quad \text{(A.2.4)}$$

The thermal force matrix for element no. 2 with the heat flux across side 2–3 was computed by the second term of (2.54):

$$\{F_e''\}_2 = 2\pi \times 6000 \times 0.02 \left\{ \begin{array}{c} 0 \\ \dfrac{1}{3} + \dfrac{1}{6} \\ \dfrac{1}{6} + \dfrac{1}{3} \end{array} \right\} = 2\pi \left\{ \begin{array}{c} 0 \\ 600 \\ 600 \end{array} \right\} \quad \text{(A.2.5a)}$$

and the thermal force matrix for element no. 3 could be evaluated by the second term of (2.53):

$$\{F_e'\}_3 = 2\pi(-6) \times 20 \times 0.2 \left\{ \begin{array}{c} 0 \\ \dfrac{1.4}{3} \\ \dfrac{1.3}{3} \end{array} \right\} = 2\pi \left\{ \begin{array}{c} 0 \\ -11.2 \\ -10.8 \end{array} \right\} \quad \text{(A.2.5b)}$$

The assembled thermal force matrix is therefore equal to

$$\{F\} = 2\pi \left\{ \begin{array}{c} 0 \\ 600 \\ 588.8 \\ -10.8 \\ 0 \end{array} \right\} \quad \text{(A.2.6)}$$

For an assumed time increment $\Delta t = 300$ seconds, the temperature at the five nodes at $t = t + \Delta t = 0 + 300 = 300$ seconds were computed by the two time-integration schemes shown below.

By the two-level explicit method (Section 2.8.1)

The equivalent thermal conductivity is found from (2.48a):

$$[K^*] = 2\pi \begin{bmatrix} 87.19 & 8.50 & 0 & 7.74 & -39.23 \\ & 96.00 & 9.25 & 0 & -41.98 \\ & & 96.37 & 8.68 & -41.98 \\ & & & 87.53 & -39.23 \\ \text{SYM} & & & & 298.38 \end{bmatrix} \quad \text{(A.2.7a)}$$

and the equivalent thermal force matrix from (2.48b):

$$\{F^*\} = 2\pi \begin{Bmatrix} 0 \\ 600 \\ 588.8 \\ -10.8 \\ 0 \end{Bmatrix} \quad \text{(A.2.7b)}$$

The prescribed nodal temperature boundary condition could be implemented by assigning the specified numerical values to the appropriate nodes followed by proper modifications of $[K^*]$ and $\{F^*\}$, a procedure which is quite standard in finite element analysis (see e.g. on p. 8 of Ref. [3], Chapter 1). The resultant temperature increments were

$$\{\Delta T\}^\mathrm{T} = \{0 \quad 6.28 \quad 6.28 \quad 0 \quad 1.79\}^\mathrm{T} \quad \text{(A.2.8)}$$

By the mid-interval scheme

By referring to (2.51c), the equivalent thermal conductivity matrix was determined according to (2.51d):

$$[K^*] = 2\pi \begin{bmatrix} 118.92 & 17.00 & 0 & 15.49 & -22.99 \\ & 132.25 & 18.51 & 0 & -24.23 \\ & & 132.63 & 17.18 & -24.23 \\ & & & 119.26 & -22.99 \\ \text{SYM} & & & & 366.36 \end{bmatrix} \quad \text{(A.2.9a)}$$

and the equivalent thermal force matrix was calculated from (2.51g) to be

$$\{F^{***}\} = -2[K]\{T_i\} + 2\{F\} = 2\pi \left\{ \begin{array}{c} 0 \\ 1200 \\ 1155 \\ -35.2 \\ 0 \end{array} \right\} \qquad \text{(A.2.9b)}$$

with $[K]$ from (A.2.2), $\{T_i\} = \{20\}$ and $\{F\}$ from (A.2.6).

Thus by substituting (A.2.9a,b) into (2.51c) and after implementing $\Delta T = 0$ at nodes 1 and 4, the temperature increments at the nodes could be solved to give:

$$\{\Delta T\}^T = \{0 \quad 8.18 \quad 7.76 \quad 0 \quad 1.054\}^T \qquad \text{(A.2.10)}$$

As can be seen by comparing the results in (A.2.8) and (A.2.10), the two time-integration schemes yield almost the same results.

A.2.4 Solution with boundary conditions imposed on the boundary elements—TEPSA approach

Following the description in Section 2.9.2, this approach treats all elements as "interior" elements with assembled overall thermal conductivity matrix in (A.2.2), the capacitance matrix in (A.2.3) and an all-zero thermal force matrix, $\{F\} = \{0\}$.

Both $[K]$ and $\{F\}$ matrices are then modified to take into account the convective boundary condition in element no. 3 with nodes 3 and 4 to be taken respectively as nodes i and j in (2.57) with the index $\alpha = 0$. The equivalent nodal heat flux thus becomes

$$b_c = \frac{h_c a}{2} = \frac{h}{2}(2\pi)\left(\frac{r_i + r_j}{2}\right)L_{ij} = 0.54(2\pi)$$

Modification to $[K]$ is carried out by (2.58), i.e.

$$K'_{ii} = K_{33} + 0.54(2\pi) \qquad \text{(A.2.11a)}$$

and

$$K'_{jj} = K_{44} + 0.54(2\pi) \qquad \text{(A.2.11b)}$$

where K_{33} and K_{44} were given in (A.2.2).

The modified matrix has the form:

$$[K'] = 2\pi \begin{bmatrix} 55.47 & 0 & 0 & 0 & -55.47 \\ & 59.73 & 0 & 0 & -59.73 \\ & & 60.27 & 0 & -59.73 \\ & & & 56.0 & -55.47 \\ \text{SYM} & & & & 230.4 \end{bmatrix} \qquad \text{(A.2.12)}$$

The modified thermal force matrix for the convective boundary condition on the side linking nodes 3 and 4 is

$$\{F\} = \{0\} + \left\{ \begin{array}{c} 0 \\ 0 \\ -0.54(2\pi) \times 20 \\ -0.54(2\pi) \times 20 \\ 0 \end{array} \right\} = 2\pi \left\{ \begin{array}{c} 0 \\ 0 \\ -10.8 \\ -10.8 \\ 0 \end{array} \right\}$$

An additional modification to this matrix for the heat flux across nodes 2 and 3 in element no. 2 is:

$$F_2 = F_3 = \frac{aq}{2} = \frac{2\pi R_0}{2}(0.2) \times 6000 = 2\pi(600) \text{ W/m}^2$$

Hence the modified thermal force matrix is

$$\{F'\} = 2\pi \left\{ \begin{array}{c} 0 \\ 0 \\ -10.8 \\ -10.8 \\ 0 \end{array} \right\} + 2\pi \left\{ \begin{array}{c} 0 \\ 600 \\ 600 \\ 0 \\ 0 \end{array} \right\} = 2\pi \left\{ \begin{array}{c} 0 \\ 600 \\ 589.20 \\ -10.8 \\ 0 \end{array} \right\}$$

$$(A.2.13)$$

By the two-level explicit method (Section 2.8.1)

The equivalent conductivity matrix is again determined by (2.48a) by summing $[K']$ in (A.2.12) and $[C]$ in (A.2.3):

$$[K^*] = 2\pi \begin{bmatrix} 87.19 & 8.50 & 0 & 7.74 & -39.23 \\ & 96.0 & 9.25 & 0 & -41.98 \\ & & 96.53 & 8.50 & -41.98 \\ & & & 87.73 & -39.23 \\ \text{SYM} & & & & 298.38 \end{bmatrix} \quad (A.2.14a)$$

The thermal force matrix is identical to that shown in (A.2.13) according to (2.48b):

$$\{F^*\} = \{F'\} = \left\{ \begin{array}{c} 0 \\ 600 \\ 589.2 \\ -10.8 \\ 0 \end{array} \right\} \quad (A.2.14b)$$

There is yet another set of boundary conditions to be taken into account before establishing the final equilibrium equation, i.e. the prescribed nodal temperatures at nodes 1 and 4. Thus, by setting $\beta = 10^6$ and following the procedure described in Section 2.9.2(a), we obtain

$$e = \beta(K_{11}^* + K_{44}^*) = 175 \times 10^6$$

and

$$F_1^* = F_4^* = e\,\Delta T_1 = e\,\Delta T_2 = 0$$

The resultant equilibrium equations are modified to the form

$$
\begin{bmatrix}
175 \times 10^6 & 8.50 & 0 & 7.74 & -39.23 \\
 & 96.00 & 9.25 & 0 & -41.98 \\
 & & 96.53 & 8.50 & -41.98 \\
 & & & 175 \times 10^6 & -39.23 \\
\text{SYM} & & & & 298.38
\end{bmatrix}
\begin{Bmatrix}
\Delta T_1 \\ \Delta T_2 \\ \Delta T_3 \\ \Delta T_4 \\ \Delta T_5
\end{Bmatrix}
=
\begin{Bmatrix}
0 \\ 600 \\ 589.2 \\ 0 \\ 0
\end{Bmatrix}
$$

$$\text{(A.2.15)}$$

Solving for $\{\Delta T\}^{\mathrm{T}} = \{0 \quad 6.4272 \quad 6.2645 \quad 0 \quad 1.7858\}^{\mathrm{T}}$

By the mid-interval scheme

Although the conductivity matrix $[K^*]$ in (A.2.9a) remained unchanged, proper modifications of $\{F\}$ in (A.2.6) have to be carried out. The equivalent thermal force matrix before implementing the boundary conditions is given in (2.51g):

$$\{F^{***}\} = -2[K]\{T_t\} + 2\{F\} = 2
\begin{Bmatrix}
0 \\ 1200 \\ 1156.8 \\ -43.2 \\ 0
\end{Bmatrix}
$$

The convective boundary condition in element no. 3 was implemented first by modifying K_{33}^* and K_{44}^* according to (A.2.11a,b) for the conductivity matrix, resulting in

$$[K^*] = 2\pi
\begin{bmatrix}
118.92 & 17 & 0 & 15.49 & -22.99 \\
 & 132.25 & 18.51 & 0 & -24.23 \\
 & & 133.17 & 17.18 & -24.23 \\
 & & & 119.80 & -22.99 \\
\text{SYM} & & & & 366.36
\end{bmatrix}
$$

The above thermal force matrix was modified according to (2.58):

$$F_3^{***} = F_3^{***} + b_c T_0 = 2\pi(1156.8 + 0.54 \times 20) = 2\pi(1167.6)$$

$$F_4^{***} = F_4^{***} + b_c T_0 = 2\pi(-43.2 + 0.54 \times 20) = 2\pi(-32.4)$$

Hence the modified thermal force matrix with convective boundary conditions becomes

$$\{F^{***}\}^T = 2\pi\{0 \quad 1200 \quad 1167.60 \quad -32.4 \quad 0\}^T \qquad \text{(A.2.16b)}$$

The next step is to incorporate the prescribed temperature condition for side 1–4 in Figure A.2.1. Again, by using the procedure outlined in Section 2.9.2(a) with $\beta = 10^6$ and $[K^*]$ in (A.2.16a), we obtain

$$e = \beta(K_{11}^* + K_{44}^*) = 238.72 \times 10^6$$

and

$$F_1^{***} = F_4^{***} = e\,\Delta T_1 = e\,\Delta T_4 = 0.$$

The final form of the equilibrium equation becomes

$$
\begin{bmatrix}
238.72 \times 10^6 & 17 & 0 & 15.49 & -22.99 \\
 & 132.25 & 18.51 & 0 & -24.23 \\
 & & 133.17 & 17.18 & -24.23 \\
 & & & 238.72 \times 10^6 & -22.99 \\
\text{SYM} & & & & 366.36
\end{bmatrix}
\begin{Bmatrix}
\Delta T_1 \\ \Delta T_2 \\ \Delta T_3 \\ \Delta T_4 \\ \Delta T_5
\end{Bmatrix}
=
\begin{Bmatrix}
0 \\ 1200 \\ 1167.6 \\ 0 \\ 0
\end{Bmatrix}
$$

$$\text{(A.2.17)}$$

Solving (A.2.17) for the temperature increments at the nodes, we find

$$\{\Delta T\}^T = \{0 \quad 8.17 \quad 7.824 \quad 0 \quad 1.058\}^T \qquad \text{(A.2.18)}$$

The numerical results obtained by these four different approaches are summarized in the following table. While there are differences in results by the two time-integration schemes described in Section 2.8, the results derived by different methods of implementing boundary conditions appear to be trivial.

Table A.2.4   Summary of the numerical results.

	Temperature increments (°C)			
	B.C. in F.E. formulation		B.C. in boundary elements	
Node no.	2-level	Mid-interval	2-level	Mid-interval
1	0	0	0	0
2	6.28	8.18	6.4272	8.17
3	6.28	7.76	6.2645	7.824
4	0	0	0	0
5	1.79	1.054	1.7858	1.058
Eqn. no.	(A.2.8)	(A.2.10)	(A.2.15)	(A.2.18)

APPENDIX 3

INTEGRANDS OF THE MODE-MIXING STIFFNESS MATRIX (MODE ZERO AND MODE ONE ONLY)

Note:
- (a) Because of symmetry, only the upper triangular part of the matrix is printed.
- (b) A = COS(θ);
     B = SIN(θ).
- (c) XI(1) = 1;
     XI(2) = 1/R;
     XI(3) = 1/R**2;
     XI(4) = Z/R;
     XI(5) = Z/R**2;
     XI(6) = Z**2/R**2.
- (d) DS(I, J) for I, J = 1, 2, 3, 4, 5, 6 are the elastoplastic matrix, $[C^{ep}]$.

(1)   1st row elements

```
101   FCT=DS(3,3)*XI(3)
102   FCT=(DS(3,1)+DS(3,3))*XI(2)
103   FCT=DS(3,3)*XI(5)+DS(3,4)*XI(2)
104   FCT=0.0
105   FCT=DS(3,4)*XI(2)
106   FCT=DS(3,2)*XI(2)
107   FCT=-DS(3,5)*XI(3)
108   FCT=0.0
109   FCT=-DS(3,5)*XI(5)+DS(3,6)*XI(2)
110   FCT=DS(3,3)*XI(3)*A-DS(3,5)*XI(3)*B
111   FCT=(DS(3,1)+DS(3,3))*XI(2)*A-DS(3,5)*XI(2)*B
112   FCT=DS(3,3)*XI(5)*A+DS(3,4)*XI(2)*A-DS(3,5)*XI(5)*B
113   FCT=-DS(3,6)*XI(3)*B
114   FCT=DS(3,4)*XI(2)*A-DS(3,6)*XI(2)*B
115   FCT=DS(3,2)*XI(2)*A-DS(3,6)*XI(5)*B
116   FCT=DS(3,3)*XI(3)*A-DS(3,5)*XI(3)*B
117   FCT=DS(3,3)*XI(2)*A
118   FCT=DS(3,3)*XI(5)*A-DS(3,5)*XI(5)*B+DS(3,6)*XI(2)*B
```

(2) 2nd row elements

```
202   FCT=(DS(1,1)+DS(1,3)+DS(3,1)+DS(3,3))*X1(1)
203   FCT=(DS(1,3)+DS(3,3))*XI(4)+(DS(1,4)+DS(3,4))*XI(1)
204   FCT=0.0
205   FCT=(DS(1,4)+DS(3,4))*XI(1)
206   FCT=(DS(1,2)+DS(3,2))*XI(1)
207   FCT=-DS(1,5)*XI(2)-DS(3,5)*XI(2)
208   FCT=0.0
209   FCT=-(DS(1,5)+DS(3,5))*XI(4)+(DS(1,6)+DS(3,6))*XI(1)
210   FCT=(DS(1,3)+DS(3,3))*XI(2)*A-(DS(1,5)+DS(3,5))*XI(2)*B
211   FCT=((DS(1,1)+DS(1,3)+DS(3,1)+DS(3,3))*A-(DS(3,5)+DS(1,5))*B)*XI(1)
212   FCT=(DS(1,3)+DS(3,3))*XI(4)*A+(DS(1,4)+DS(3,4))*A*XI(1)-(DS(1,5)
      1+DS(3,5))*XI(4)*B
213   FCT=-(DS(1,6)+DS(3,6))*XI(2)*B
214   FCT=((DS(1,4)+DS(3,4))*A-(DS(1,6)+DS(3,6))*B)*XI(1)
215   FCT=(DS(1,2)+DS(3,2))*A*XI(1)-(DS(1,6)+DS(3,6))*XI(4)*B
216   FCT=(DS(1,3)+DS(3,3))*XI(2)*A-(DS(1,5)+DS(3,5))*XI(2)*B
217   FCT=(DS(1,3)+DS(3,3))*A*XI(1)
218    FCT=(DS(1,3)+DS(3,3))*XI(4)*A-(DS(1,5)+DS(3,5))*XI(4)*B+(DS(1,6)
      1+DS(3,6))*B*XI(1)
```

(3) 3rd row elements

```
303   FCT=DS(3,3)*XI(6)+DS(3,4)*XI(4)+DS(4,3)*XI(4)+DS(4,4)*XI(1)
304   FCT=0.0
305   FCT=DS(3,4)*XI(4)+DS(4,4)*XI(1)
306   FCT=DS(3,2)*XI(4)+DS(4,2)*XI(1)
307   FCT=-DS(3,5)*XI(5)-DS(4,5)*XI(2)
308   FCT=0.0
309   FCT=-DS(3,5)*XI(6)+DS(3,6)*XI(4)-DS(4,5)*XI(4)+DS(4,6)*XI(1)
310   FCT=(DS(3,3)*A-DS(3,5)*B)*XI(5)+(DS(4,3)*A-DS(4,5)*B)*XI(2)
311   FCT=(DS(3,1)*A+DS(3,3)*A-DS(3,5)*B)*XI(4)+(DS(4,1)*A+DS(4,3)
      1*A-DS(4,5)*B)*XI(1)
312    FCT=DS(3,3)*XI(6)*A+DS(3,4)*XI(4)*A-DS(3,5)*XI(6)*B+DS(4,3)
      1*XI(4)*A+DS(4,4)*A*XI(1)-DS(4,5)*XI(4)*B
313   FCT=-DS(3,6)*XI(5)*B-DS(4,6)*XI(2)*B
314   FCT=(DS(3,4)*A-DS(3,6)*B)*XI(4)+(DS(4,4)*A-DS(4,6)*B)*XI(1)
315   FCT=DS(3,2)*XI(4)*A-DS(3,6)*XI(6)*B+DS(4,2)*A*XI(1)-DS(4,6)
      1*XI(4)*B
316   FCT=(DS(3,3)*A-DS(3,5)*B)*XI(5)+(DS(4,3)*A-DS(4,5)*B)*XI(2)
317   FCT=DS(3,3)*XI(4)*A+DS(4,3)*A*XI(1)
318   FCT=(DS(3,3)*A-DS(3,5)*B)*XI(6)+DS(3,6)*XI(4)*B+(DS(4,3)
      1*A-DS(4,5)*B)*XI(4)+DS(4,6)*XI(1)*B
```

(4) 4th row elements

```
400   FCT=0.0
  .
  .
  .
418   FCT=0.0
```

309

(5) 5th row elements

```
505    FCT=DS(4,4)*XI(1)
506    FCT=DS(4,2)*XI(1)
507    FCT=-DS(4,5)*XI(2)
508    FCT=0.0
509    FCT=-DS(4,5)*XI(4)+DS(4,6)*XI(1)
510    FCT=(DS(4,3)*A-DS(4,5)*B)*XI(2)
511    FCT=(DS(4,1)*A+DS(4,3)*A-DS(4,5)*B)*XI(1)
512    FCT=(DS(4,3)*A-DS(4,5)*B)*XI(4)+DS(4,4)*XI(1)*A
513    FCT=-DS(4,6)*XI(2)*B
514    FCT=(DS(4,4)*A-DS(4,6)*B)*XI(1)
515    FCT=DS(4,2)*A*XI(1)-DS(4,6)*XI(4)*B
516    FCT=(DS(4,3)*A-DS(4,5)*B)*XI(2)
517    FCT=DS(4,3)*XI(1)*A
518    FCT=(DS(4,3)*A-DS(4,5)*B)*XI(4)+DS(4,6)*XI(1)*B
```

(6) 6th row elements

```
606    FCT=DS(2,2)*XI(1)
607    FCT=-DS(2,5)*XI(2)
608    FCT=0.0
609    FCT=-DS(2,5)*XI(4)+DS(2,6)*XI(1)
610    FCT=(DS(2,3)*A-DS(2,5)*B)*XI(2)
611    FCT=(DS(2,1)*A+DS(2,3)*A-DS(2,5)*B)*XI(1)
612    FCT=(DS(2,3)*A-DS(2,5)*B)*XI(4)+DS(2,4)*A*XI(1)
613    FCT=-DS(2,6)*B*XI(2)
614    FCT=(DS(2,4)*A-DS(2,6)*B)*XI(1)
615    FCT=DS(2,2)*A*XI(1)-DS(2,6)*B*XI(4)
616    FCT=(DS(2,3)*A-DS(2,5)*B)*XI(2)
617    FCT=DS(2,3)*A*XI(1)
618    FCT=(DS(2,3)*A-DS(2,5)*B)*XI(4)+DS(2,6)*B*XI(1)
```

(7) 7th row elements

```
707    FCT=DS(5,5)*XI(3)
708    FCT=0.0
709    FCT=DS(5,5)*XI(5)-DS(5,6)*XI(2)
710    FCT=(DS(5,5)*B-DS(5,3)*A)*XI(3)
711    FCT=(DS(5,5)*B-DS(5,1)*A-DS(5,3)*A)*XI(2)
712    FCT=-(DS(5,3)*A+DS(5,5)*B)*XI(5)-DS(5,4)*A*XI(2)
713    FCT=DS(5,6)*B*XI(3)
714    FCT=(DS(5,6)*B-DS(5,4)*A)*XI(2)
715    FCT=DS(5,6)*B*XI(5)-DS(5,2)*A*XI(2)
716    FCT=(DS(5,5)*B-DS(5,3)*A)*XI(3)
717    FCT=-DS(5,3)*A*XI(2)
718    FCT=(DS(5,5)*B-DS(5,3)*A)*XI(5)-DS(5,6)*B*XI(2)
```

(8) 8th row elements

```
800    FCT=0.0
 .
 .
 .
818    FCT=0.0
```

310

(9)   9th row elements

```
909    FCT=DS(5,5)*XI(6)-(DS(5,6)+DS(6,5))*XI(4)+DS(6,6)*XI(1)
910    FCT=(-DS(5,3)*A+DS(5,5)*B)*XI(5)+(DS(6,3)*A-DS(6,5)*B)*XI(2)
911    FCT=-(DS(5,1)*A+DS(5,3)*A-DS(5,5)*B)*XI(4)+(DS(6,1)*A+DS
       1(6,3)*A-DS(6,5)*B)*XI(1)
912    FCT=-(DS(5,3)*A-DS(5,5)*B)*XI(6)+(DS(6,3)*A-DS(6,5)*B-DS
       1(5,4)*A)*XI(4)+DS(6,4)*A*XI(1)
913    FCT=DS(5,6)*B*XI(5)-DS(6,6)*B*XI(2)
914    FCT=-(DS(5,4)*A-DS(5,6)*B)*XI(4)+(DS(6,4)*A-DS(6,6)*B)*XI(1)
915    FCT=DS(5,6)*B*XI(6)-(DS(5,2)*A+DS(6,6)*B)*XI(4)+DS(6,2)
       1*A*XI(1)
916    FCT=(-DS(5,3)*A+DS(5,5)*B)*XI(5)+(DS(6,3)*A-DS(6,5)*B)*XI(2)
917    FCT=-DS(5,3)*A*XI(4)+DS(6,3)*A*XI(1)
918    FCT=(DS(5,5)*B-DS(5,3)*A)*XI(6)+(DS(6,3)*A-DS(6,5)*B-DS
       1(5,6)*B)*XI(4)+DS(6,6)*B*XI(1)
```

(10)   10th row elements

```
1010    FCT=(DS(3,3)*A-DS(3,5)*B)*A*XI(3)-(DS(5,3)*A-DS(5,5)
        1*B)*B*XI(3)
1011    FCT=(DS(3,1)*A+DS(3,3)*A-DS(3,5)*B)*A*XI(2)-(DS(5,1)
        1*A+DS(5,3)*A-DS(5,5)*B)*B*XI(2)
1012    FCT=(DS(3,3)*A**2-DS(3,5)*A*B-DS(5,3)*A*B+DS(5,5)*B**2)
        1*XI(5)+(DS(3,4)*A-DS(5,4)*B)*A*XI(2)
1013    FCT=(DS(5,6)*B-DS(3,6)*A)*B*XI(3)
1014    FCT=(DS(3,4)*A**2-DS(3,6)*A*B-DS(5,4)*A*B+DS(5,6)*B**2)
        1*XI(2)
1015    FCT=(DS(5,6)*B-DS(3,6)*A)*B*XI(5)+(DS(3,2)*A-DS(5,2)*B)
        1*A*XI(2)
1016    FCT=(DS(3,3)*A**2-DS(3,5)*A*B-DS(5,3)*A*B+DS(5,5)*B**2)
        1*XI(3)
1017    FCT=(DS(3,3)*A-DS(5,3)*B)*A*XI(2)
1018    FCT=(DS(3,3)*A**2-DS(3,5)*A*B-DS(5,3)*A*B+DS(5,5)*B**2)
        1*XI(5)+(DS(3,6)*A-DS(5,6)*B)*B*XI(2)
```

(11)   11th row elements

```
1111    FCT=(((DS(1,1)+DS(1,3)+DS(3,1))*A-(DS(1,5)+DS(3,5))
        1*B)*A-(DS(5,1)*A+DS(5,3)*A-DS(5,5)*B)*B)*XI(1)
1112    FCT=((DS(1,3)+DS(3,3))*A**2-(DS(1,5)+DS(3,5))*A*B+DS
        1(5,5)*B**2-DS(5,3)*A*B)*XI(4)+((DS(1,4)+DS(3,4))*A-DS
        2(5,4)*B)*A*XI(1)
1113    FCT=(DS(5,6)*B**2-(DS(1,6)+DS(3,6))*A*B)*XI(2)
1114    FCT=((DS(1,4)*A-DS(1,6)*B+DS(3,4)*A-DS(3,6)*B)*A-(DS
        1(5,4)*A-DS(5,6)*B)*B)*XI(1)
1115    FCT=((DS(1,2)+DS(3,2))*A-DS(5,2)*B)*A*XI(1)+(DS(5,6)
        1*B-(DS(1,6)+DS(3,6))*A)*B*XI(4)
1116    FCT=((DS(1,3)*A-DS(1,5)*B+DS(3,3)*A-DS(3,5)*B)*A-(DS
        1(5,3)*A-DS(5,5)*B)*B)*XI(2)
1117    FCT=((DS(1,3)+DS(3,3))*A-DS(5,3)*B)*A*XI(1)
1118    FCT=((DS(1,3)+DS(3,3))*A**2-(DS(1,5)+DS(3,5)+DS(5,3))
        1*A*B+DS(5,5)*B**2)*XI(4)+((DS(1,6)+DS(3,6))*A-DS(5,6)
        2*B)*B*XI(1)
```

311

(12)   12th row elements

```
1212 FCT=(DS(3,3)*A**2-DS(3,5)*A*B-DS(5,3)*A*B+DS(5,5)
     1*B**2)*XI(6)+(DS(3,4)*A**2-DS(5,4)*A*B+DS(4,3)*A**2-DS
     2(4,5)*A*B)*XI(4)+DS(4,4)*A**2*XI(1)
1213 FCT=(DS(5,6)*B-DS(3,6)*A)*B*XI(5)-DS(4,6)*A*B*XI(2)
1214 FCT=(DS(3,4)*A**2-DS(3,6)*A*B-DS(5,4)*A*B+DS(5,6)*B**2)
     1*XI(4)+(DS(4,4)*A-DS(4,6)*B)*A*XI(1)
1215 FCT=(DS(5,6)*B-DS(3,6)*A)*B*XI(6)+(DS(3,2)*A-DS(4,6)
     1*B-DS(5,2)*B)*A*XI(4)+DS(4,2)*A**2*XI(1)
1216 FCT=(DS(3,3)*A**2-DS(3,5)*A*B-DS(5,3)*A*B+DS(5,5)*B**2)
     1*XI(5)+(DS(4,3)*A-DS(4,5)*B)*A*XI(2)
1217 FCT=(DS(3,3)*A-DS(5,3)*B)*A*XI(4)+DS(4,3)*A**2*XI(1)
1218 FCT=(DS(3,3)*A**2-DS(3,5)*A*B-DS(5,3)*A*B+DS(5,5)*B**2)
     1*XI(6)+(DS(3,6)*A*B-DS(5,6)*B**2+DS(4,3)*A**2-DS(4,5)
     2*A*B)*XI(4)+DS(4,6)*A*B*XI(1)
```

(13)   13th row elements

```
1313 FCT=DS(6,6)*B**2*XI(3)
1314 FCT=(DS(6,6)*B-DS(6,4)*A)*B*XI(2)
1315 FCT=(DS(6,6)*B*XI(5)-DS(6,2)*A*XI(2))*B
1316 FCT=(DS(6,5)*B-DS(6,3)*A)*B*XI(3)
1317 FCT=-DS(6,3)*A*B*XI(2)
1318 FCT=-(DS(6,3)*A-DS(6,5)*B)*XI(5)*B-DS(6,6)*B**2*XI(2)
```

(14)   14th row elements

```
1414 FCT=(DS(4,4)*A**2-DS(4,6)*A*B-DS(6,4)*A*B+DS(6,6)*B**2)
     1*XI(1)
1415 FCT=(DS(4,2)*A-DS(6,2)*B)*A*XI(1)+(DS(6,6)*B-DS(4,6)
     1*A)*B*XI(4)
1416 FCT=(DS(4,3)*A**2-DS(4,5)*A*B-DS(6,3)*A*B-DS(6,5)*B**2)
     1*XI(2)
1417 FCT=(DS(4,3)*A-DS(6,3)*B)*A*XI(1)
1418 FCT=(DS(4,3)*A**2-DS(4,5)*A*B-DS(6,3)*A*B+DS(6,5)*B**2)
     1*XI(4)+(DS(4,6)*A-DS(6,6)*B)*B*XI(1)
```

(15)   15th row elements

```
1515 FCT=DS(2,2)*A**2*XI(1)-(DS(2,6)+DS(6,2))*A*B*XI(4)+DS
     1(6,6)*B**2*XI(6)
1516 FCT=(DS(2,3)*A-DS(2,5)*B)*A*XI(2)-(DS(6,3)*A-DS(6,5)
     1*B)*B*XI(5)
1517 FCT=DS(2,3)*A**2*XI(1)-DS(6,3)*A*B*XI(4)
1518 FCT=(DS(2,3)*A**2-DS(2,5)*B*A-DS(6,6)*B**2)*XI(4)+DS
     1(2,6)*A*B*XI(1)+(DS(6,5)*B-DS(6,3)*A)*B*XI(6)
```

(16) 16th row elements

```
1616  FCT=(DS(3,3)*A**2-DS(3,5)*A*B-DS(5,3)*A*B+DS(5,5)*B**2)
      1*XI(3)
1617  FCT=(DS(3,3)*A**2-DS(5,3)*A*B)*XI(2)
1618  FCT=(DS(3,3)*A**2-DS(3,5)*A*B-DS(5,3)*A*B+DS(5,5)*B**2)
      1*XI(5)+(DS(3,6)*A-DS(5,6)*B)*B*XI(2)
```

(17) 17th row elements

```
1717  FCT=DS(3,3)*A**2*XI(1)
1718  FCT=(DS(3,3)*A**2-DS(3,5)*A*B)*XI(4)+DS(3,6)*A*B*XI(1)
```

(18) 18th row elements

```
1800  FCT=(DS(3,3)*A**2-DS(3,5)*A*B-DS(5,3)*A*B+DS(5,5)*B**2)
      1*XI(6)+(DS(3,6)*A*B-DS(5,6)*B**2+DS(6,3)*A*B-DS(6,5)
      2*B**2)*XI(4)+DS(6,6)*B**2*XI(1)
```

313

APPENDIX 4

USER'S GUIDE FOR TEPSAC

A.4.1 Introduction

The finite element formulations described in Chapters 2, 3 and 4 have been incorporated into a computer program called TEPSAC (Thermo Elastic–Plastic Stress Analysis with Creep). This code was constructed to handle quasi-coupled thermal and/or mechanical stress analysis of structures of two-dimensional planar or three-dimensional axisymmetrical geometries. Some of its main features can be outlined as follows:

(1) Simplex elements of constant stresses and strains are used. The shape of the elements can be either:
   (a) triangular and/or quadrilateral plates for planar structures; or
   (b) triangular and/or quadrilateral torus rings for axisymmetric structures.
(2) Up to six different materials with their mechanical properties at up to five different temperatures are allowed.
(3) The code accepts temperature and strain-rate dependent material properties.
(4) Incremental plasticity theory with either isotropic or kinematic hardening rule can be used in the elastoplastic analysis.
(5) Stresses/strains due to creep deformations of materials following Norton's creep law can be computed in conjunction with the thermo-elastic–plastic analysis.
(6) Thermal analysis with temperature–dependent thermophysical properties can be used.
(7) Both thermal and mechanical analyses are quasi-coupled; i.e. the thermal analysis is performed on the structural geometries at the immediate last load step.
(8) A special continuous function is used to describe true stress *vs.* true strain curve of the material up to the rupture strength, thus avoiding the need for distinct treatments for the elastic and plastic material behaviors in the analysis.
(9) The structure can be modelled in regions. Local alteration of the model can be made without having to re-number the nodes and elements in other regions.

314

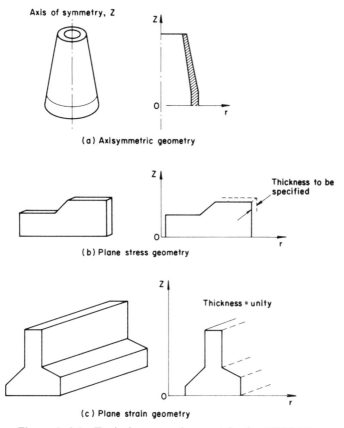

Axis of symmetry, Z

(a) Axisymmetric geometry

Thickness to be specified

(b) Plane stress geometry

Thickness = unity

(c) Plane strain geometry

Figure A.4.1 Typical structural geometries for TEPSAC.

A.4.2 Coordinate systems

The code can analyze axisymmetric, plane stress, and plane strain structural geometries. The coordinate systems used for these structures are illustrated in Figure A.4.1. The origin of the coordinate system may be chosen arbitrarily at any convenient location.

A.4.3 Element types and restrictions

Only simplex elements with triangular and quadrilateral configurations are allowed. Plates and rings are used for the respective planar and axisymmetric structural geometries.

The element aspect ratio, which is defined as the ratio of the length of the longest side to the height of the triangle with respect to the same side for a triangular element; and the ratio of the diameter of the smallest circum-scribed circle to the minimum perpendicular distance between opposite sides for a quadrilateral element, is to be kept below five. This aspect ratio limitation does not need to be followed when the state of stress is known to be essentially uniaxial. In this case, the length of the element in the direction of the applied stress may be large compared to the width.

A.4.4 Subdivision of models by regions

The subdivision of a structure into regions has the principal advantages of subsequent revision or refinement of a small portion of the finite element model without having to modify the remainder of the structure. In general, the subdivision should be done bearing in mind the likely distribution of the solution values, e.g. stresses or strains under consideration. For example, it is suggested that separate regions, where modifications to the structure discretization can be done with relative ease, be used for areas of high stress concentration.

There are certain restrictions on the manner in which the subdivision can be done. These restrictions are summarized here, with further illustrations in Figures A.4.2 and A.4.3:

(a) no node point may be common to more than two regions;
(b) a region may not have interfaces with more than two other regions;
(c) the regions must be numbered consecutively in the order that they occur in the structure;

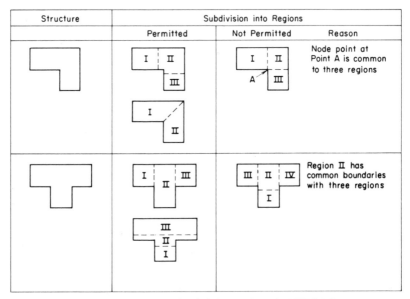

Figure A.4.2 Rules for joining regions in TEPSAC.

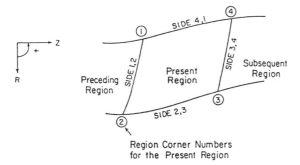

Region Corner Numbers
for the Present Region

Figure A.4.3 Common nodes for regions.

(d) interfaces between two regions may be of any shape, but the restrictions on node numbering must be followed;

(e) orientation of the regions is prescribed by the region corner numbering sequence which requires that region corner numbers be assigned.

The numbering of elements within a region is independent of element numbers in any other region. The total number of elements within a region is restricted only by the limit of a maximum of 378 elements for the entire finite element model. Element numbers within a region are termed region element numbers and must be assigned consecutively starting with the value one.

The region node numbers are also independent of the values for other regions, but to insure compatibility at the interfaces between regions, certain restrictions are necessary. The restrictions are best illustrated by reference to the examples given in Figures A.4.3 and A.4.4.

(a) The initial node points of the region must lie on the interface (side 1,2) with the preceding region, and must be numbered sequentially starting at a corner number.

(b) The final node points of the region must lie on the interface with the subsequent region (side 3,4) and must be numbered sequentially. These final node points "belong" to the subsequent region, and in fact are input (r and z coordinates) with the rest of the nodes for that region. To emphasize this, they have been enclosed by quotation marks in Figure A.4.4. Note, however, that the description of elements for the current region includes these boundary nodes (input in the subsequent region), and identifies them as node numbers in the current region. Since these node points "belong" to the subsequent region, they are *not* included in the node point count for the region, since node point data is not required for these nodes for the current region. Accordingly, the number of node points in the regions illustrated is 16 and region node number 16 is the last one for which node point data would be supplied in the input for this region.

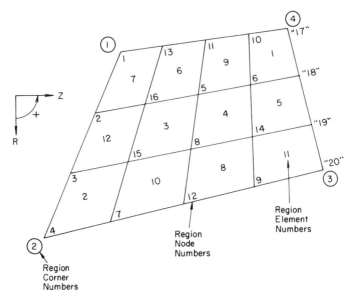

Figure A.4.4 Node numbers for finite element idealization in regions.

(c) As illustrated in Figure A.4.4, the numbering of the other node points must be sequential, but they need not be in any particular order within the region. To reduce errors and to satisfy the bandwidth limitations, it is best to use an ordered numbering system. Judicious choice of a discretization scheme permits the user to take advantage of the limited built-in automatic node and mesh generating features of the code.

(d) In order to reduce required computer storage requirements and computer running times, the nonzero elements of the stiffness matrix are restricted to a relatively narrow band adjacent to the main diagonal of the stiffness matrix. For the user, this requires that the maximum difference in node point numbers around a single element may not exceed 26.

(e) The maximum number of node points in a given region is restricted only by the requirement that the total number of node points in the complete analytical model is limited to 378.

A.4.5   Mechanical loadings and boundary conditions

Nodal forces and displacements input

Mechanical loads may be input either as concentrated nodal forces or loads, and/or nodal displacements, and/or surface pressures. Nodal load and/or displacement inputs are input along with the nodal coordinate specification

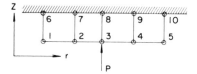

Figure A.4.5 Specifications of forces and displacements at nodes.

(Group III-(3)), whereas surface pressure inputs are done using Group III-(5) which specifies the appropriate nodes subject to pressure, and Group VII-(1) which specifies the pressure value for the particular load step. The direction of the pertinent nodal loads and/or displacement values are specified by using the appropriate value of the UNC variable chosen from the specified code numbers in (Group III-(3)).

For example, if one wants to apply a 1000 lb load to node 3 of the simple structure illustrated in Figure A.4.5, the appropriate entry for the node would be (following Group III-(3)):

node ... r(force)	z(force)	code
3	1000.0	0

By specifying code 0, values entered in columns 41 to 50 and 51 to 60 will be considered to be force inputs: blanks are considered as zeros.

If instead of a concentrated load at node 3, a constant displacement, say 0.0001, in the same direction as P is desired, the appropriate entry would be:

node ... r(force)	z(displ.)	code
3	0.0001	2

For analyses in which either concentrated loads and/or displacements is the desired mode of loading only (i.e. no pressure loads), the NBC variable in Group III-(1) must be either zero or left blank. It then no longer becomes necessary to specify Groups III-(5) and VII-(1). The user simply repeats Group V-(1) for as many load (or time) steps as desired. Note that at each step the incremental value of load or displacement will of course be that specified in the appropriate columns of Group III-(3). This means that the user cannot vary the size of the increment from step to step. The load size can be varied only when using pressure inputs.

Pressure boundary input

For an element subject to a pressure loading, it is first necessary to identify the particular side of the element over which the pressure acts. Identification is done by indicating the two nodes that define the side. The order of specification of these two nodes (i, j) is such that node i must precede node j

319

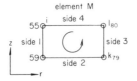

Figure A.4.6 Element description for positive pressure loading.

when following a counterclockwise path around the element (Group III-(5)). With reference to Figure A.4.6, this means that, for example:

IP(1)	IP(2)	
55	59	for side 1
59	79	for side 2
79	80	for side 3
80	55	for side 4

Finally, it is necessary to specify the direction (or sign) of the pressure loading for the input to Group VII-(1). The sign is specified in consideration of the effect of the pressure on the element, with pressures that produce tensile stresses input as negative pressures and those that produce compressive stresses input as positive pressures.

If, after all of the above exercise, the sign of the stresses as output from the program indicates clearly that the pressure is in the wrong direction, one may set the IOREVL variable equal to one in Group I-(3). This will effectively reverse the direction (sign) of the pressure load.

A.4.6   Input information

There are seven groups of input required for use of the TEPSAC code. All input values are to be right justified in the field of the specified column numbers. Integer numbers are to be used for those variable names starting with the letters I to N. Exponential formats are allowed for variables using floating points.

Group I: case identification and job control information

			Column nos.
I-(1)	HED	– heading; job description up to 72 characters	(1–72)
I-(2)	NUMMAT	– total number of materials used in the analysis (max. = 6)	(1–5)
	NRMAX	– total number of regions used in the analysis (max. = 10)	(6–10)

320

<table>
<thead>
<tr><th></th><th></th><th></th><th>Column
nos.</th></tr>
</thead>
<tbody>
<tr><td></td><td>NPP</td><td>– structure geometry
= 0 for axisymmetric
= 1 for plane stress
= 2 for plane strain</td><td>(11–15)</td></tr>
<tr><td></td><td>NOLODS</td><td>– number of load (time) steps</td><td>(16–20)</td></tr>
<tr><td></td><td>JOBTYP</td><td>– type of analysis
= 0 stress only
= 1 steady-state thermal only
= 2 transient thermal only with explicit
  time integration scheme
= 3 steady-state thermal and stress analysis
= 4 transient thermal (explicit time
  integration) and stress analysis
= 51 stress and creep analysis
= 52 steady state thermal, stress and creep
  analysis
= 53 transient thermal, stress and creep
  analysis</td><td>(21–25)</td></tr>
<tr><td></td><td>Q</td><td>– reference temperature</td><td>(24–35)</td></tr>
<tr><td>I-(3)</td><td>IOPRNT</td><td>– print control parameter
= n, print computed information at every
  nth step starting at step n</td><td>(1–5)</td></tr>
<tr><td></td><td>IOPRSP</td><td>– print suppression indicator
= 1, print suppression parameters to follow
  (I-(5))</td><td>(6–10)</td></tr>
<tr><td></td><td>IOGEOM</td><td>– geometry update parameter
= 1, update nodal coordinates after each
  load (time) step</td><td>(11–15)</td></tr>
<tr><td></td><td>IOREVL</td><td>– pressure sign change parameter
= 1, reverse sign of pressure load</td><td>(16–20)</td></tr>
<tr><td></td><td>KINHAR</td><td>– kinematic hardening parameter
= 0, isotropic hardening rule
= 1, kinematic hardening rule</td><td>(21–25)</td></tr>
<tr><td>I-(4)</td><td>[Optional</td><td>– see I-(2) JOBTYP variable]</td><td></td></tr>
<tr><td></td><td>COEF</td><td>– creep law coefficient</td><td>(1–10)</td></tr>
<tr><td></td><td>ENC</td><td>– stress exponent</td><td>(11–20)</td></tr>
<tr><td></td><td>ACTQ</td><td>– activation energy</td><td>(21–30)</td></tr>
<tr><td></td><td>RCONT</td><td>– gas constant</td><td>(31–40)</td></tr>
<tr><td></td><td>DTOO</td><td>– for automatic time control</td><td>(41–50)</td></tr>
</tbody>
</table>

					Column nos.
T11	– start time (time step number) for creep analysis				(11–20)
T12	– end time for creep analysis				(21–30)

I-(5) [Optional – see I-(3) IOPRSP variable]

IPSNDD	– nodal displacement results print suppress flag	(1–5)
	= 1, do not print nodal displacement results	
IPSELR	– element strain results print suppress flag	(6–10)
	= 1, do not print element strain results	
IPSELS	– element stress results print suppress flag	(11–15)
	= 1, do not print element stress results	

Group II: mechanical properties

[Skip this group for thermal analysis only (I-(2) JOBTYP variable)]

II-(1)	MTYPE	– material identification number (up to 6)	(1–5)
	NUMTC	– number of temperature conditions for which properties are given (up to 5)	(6–10)
	NMSTRC	– number of strain rates for which properties are given (up to 5)	(11–15)
	DENSTY	– mass density of material	(16–25)

II-(2) RER – dummy array for temperature and strain-rate dependent properties specified as follows:

Column nos. (1–10) (11–20) ... (41–50) – for NUMTC > 1 as required

II-(2)		(1-10)	(11-20)		(41-50)	
– 1	T_1	T_2	...	T_5	temperature (max. = 5)	
– 2	E_1	E_2	...	E_5	Young's modulus	
– 3	v_1	v_2	...	v_5	Poisson ratio	
– 4	G_1	G_2	...	G_5	shear modulus	
– 5	a_1	a_2	...	a_5	therm. exp. coeff.	
– 6	σ_{y1}	σ_{y2}	...	σ_{y5}	initial yield strength	
– 7	σ_{ult1}	σ_{ult2}	...	σ_{ult5}	rupture stress	
– 8	E_1'	E_2'	...	E_5'	plastic modulus	
– 9	n_1	n_2	...	n_5	stress–strain power	
– 10	σ_{kink1}	σ_{kink2}	...	σ_{kink5}	elastic–plastic interface (kink) stress	
– 11	$\dot{\varepsilon}_1$	$\dot{\varepsilon}_2$	...	$\dot{\varepsilon}_5$	strain rate ($\dot{\varepsilon}_1$)	for $\dot{\varepsilon}_1$
– 12	E_1	E_2	...	E_5	Young's modulus	for $\dot{\varepsilon}_2$

-13 v_1	v_5	...	v_5	Poisson ratio
-14 G_1	G_2	...	G_5	shear modulus
-15 σ_{y1}	σ_{y2}	...	σ_{y5}	initial yield strength
-16 σ_{ult1}	σ_{ult2}	...	σ_{ult5}	rupture stress
-17 E_1'	E_2'	...	E_5'	plastic modulus
-18 n_1	n_2	...	n_5	stress–strain power
-19 σ_{kink1}	σ_{kink2}	...	σ_{kink5}	elastic–plastic interface (kink) stress
-20 $\dot{\varepsilon}_1$	$\dot{\varepsilon}_2$	...	$\dot{\varepsilon}_5$	strain rate ($\dot{\varepsilon}_2$)

for NMSTRC > 1 as required (maximum $= 5$, i.e. up to $\dot{\varepsilon}_5$).

Group III: structure description

			Column nos.
III-(1)	NREJUN	– region number (up to 10)	(1–5)
	NNP12	– number of nodal points along side 1,2 of this region	(6–10)
	NNP34	– number of nodal points along side 3,4 of this region	(11–15)
	NNPR	– number of nodal points in this region	(16–20)
	NELR	– number of elements in this region	(21–25)
	NBC	– number of element sides in this region subject to pressure loading	(26–30)
III-(2)	Node inputs (optional – for thermal analysis only)		
	N	– node number for this region	(1–5)
	UN(1)	– r-coordinate of node	(6–15)
	UN(2)	– z-coordinate of node	(16–25)
	UN(3)	– nodal temperature	(26–35)
	ITEMP	– prescribed nodal temperature identification	(36–40)
		= 0, temperature not prescribed	
		= n, see Group VI	

III-(3)   Node inputs
[Skip this for thermal analysis only]

N	– ⎫		
UN(1)	– ⎪		
UN(2)	– ⎬ as III-(2)		
UN(3)	– ⎪		
ITEMP	– ⎭		
UN(4)	– initial r force or displacement		(41–50)

UN(5) – initial z force or displacement   (51–60)
UN(6) – element thickness at this node   (61–70)
UNC – force or displacement code   (71–72)

Code table

UNC	Value for UN(4) considered to be	Value for UN(5) considered to be
0	r-force	z-force
1	r-displacement	z-force
2	r-force	z-displacement
3	r-displacement	z-displacement

Input node points should be in an ascending order. Omitted node points will be generated between the defined nodes. Code will be set to zero. Repeat III-(2) or III-(3) for the required NNPR specified for the region.

III-(4) Element inputs
MM – element number for this region   (1–5)
JX(1) – node I of element   (6–10)
JX(2) – node J of element   (11–15)
JX(3) – node K of element   (16–20)
JX(4) – node L of element   (21–25)
      (= node K for triangular elements)
JX(5) – material number for element   (26–30)

Input elements in ascending order. Omitted elements will be generated automatically between defined elements with unity incrementing of node numbers. Specify I, J, K, L nodes in counter-clockwise fashion. Repeat III-(4) for the required NELR specified for the region.

III-(5) Pressure inputs (optional – only for NBC $\neq$ 0 for the region)

IP(1) – surface node i of element side under pressure   (1–5)
IP(2) – surface node j of element side under pressure   (6–10)

Node i must precede node j when following a counterclockwise path around an element. Repeat III-(5) for the required NBC specified for the region. Repeat III-(2) (or III-(3)), III-(4), and III-(5) where appropriate, for the required NRMAX (from I-(2)). Specify regions in increasing number, commencing with one.

Group IV: output control

Column
nos.

IV-(1) NODPR – number of nodes for which nodal   (1–5)
information will be printed
= 0, all nodal information printed

        NELPR – number of nodes for which element   (6–10)
information will be printed
= 0, all element information printed

IV-(2) (optional – only for NODPR ≠ 0)
IDNDPR (NODPR) – nodal numbers of nodes for which   (1–70)
information is to be printed
(14I5)

IV-(3) (optional – only for NELPR ≠ 0)
IDELPR (NELPR) – element numbers for which   (1–70)
information is to be printed
(14I5)

Group V: time (load) step input

V-(1) DTIME (ITME) – time (load) step increment   (1–10)

Group VI: (for thermal and/or thermal/stress analysis)

VI-(1)	HED	– heading for thermal analysis	(1–48)
	NMAT	– total number of materials (max. = 6)	(1–5)
	NBCT	– total number of element sides for which surface heat transfer boundary conditions are given	(6–10)
	NTEM	– number of temperatures for which thermo-physical properties of materials are given	(11–15)
	NTIM	– number of time points at which time-varying functions are given	(16–20)
	NTF	– number of time-varying functions are given	(21–25)
VI-(2)	MTYPE	– material number	(1–5)
	XCON	– thermal conductivity of material	(6–15)
	CP	– specific heat of material	(16–25)
	RO	– density of material	(26–35)

[If NTEM ≠ 0, set properties = 0.0]
Repeat VI-(2) for the required NMAT.

325

VI-(3) TIF(*j*, *i*) – array of time function values used in
                        specifying the various time varying
                        functions

    TIF(1, *i*) – time point; *i* = 1 ... NTIM (1–10)

    (*j*, *i*) – value of function *j* at time *i*; (11–80)
                 j = 1 ... NTF

VI-(4) IBCT(*i*) – surface node *i* of specified boundary (1–5)

    JBCT(*j*) – surface node *j* of specified boundary (6–10)

    IDB(1, *k*) – identification number for bulk fluid (11–15)
                  temperature

    IDB(2, *k*) – identification number for free convective (16–20)
                  heat transfer coefficient

    IDB(3, *k*) – identification number for forced convective (21–25)
                  heat transfer coefficient

    IDB(4, *k*) – identification number for radiation factor (26–30)

    IDB(5, *k*) – identification number for exponent of forced (31–35)
                  or free convective boundary condition

    IDB(6, *k*) – identification for heat flux boundary (36–40)
                  condition

Repeat VI-(4) for the appropriate number of NBCT sides (*k* = 1
to NBCT).

VI-(5) (optional – for NTEM ≠ 0)

    TFUN(*i*, *j*, *k*) – temperature dependent thermal (1–60)
                        conductivities (*k* = 1), specific heats
                        (*k* = 2) and densities (*k* = 3) of the
                        materials

Repeat for appropriate number of materials (*j* = 2, ..., NMAT + 1)
and temperature points specified (*i* = 1, ..., NTEM) for each
property.

VII-(1) [optional – only for NBC ≠ 0 (III-(1))]

    PRRR – pressure increment for this time (load) step (1–10)

A.4.7 Output information

Among a number of outputs from the program, which include the element
description, nodal coordinates and material properties, the computed nodal
displacement components and stresses and strains in elements are of great
interest to the user. These results are output in the following formats.

Displacements at nodes

Node	Code	R-coord.	Z-coord.	UR	UZ	TUR	TUZ
.	.	.	.	.	.	.	.
.	.	.	.	.	.	.	.
.	.	.	.	.	.	.	.

where $Node$ = node number
        $Code$ = nodal conditions (see input group III)
        UR = incremental displacement component in r-direction
        UZ = incremental displacement component in z-direction
        TUR = total displacement in r-direction
        TUZ = total displacement in z-direction

Stresses in element

EL.	coordinates		*	*	*	*	*	STRESS	*		*	*	*	*********
no.	R Z		R	Z	T	RZ	MAX	MIN	ANGLE	SHEAR	EFFECTIVE

where   EL. no.   = element number
        R, Z   = coordinates of the centroid of the element at
                       which the stress components act
        R   = stress component along r-direction
        Z   = stress component along z-direction
        T   = hoop stress component in θ-direction
        RZ   = shear stress component acting on the r–z plane
        MAX, MIN   = respective maximum and minimum principal
                       stress components
        ANGLE   = the angular orientation (in degrees) of the
                       maximum principal stress with respect to the
                       r-direction
        SHEAR   = the shear stress in the r–z plane and tangent to a
                       plane parallel to I, J side of the element
        EFFECTIVE = the effective stress at the centroid of the element
                       computed according to Eqn. (3.21)

Strains and plastic zone indicator in elements

EL. no.	RATS1	RATS2	EPSR	EPST	EPSRZ	EFFEPS	EPSDOT	ALF

where   EL. no.   = element number
        RATS1   = current effective stress/initial yield strength of
                     material (> 1 for plastic deformation)

327

RATS2 = computed effective stress/input effective stress estimated by the computed effective strain ($= 1.0$ for complete agreement, $\neq 1$ means deviation from input value)

EPSR = element strain in r-direction
EPSZ = element strain in z-direction
EPST = element hoop strain in θ-direction
EPSRZ = shear strain on r-z plane
EFFEPS = effective strain computed by Eqn. (3.22)
EPSDOT = rate of change of the effective strain (or strain rate)
ALF = shift of yield surfaces when using kinematic hardening rule.

The value RATS1 is usually used as a measure of the elastoplastic boundary in a structure and RATS2 indicates the degree of accuracy achieved by the assigned loading steps for an element in the plastic range.

APPENDIX 5

LISTING OF TEPSAC CODE

The following listing of the TEPSAC code was compiled and tested on a VAX-11/750 computer using a VAX/VMS operating system. The program is written in FORTRAN 77 language. Implementation of other operating systems may require minor modifications, for example in the SYNTAX of OPEN statements.

This listing is prepared according to the theoretical formulations presented in Chapters 2, 3 and 4 in the text. It is offered to readers primarily as a research and development tool.

Despite sincere efforts made by the author and his research associate to achieve accuracy and integrity of the code, they shall not be held liable for any consequences resulting from the use of this listing.

```
0001    C
0002    C
0003    C***********************************************************************
0004    C
0005    C
0006    C         THIS PROGRAM MAY BE USED FOR : STATIC STRESS ANALYSIS; HEAT
0007    C    TRANSFER ANALYSIS (TRANSIENT OR STEADY STATE); THERMAL STRESS
0008    C    ANALYSIS; CREEP ANALYSIS; AND COMBINATIONS OF THESE ANALYSES.
0009    C
0010    C         THE CODE IS BASED ON THE FINITE ELEMENT VARIATIONAL TECHNIQUE,
0011    C    USING LINEAR CONSTANT-STRAIN ELEMENTS (TRIANGULAR AND/OR QUADRILATERAL).
0012    C
0013    C         FEATURES OF THIS IMPLEMENTATION INCLUDE  SPECIFICATION OF
0014    C    MECHANICAL PROPERTIES FOR UP TO SIX DIFFERENT MATERIALS, AT UP TO FIVE
0015    C    DIFFERENT TEMPERATURE AND STRAIN RATE CONDITIONS FOR EACH MATERIAL.
0016    C    LOADING MAY BE INPUT AS AN APPLIED PRESSURE, VARYING FROM LOAD STEP TO
0017    C    LOAD STEP, AND/OR CONSTANT INCREMENT CONCENTRATED NODAL LOADS AND/OR
0018    C    NODAL DISPLACEMENTS.  CURRENT CAPACITY CAN ACCOMODATE UP TO 378 NODES
0019    C    AND 378 ELEMENTS TO DISCRETIZE THE STRUCTURE UNDER CONSIDERATION.
0020    C    EITHER KINEMATIC OR ISOTROPIC HARDENING MAY BE SPECIFIED FOR MATERIAL
0021    C    BEHAVIOUR IN THE PLASTIC RANGE.
0022    C
0023    C
0024    C***********************************************************************
0025    C
0026    C
0027          COMMON/TEMP1/NUMMAT,NUMTC,DENSTY(6)
0028          COMMON/MECH1/NMSTRC,Q,QT(378)
0029          COMMON/PRINT1/IPRINT,NELPR,NODPR,IOPRNT,IDELPR(378),IDNDPR(378)
0030          COMMON/PRINT2/IPSNDD,IPSELR,IPSELS
0031          COMMON/STRAIN/EPSDI(378),EPSDOT(378),SIGAVE(378),DEVSIG(378,5),
0032         *       EPSDTC(378),EPSDTP(378)
0033          COMMON/DISP1/UR(378),UZ(378),CODE(378)
0034          COMMON/BOUND/IBC(200),JBC(200),NPBCR(10),NUMPC,PR(200)
0035          COMMON/SYSMSH/NUMEL,NUMNP,NRMAX
0036          COMMON/COOR/R(378),Z(378),T(378),IX(378,5),TK(378)
0037          COMMON/TIMSTP/ITIME
0038          COMMON/DSTRES/RATS2(378),DSIG(378,4),DIVEPS(378,4)
0039          COMMON/STREZZ/RATS1(378),TSIG(378,5),TOTEPS(378,4),EPS(378),
0040         *       DLAM(378),RATST(378),DEPS(378,4)
0041          COMMON/DISPL/BB(756),BBT(756)
```

329

```
0042          COMMON/PLANE/NPP
0043          COMMON/FLAGS/PLAST
0044          COMMON/RESTR/MBANX,DTIME(1000)
0045          COMMON/HEAT3/CFX(378),XCONX(6),CPX(6),ROX(6),
0046        *     TFUNX(9,6,3),TIFX(21,99),XLX(378)
0047          COMMON/HEAT1/NMATX,NBCX,NTEMX,NTIMX,NTFX,MBANTX,NMBX
0048          COMMON/SWITCH/IXIX(5,378)
0049          COMMON/BIGR/RBIGRX(378)
0050          COMMON/HTBOND/ITCT(75),JBCT(75),IDB(6,75)
0051          COMMON/PAR/IDTEMP(378)
0052          COMMON/HEATR/XA,DTI,TIME,TIMECR
0053          COMMON/MATRI/III
0054          COMMON/CCRP/SCP(378,14),SSCP(378,2)
0055          COMMON/ATEM/ACTQ,RCONT,COEF,ENC
0056          COMMON/DOT/DTOO
0057          COMMON/RITBE/T11,T12
0058          COMMON/TIMKEP/TIMEK
0059          COMMON/MECH2/TSIP(378),AEPS(378),EEPS(378),ANP(378),ANG(378),
0060        *     STP(378),EFFCLF(378)
0061          COMMON/MECH3/BLF(378,4),BEPS(378,4),CLF(378,4),CEPS(378,4),NEU
0062          COMMON/KINH/IRE(378),ICK(378),LCY(378),EFFBLF(378),RATSP(378)
0063          COMMON/PLASTA/D2(378),EFSLOP(378),EPSPL(378,4)
0064          COMMON/PLASTB/AVDEVI(378),TRDEVI(378,4)
0065          COMMON/PLASTC/TRLSIG(378,4),PYS(378)
0066          COMMON/ALFA/ALF(378,4),EFFALF(378)
0067          DOUBLE PRECISION QDEPCJ,QP,QDSIG,QH
0068          LOGICAL PLAST(378)
0069          DIMENSION TOTPR(378)
0070          INTEGER HED(18)
0071   C
0072   C**** FILE SPECIFICATION (PERTAINING TO THE VAX)
0073   C
0074          OPEN (1,STATUS='SCRATCH',FILE='STOR1',FORM='UNFORMATTED',
0075        *     RECL=8100)
0076          OPEN (2,STATUS='SCRATCH',FILE='STOR2',FORM='UNFORMATTED',
0077        *     RECL=8100)
0078          OPEN (8,ACCESS='DIRECT',STATUS='SCRATCH',INITIALSIZE=100,RECL=800)
0079   C
0080   C**** INPUT AND OUTPUT GENERAL PROBLEM DESCRIPTION
0081   C
0082          READ (5,1000,END=60) HED
0083          READ (5,1001) NUMMAT,NRMAX,NPP,NOLODS,JOBTYP,Q
0084          READ (5,1002) IOPRNT,IOPRSP,IOGEOM,IOREVL,KINHAR
0085          IF (JOBTYP.EQ.51 .OR. JOBTYP.EQ.52 .OR. JOBTYP.EQ.53)
0086        *     READ (5,1003) COEF,ENC,ACTQ,RCONT,DTOO,T11,T12
0087          IF (JOBTYP.EQ.51 .OR. JOBTYP.EQ.52 .OR. JOBTYP.EQ.53)
0088        *     WRITE (6,2000) COEF,ENC,ACTQ,RCONT,DTOO,T11,T12
0089          IF (IOPRSP.NE.1) GO TO 5
0090          READ (5,1004) IPSNDD,IPSELR,IPSELS
0091        5 WRITE (6,2001) HED,NUMMAT,Q,JOBTYP,NRMAX
0092          IF (NPP-1) 10,15,20
0093       10 WRITE (6,2002)
0094          GO TO 25
0095       15 WRITE (6,2003)
0096          GO TO 25
0097       20 WRITE (6,2004)
0098       25 WRITE (6,2005)
0099          WRITE (6,2006) IOGEOM,IOREVL,IOPRNT
0100   C
0101   C**** GENERAL INITIALIZATION
0102   C
0103          IF (KINHAR.EQ.1) WRITE (6,2020)
0104          IF (KINHAR.EQ.0) WRITE (6,2021)
0105          TIME=0.0
0106          ITIME=0
0107          UNL=1.0
0108          LODCT=0
0109          RDUM=0.0
0110          RMAX=0.0
0111          TIMEK=.0
```

```
0112          TIMECR=.0
0113          III=1
0114          NEU=0
0115          IF (JOBTYP.EQ.1 .OR. JOBTYP.EQ.2) GO TO 30
0116    C
0117    C**** INPUT MATERIAL PROPERTIES
0118    C
0119          CALL PRPRTY
0120    C
0121    C**** INITIALIZATION FOR STRESS ANALYSIS
0122    C
0123          DO 500 I=1,1000
0124      500 DTIME(I)=0.0
0125          DO 510 I=1,378
0126          DO 520 K14=1,14
0127      520 SCP(I,K14)=0.0
0128          DLAM(I)=1.0E-11
0129          RATS1(I)=.000000001
0130          PLAST(I)=.FALSE.
0131          EPSDTP(I)=1.E-20
0132          EPSDOT(I)=1.E-20
0133          EPS(I)=1.E-20
0134          EFFALF(I)=0.0
0135          LCY(I)=1
0136          IRE(I)=0
0137          EFFBLF(I)=0.0
0138          RATST(I)=0.0
0139          TSIP(I)=0.0
0140          D2(I)=10.E-06
0141          EFSLOP(I)=10.E+20
0142          AVDEVI(I)=0.0
0143          RATS2(I)=0.0
0144          DO 530 JK=1,4
0145          TSIG(I,JK)=0.0
0146          TRDEVI(I,JK)=0.0
0147          TRLSIG(I,JK)=0.0
0148          ALF(I,JK)=0.0
0149          EPSPL(I,JK)=0.0
0150          DSIG(I,JK)=0.0
0151          DIVEPS(I,JK)=0.0
0152      530 DEVSIG(I,JK)=.0
0153          TSIG(I,5)=0.0
0154      510 DEVSIG(I,5)=10.E-10
0155    C
0156    C**** INPUT STRUCTURE DISCRETIZATION
0157    C
0158       30 CALL MESH(JOBTYP)
0159          DO 540 I=1,NUMNP
0160      540 RMAX=AMAX1(RMAX,R(I))
0161          IF (NPP.NE.0) RDUM=RMAX*1.E+03
0162    C
0163    C**** INPUT NODE AND ELEMENT PRINT SPECIFICATION
0164    C
0165          READ (5,1005) NODPR,NELPR
0166          IF (NODPR.EQ.0) GO TO 35
0167    C
0168    C**** INFORMATION FOR THE FOLLOWING NODE NOS. WILL BE PRINTED
0169    C
0170          DO 550 I=1,NODPR,14
0171          II=I+13
0172      550 READ (5,1006) (IDNDPR(J),J=I,II)
0173       35 IF (NELPR.EQ.0) GO TO 40
0174    C
0175    C**** INFORMATION FOR THE FOLLOWING ELEMENT NOS. WILL BE PRINTED
0176    C
0177          DO 560 I=1,NELPR,14
0178          II=I+13
0179      560 READ (5,1006) (IDELPR(J),J=I,II)
0180       40 J=0
0181          IF (JOBTYP.EQ.1 .OR. JOBTYP.EQ.2) GO TO 45
0182    C
```

```
0183    C**** CHECK BANDWIDTH
0184    C
0185            DO 570 N=1,NUMEL
0186            DO 570 I=1,4
0187            DO 580 L=1,4
0188            KK=IABS(IX(N,I)-IX(N,L))
0189            IF (26-KK) 50,55,55
0190         50 WRITE (6,2007) N
0191            GO TO 60
0192         55 IF (KK-J) 580,580,65
0193         65 J=KK
0194        580 CONTINUE
0195        570 CONTINUE
0196            MBAND=2*J+2
0197            MBANX=MBAND
0198    C
0199    C**** END OF INITIALIZATION
0200    C
0201    C
0202    C**** BEGIN ANALYSIS LOOP FOR TIME (LOAD) STEPS
0203    C
0204         45 IPRINT=1
0205        145 CONTINUE
0206    C
0207    C**** CHECK STEP NUMBER
0208    C
0209            IF (LODCT.EQ.NOLODS) GO TO 70
0210    C
0211    C**** INPUT TIME (LOAD) STEP INFORMATION
0212    C
0213            IF (JOBTYP.EQ.51 .AND. TIME.GT.T11) DTIME(ITIME+1)=DTIME(ITIME)
0214            IF (JOBTYP.EQ.52 .AND. TIME.GT.T11) DTIME(ITIME+1)=DTIME(ITIME)
0215            IF (JOBTYP.EQ.53 .AND. TIME.GT.T11) DTIME(ITIME+1)=DTIME(ITIME)
0216            IF ((III.GT.1).AND.(DTOO.GT..1)) DTIME(ITIME+1)=DTIME(ITIME)
0217            ITIME=ITIME+1
0218            IF (DTIME(ITIME).NE.0.0) GO TO 75
0219            READ (5,1007) DTIME(ITIME)
0220         75 IF (TIME.EQ.T11) TIME=TIME+.1
0221            TIME=TIME+DTIME(ITIME)
0222            IF (JOBTYP.EQ.0 .OR. JOBTYP.EQ.1 .OR. JOBTYP.EQ.2) GO TO 80
0223            IF (JOBTYP.EQ.3 .OR. JOBTYP.EQ.4) GO TO 80
0224            IF (TIME.GT.T11) TIMECR=TIMECR+DTIME(ITIME)
0225    C
0226    C**** OUTPUT TIME STEP INFORMATION (CREEP ONLY)
0227    C
0228            WRITE (6,2008)
0229            WRITE (6,2009) ITIME,DTIME(ITIME),TIME,TIMECR
0230    C
0231    C**** BRANCH TO THERMAL ANALYSIS
0232    C
0233         80 IF (JOBTYP.EQ.1.OR.JOBTYP.EQ.2.OR.JOBTYP.EQ.3.OR.JOBTYP.EQ.4)
0234          *    CALL THERM(DTIME(ITIME),TIME,RDUM,JOBTYP)
0235            IF (JOBTYP.EQ.52 .AND. TIME.LE.T11)
0236          *    CALL THERM(DTIME(ITIME),TIME,RDUM,JOBTYP)
0237            IF (JOBTYP.EQ.53 .AND. TIME.LE.T11)
0238          *    CALL THERM(DTIME(ITIME),TIME,RDUM,JOBTYP)
0239            IF (JOBTYP.EQ.1 .OR. JOBTYP.EQ.2) GO TO 85
0240            IF (IOREVL.EQ.1) UNL=-1.0
0241            IF (NUMPC.EQ.0) GO TO 85
0242    C
0243    C**** INPUT PRESSURE LOAD
0244    C
0245            DO 590 L=1,NUMPC
0246            IF (JOBTYP.EQ.51 .AND. TIME.GT.T11) GO TO 90
0247            IF (JOBTYP.EQ.52 .AND. TIME.GT.T11) GO TO 90
0248            IF (JOBTYP.EQ.53 .AND. TIME.GT.T11) GO TO 90
0249            IF (L.EQ.1) READ (5,1008) PRRR
0250            PR(L)=PRRR*UNL
0251            PRT=TOTPR(L)
0252            TOTPR(L)=TOTPR(L)+PR(L)
```

```
0253              GO TO 590
0254         90 PR(L)=0.0
0255        590 CONTINUE
0256         85 LODCT=LODCT+1
0257              IF (JOBTYP.EQ.1 .OR. JOBTYP.EQ.2) GO TO 95
0258              IF (JOBTYP.EQ.51 .OR. JOBTYP.EQ.52 .OR. JOBTYP.EQ.53) GO TO 100
0259      C
0260      C**** WRITE TIME (LOAD) STEP INFORMATION (NO CREEP)
0261      C
0262              WRITE (6,2008)
0263              WRITE (6,2010) ITIME,DTIME(ITIME),TIME
0264        100 LBC=0
0265              IF (JOBTYP.EQ.51 .AND. TIME.GT.T11) GO TO 105
0266              IF (JOBTYP.EQ.52 .AND. TIME.GT.T11) GO TO 105
0267              IF (JOBTYP.EQ.53 .AND. TIME.GT.T11) GO TO 105
0268              IF (NUMPC.EQ.0) GO TO 105
0269              WRITE (6,2011)
0270              DO 600 I=1,NRMAX
0271              IF (I.GT.1) GO TO 105
0272              NR=NPBCR(I)
0273              DO 610 LR=1,NR
0274              L=LBC+LR
0275              IF (L.EQ.1) GO TO 110
0276              IF (L.GT.1.AND.PR(L).EQ.PR(L-1)) GO TO 610
0277        110 WRITE (6,2012) IBC(L),JBC(L),PR(L),TOTPR(L)
0278        610 CONTINUE
0279              LBC=L
0280        600 CONTINUE
0281      C
0282      C**** ASSEMBLE STRUCTURAL STIFNESS MATRIX
0283      C
0284        105 IF (NUMPC.EQ.0) GO TO 6213
0285              PRS=0.0
0286              DO 6211 IN=1,NUMPC
0287       6211 PRS=PRS+TOTPR(IN)
0288              PRS=PRS/FLOAT(NUMPC)
0289              IF (PRS.EQ.0) NEU=2
0290       6213 CALL STIFF(MBAND,JOBTYP,KINHAR)
0291      C
0292      C**** SOLVE FOR NODAL DISPLACEMENTS
0293      C
0294              CALL BANSOL
0295              DO 6210 N=1,NUMEL
0296       6210 RATSP(N)=RATST(N)
0297      C
0298      C**** OUTPUT NODAL DISPLACEMENT RESULTS
0299      C
0300              IF (IPRINT-IOPRNT) 115,120,115
0301        120 IF (IPSNDD.EQ.1) GO TO 115
0302              IF (NODPR.NE.0) GO TO 125
0303              WRITE (6,2013) (N,CODE(N),R(N),Z(N),T(N),BB(2*N-1),BB(2*N),
0304          *       BBT(2*N-1),BBT(2*N),N=1,NUMNP)
0305              GO TO 115
0306        125 WRITE (6,2013) (IDNDPR(J),CODE(IDNDPR(J)),R(IDNDPR(J)),
0307          * Z(IDNDPR(J)),T(IDNDPR(J)),BB(2*(IDNDPR(J))-1),BB(2*IDNDPR(J)),
0308          * BBT(2*(IDNDPR(J))-1),BBT(2*IDNDPR(J)),J=1,NODPR)
0309      C
0310      C**** CALCULATE ELEMENT STRESSES
0311      C
0312        115 CALL STRESS(DTIME(ITIME),TIME,JOBTYP,KINHAR)
0313      C
0314      C**** FORM NEW GEOMETRY
0315      C
0316              IF (IOGEOM.EQ.0) GO TO 130
0317              DO 620 IKL=1,NUMNP
0318              R(IKL)=R(IKL)+BB(2*IKL-1)
0319        620 Z(IKL)=Z(IKL)+BB(2*IKL)
0320      C
0321      C**** OUTPUT ELEMENT STRAIN RESULTS
0322      C
```

333

```
0323      130 IF (IPRINT.NE.IOPRNT) GO TO 135
0324          IF (IPSELR.EQ.1) GO TO 135
0325          WRITE (6,2014)
0326          IF (NELPR.NE.O) GO TO 140
0327          DO 630 N=1,NUMEL
0328      630 WRITE (6,2015) N,RATS1(N),RATS2(N),TOTEPS(N,1),
0329        *    TOTEPS(N,2),TOTEPS(N,3),TOTEPS(N,4),EPS(N),EPSDOT(N)
0330          GO TO 135
0331      140 DO 640 N=1,NELPR
0332      640 WRITE (6,2015) IDELPR(N),RATS1(IDELPR(N)),RATS2(IDELPR(N)),
0333        * TOTEPS(IDELPR(N),1),TOTEPS(IDELPR(N),2),TOTEPS(IDELPR(N),3),
0334        * TOTEPS(IDELPR(N),4),EPS(IDELPR(N)),EPSDOT(IDELPR(N))
0335      135 IPRINT=IPRINT+1
0336          IF (JOBTYP.EQ.O .OR. JOBTYP.EQ.1 .OR. JOBTYP.EQ.2) GO TO 95
0337          IF (JOBTYP.EQ.3 .OR. JOBTYP.EQ.4) GO TO 95
0338          IF (TIME.LT.T11) GO TO 95
0339    C
0340    C**** OUTPUT ELEMENT CREEP RESULTS
0341    C
0342          WRITE (6,2016)
0343          WRITE (6,2017)
0344          IF (NELPR.NE.O) GO TO 95
0345          DO 650 N=1,NUMEL
0346      650 WRITE (6,2018) N, (SCP(N,II),II=1,9)
0347    C
0348    C**** END OF ANALYSIS LOOP:  BRANCH TO TOP OF LOOP FOR NEXT STEP
0349    C
0350       95 IF (IPRINT-1.EQ.IOPRNT) GO TO 45
0351          GO TO 145
0352    C
0353    C**** PROGRAM TERMINATION
0354    C
0355       70 WRITE (6,2019)
0356       60 STOP
0357    C
0358    C**** INPUT FORMATS
0359    C
0360     1000 FORMAT (18A4)
0361     1001 FORMAT (5I5,E10.0)
0362     1002 FORMAT (5I5)
0363     1003 FORMAT (7E10.0)
0364     1004 FORMAT (3I5)
0365     1005 FORMAT (2I5)
0366     1006 FORMAT (14I5)
0367     1007 FORMAT (E10.0)
0368     1008 FORMAT (1E10.0)
0369    C
0370    C**** OUTPUT FORMATS
0371    C
0372     2000 FORMAT (' COEF. OF CREEP LAW-------------- COEF=',E10.4//
0373        *' EXPONENT OF STRESS---------------ENC=',E10.4//
0374        *' ACTIVATION ENERGY --------------ACTQ=',E10.4//
0375        *' GAS CONSTANT       ------------ RCONT=',E10.4//
0376        *' AUTO-TIME CONTROL ------------- DTOO=',E10.4//
0377        *' START TIME OF CREEP--------------T11=',E10.4//
0378        *' END TIME OF CREEP  --------------T12=',E10.4//)
0379     2001 FORMAT (1H1,18A4/
0380        * 30H  NUMBER OF MATERIALS--------¬--,I3 /
0381        * 30H  REFERENCE TEMPERATURE-------,E12.4/
0382        * 30H  TYPE OF ANALYSIS------------,I3/
0383        * '          (=0,  STRESS ONLY)'/
0384        * '          (=1,  STEADY STATE THERMAL)'/
0385        * '          (=2,  TRANSIENT THERMAL)'/
0386        * '          (=3,  S. S. THERM./STRESS)'/
0387        * '          (=4,  TRANS. THERM./STRESS)'/
0388        * '          (=51, STRESS/CREEP)'/
0389        * '          (=52, S. S. THERM./STRESS/CREEP)'/
0390        * '          (=53, TRANS. THERM./STRESS/CREEP)'/
0391        * 30H  NUMBER OF REGIONS-----------,I3)
0392     2002 FORMAT (23H AXISYMMETRIC STRUCTURE,///)
```

```
0393      2003 FORMAT(23H PLANE STRESS STRUCTURE,///)
0394      2004 FORMAT(23H PLANE STRAIN STRUCTURE,///)
0395      2005 FORMAT(////'  IOGEOM  IOREVL    IOPRNT')
0396      2006 FORMAT(1X,I5,3X,I5,5X,I5)
0397      2007 FORMAT(29HOBAND WIDTH EXCEEDS ALLOWABLE,I4)
0398      2008 FORMAT(1H1,'STEP INFORMATION',//)
0399      2009 FORMAT(' ',' TIME STEP=',I5,'   TIME INCREMENT=',E15.6,
0400         *  '   TIME=',E15.6,'   CREEP TIME=',E15.6,///)
0401      2010 FORMAT(11H TIME STEP=,I5,16H TIME INCREMENT=,E15.6,6H TIME=,E15.6,
0402         *///)
0403      2011 FORMAT(1H ,3X,'PRESSURE LOAD CHANGES'/3X,'IBC',2X,'JBC',3X,'PRESS.
0404         *.INCR.',4X,'TOTAL PRESS.'/)
0405      2012 FORMAT(2I5,2E15.6)
0406      2013 FORMAT(//10H NODE CODE,7X,1HR,11X,1HZ,11X,1HT,11X,2HUR,11X,2HUZ,
0407         *11X,3HTUR,10X,3HTUZ/(I4,F4.1,7E13.5))
0408      2014 FORMAT(///,' EL.NO.',2X,'RATS1',7X,'RATS2',7X,'EPSR',8X,'EPSZ',
0409         *  8X,'EPST',7X,'EPSRZ',8X,'EFFEPS',5X,'EPSDOT',//)
0410      2015 FORMAT(I4,8E12.4)
0411      2016 FORMAT(1H ,//,30X,'C R E E P      A N A L Y S I S ',//)
0412      2017 FORMAT(//,1H ,'EL.NO.',2X,'ECRAT','4X,'DRAT/DEC',
0413         *5X,' DEC ',7X,'ECR',8X,' ECZ',
0414         *8X,'ECT ',10X,'ECRZ ',8X,'TOTEC ',5X,'SLOPE-B',5X,'DAMAGE-PHE',//)
0415      2018 FORMAT(I4,9E12.4)
0416      2019 FORMAT(///,' TERMINATION OF ANALYSIS')
0417      2020 FORMAT(/////,' K I N E M A T I C    H A R D E N I N G ',/////)
0418      2021 FORMAT(/////,' I S O T R O P I C    H A R D E N I N G ',/////)
0419           END




0001           SUBROUTINE MESH(JOBTYP)
0002      C              (CALLED FROM "MAIN" - FOR ALL ANALYSES)
0003      C              (SEE USER MANUAL)
0004      C***********************************************************************
0005      C
0006      C
0007      C      THIS SUBROUTINE READS AND WRITES THE STRUCTURE DISCRETIZATION
0008      C      INFORMATION, NODAL COORDINATES, BOUNDARY CONDITIONS (NODAL
0009      C      FORCE/DISPLACEMENT DATA), ELEMENT MESH DESCRIPTION, AND
0010      C      IDENTIFICATION OF ELEMENT SURFACES SUBJECT TO PRESSURE LOADING
0011      C      AS SUPPLIED BY THE USER.
0012      C
0013      C
0014      C***********************************************************************
0015      C
0016      C
0017           COMMON/DISP1/UR(378),UZ(378),CODE(378)
0018           COMMON/BOUND/IBC(200),JBC(200),NPBCR(10),NUMPC,PR(200)
0019           COMMON/SYSMSH/NUMEL,NUMNP,NRMAX
0020           COMMON/COOR/R(378),Z(378),T(378),IX(378,5),TK(378)
0021           COMMON/PLANE/NPP
0022           COMMON/PAR/IDTEMP(378)
0023           DIMENSION JX(5),IXR(5),UN(7),IP(2)
0024      C
0025      C**** INITIALIZE NODE (L), ELEMENT (NLMAX), PRESSURE
0026      C     BOUNDARY (LBC), AND REGION BOUNDARY (NANP) COUNTERS
0027      C
0028           NSH=0
0029           NANP=0
0030           LBC=0
0031           NLMAX=0
0032           NUMPC=0
0033      C
0034      C**** COMMENCE LOOP FOR THIS REGION
0035      C
0036      C
0037      C**** BEGIN WITH STRUCTURE DISCRETIZATION INFORMATION
0038      C
```

335

```
0039       165 READ (5,1000) NREJUN,NNP12,NNP34,NNPR,NELR,NBC
0040           NPBCR (NREJUN) =NBC
0041           IF (NREJUN-1) 5,10,5
0042         5 IF (NNP12-NANP) 15,10,15
0043        15 NABOR=NREJUN-1
0044           WRITE (6,2000) NABOR,NREJUN
0045           CALL EXIT
0046        10 IF (JOBTYP.EQ.1 .OR.JOBTYP.EQ.2) WRITE (6,2001)
0047           IF (JOBTYP.EQ.1 .OR. JOBTYP.EQ.2) GO TO 20
0048           WRITE (6,2002)
0049        20 L=NSH
0050           NANP=NNP34
0051     C
0052     C**** INPUT AND ECHO NODAL INFORMATION
0053     C
0054        85 IF (JOBTYP.EQ.1 .OR. JOBTYP.EQ.2)
0055         *    READ (5,1001) N,UN(1),UN(2),UN(3),ITEMP
0056           IF (JOBTYP.EQ.1 .OR. JOBTYP.EQ.2) GO TO 25
0057           READ (5,1002) N,UN(1),UN(2),UN(3),ITEMP,UN(4),UN(5),UN(6),UNC
0058        25 N=N+NSH
0059           R (N) =UN (1)
0060           Z (N) =UN (2)
0061           T (N) =UN (3)
0062           UR (N) =UN (4)
0063           UZ (N) =UN (5)
0064           TK (N) =UN (6)
0065           CODE (N) =UNC
0066           IDTEMP (N) =ITEMP
0067           NL=L+1
0068           ZX=N-L
0069           IF (L) 30,30,35
0070        35 DR= (R (N) -R (L) )/ZX
0071           DZ= (Z (N) -Z (L) )/ZX
0072           DT= (T (N) -T (L) )/ZX
0073        30 IF (NPP-1) 40,45,40
0074        45 IF (L) 50,50,55
0075        55 DTK= (TK (N) -TK (L) )/ZX
0076           GO TO 50
0077        40 TK (N) =1.0
0078           DTK=0.0
0079        50 L=L+1
0080           IF (N-L) 75,65,70
0081        70 CODE (L) =0.0
0082           R (L) =R (L-1) +DR
0083           Z (L) =Z (L-1) +DZ
0084           UR (L) =0.0
0085           UZ (L) =0.0
0086           T (L) =T (L-1) +DT
0087           TK (L) =TK (L-1) +DTK
0088           GO TO 50
0089        65 DO 500 K=NL,N
0090           K1=1000*NREJUN+K-NSH
0091           IF (JOBTYP.EQ.1 .OR. JOBTYP.EQ.2)
0092         *    WRITE (6,2003) K1,K,R(K),Z(K),T(K),IDTEMP(K)
0093           IF (JOBTYP.EQ.1 .OR. JOBTYP.EQ.2) GO TO 500
0094           WRITE (6,2004) K1,K,R(K),Z(K),UR(K),UZ(K),CODE(K),T(K),TK(K),
0095         *    IDTEMP (K)
0096       500 CONTINUE
0097           IF (NNPR+NSH-N) 75,80,85
0098        75 WRITE (6,2005) NREJUN,N
0099           CALL EXIT
0100        80 NPMAX=N
0101           WRITE (6,2006)
0102           N=NLMAX
0103     C
0104     C**** INPUT AND ECHO ELEMENT INFORMATION
0105     C
0106       120 READ (5,1003) MM,JX
0107           M=MM+NLMAX
0108           M1=MM+1000*NREJUN
```

```
0109          DO 510 II=1,4
0110          IX(M,II)=JX(II)+NSH
0111      510 IXR(II)=IX(M,II)-NSH+1000*NREJUN
0112          IF (JX(5)) 90,95,90
0113       95 JX(5)=IX(M-1,5)
0114       90 IX(M,5)=JX(5)
0115      110 N=N+1
0116          IF (M-N) 100,100,105
0117      105 DO 520 II=1,4
0118          IX(N,II)=IX(N-1,II)+1
0119      520 IXR(II)=IX(N,II)-NSH+1000*NREJUN
0120          IX(N,5)=IX(N-1,5)
0121          N1=N-NLMAX+1000*NREJUN
0122          WRITE (6,2007) N1,N,(IX(N,II),II=1,4),(IXR(II),II=1,4),IX(N,5)
0123          GO TO 110
0124      100 DO 530 II=1,4
0125      530 IXR(II)=IX(M,II)-NSH+1000*NREJUN
0126          WRITE (6,2007) M1,M,(IX(M,II),II=1,4),(IXR(II),II=1,4),IX(M,5)
0127          IF (NELR-MM) 115,115,120
0128      115 NLMAX=M
0129          NUMEL=NLMAX
0130          IF (NPBCR(NREJUN)) 125,130,125
0131      125 WRITE (6,2008)
0132          ISKIP=0
0133          IP(1)=0
0134          NR=NPBCR(NREJUN)
0135          DO 540 LR=1,NR
0136          L=LBC+LR
0137          IF (ISKIP-IP(1)) 135,140,135
0138    C
0139    C**** INPUT AND ECHO PRESSURE LOADING INFORMATION
0140    C
0141      140 READ (5,1004) IP(1),IP(2)
0142          IF (IP(1)-IP(2)) 145,135,145
0143      145 ISKIP=IP(1)
0144          GO TO 150
0145      135 IP(1)=IP(1)+1
0146          IP(2)=IP(2)+1
0147          ISKIP=ISKIP+1
0148      150 IBC(L)=IP(1)+NSH
0149          JBC(L)=IP(2)+NSH
0150          IBCR=IP(1)+1000*NREJUN
0151          JBCR=IP(2)+1000*NREJUN
0152      540 WRITE (6,2009) IBC(L),JBC(L),IBCR,JBCR
0153          NUMPC=L
0154          LBC=L
0155      130 NSH=NPMAX
0156    C
0157    C**** CHECK IF ON LAST REGION
0158    C
0159          IF (NRMAX-NREJUN) 155,160,165
0160      155 WRITE (6,2010) NREJUN
0161          CALL EXIT
0162      160 NUMNP=NSH
0163          RETURN
0164    C
0165    C**** INPUT FORMATS
0166    C
0167     1000 FORMAT(615)
0168     1001 FORMAT(I5,3E10.0,I5)
0169     1002 FORMAT(I5,3E10.0,I5,3E10.0,F2.0)
0170     1003 FORMAT(615)
0171     1004 FORMAT(2I5)
0172    C
0173    C**** OUTPUT FORMATS
0174    C
0175     2000 FORMAT(23HONODE MISMATCH= REGIONS ,I3,5H  AND,I3)
0176     2001 FORMAT('0',3X,'REGION',4X,'SYSTEM',7X,'R',11X,'Z',7X,
0177         * 'TEMPERATURE',4X,'NODAL TEMP'/2X,'NODE NO.',2X,'NODE NO.',3X,
0178         * 'COORDINATE',2X,'COORDINATE',15X,' FUNCTION IDN ')
```

337

```
0179        2002 FORMAT(1H1,3X,'REGION',4X,'SYSTEM',7X,'R',11X,'Z',7X,
0180          * 'R LOAD OR',5X,'Z LOAD OR',6X,'TYPE',4X,'TEMPERATURE',
0181          * 5X,'THICKNESS',4X,'NODAL TEMP.'/2X,'NODE NO.',2X,
0182          * 'NODE NO.',3X,'COORDINATE',2X,'COORDINATE',2X,'DISPLACEMENT',2X,
0183          * 'DISPLACEMENT',40X,'FUNCTION IDN.')
0184        2003 FORMAT(3X,I5,5X,I3,4X,3F12.4,5X,I5)
0185        2004 FORMAT(3X,I5,5X,I3,4X,2E12.4,2E14.5,3X,F4.1,2X,F9.3,3X,E12.4,
0186          * 8X,I5)
0187        2005 FORMAT(24HONODE POINT ERROR,REGION I3,15H,NODE NODE NO. I3)
0188        2006 FORMAT(1H1,2X,'REGION',4X,'SYSTEM',3X,'-----SYSTEM----',1X,
0189          * '-------REGION----------',3X,'MATL '/2X,'ELEMENT',4X,
0190          * 'ELEMENT',3X,'I     J    K    L',4X,'I     J    K    L',
0191          * 4X,'CODE'/2X,'NUMBER',5X,'NUMBER')
0192        2007 FORMAT(2X,I5,7X,I3,2X,4I4,2X,4I6,4X,I2,3X,I2)
0193        2008 FORMAT(1H1,'PRESSURE BOUNDARY CONDITIONS'/'-SYSTEM-',
0194          * 4X,'---REGION--'/1X,'I     J    I    J',//)
0195        2009 FORMAT(1X,2I4,2X,2I6)
0196        2010 FORMAT(27HOREGION NUMBER ERROR,REGION I3)
0197             END




0001             SUBROUTINE BANSOL
0002        C                    (CALLED FROM "MAIN" - FOR STRESS ANALYSIS)
0003        C                    (SEE 3-16-(12))
0004        C***********************************************************************
0005        C
0006        C
0007        C    THIS SUBROUTINE SOLVES THE BANDED SYSTEM OF EQUATIONS
0008        C    AS ASSEMBLED IN STIFF SUBROUTINE BLOCK BY BLOCK.
0009        C    INCREMENTAL NODAL DISPLACEMENTS ARE CALCULATED FROM THE
0010        C    GLOBAL SYSTEM OF EQUATIONS USING GAUSSIAN ELIMINATION.
0011        C
0012        C
0013        C***********************************************************************
0014        C
0015        C
0016             COMMON/BANARG/MM,NUMBLK,B(108),A(197,54)
0017             COMMON/DISPL/BB(756),BBT(756)
0018             NN=54
0019             NL=NN+1
0020             NH=NN+NN
0021             REWIND 1
0022             REWIND 2
0023             NB=0
0024             GO TO 5
0025        C
0026        C**** REDUCE EQUATIONS BY BLOCKS
0027        C
0028        C
0029        C**** SHIFT BLOCK OF EQUATIONS
0030        C
0031         15 NB=NB+1
0032             DO 500 N=1,NN
0033             NM=NN+N
0034             B(N)=B(NM)
0035             B(NM)=0.0
0036             DO 500 M=1,MM
0037             A(N,M)=A(NM,M)
0038        500 A(NM,M)=0.0
0039        C
0040        C**** READ NEXT BLOCK OF EQUATIONS INTO CORE
0041        C
0042             IF(NUMBLK-NB) 5,10,5
0043          5 READ (2) (B(N),N=NL,NH),((A(N,M),N=NL,NH),M=1,MM)
0044             IF(NB) 10,15,10
0045        C
0046        C**** REDUCE BLOCK OF EQUATIONS
0047        C
```

338

```
0048        10 DO 510 N=1,NN
0049           IF (A(N,1)) 20,510,20
0050        20 B(N)=B(N)/A(N,1)
0051           DO 520 L=2,MM
0052           IF (A(N,L)) 25,520,25
0053        25 C=A(N,L)/A(N,1)
0054           I=N+L-1
0055           J=0
0056           DO 530 K=L,MM
0057           J=J+1
0058       530 A(I,J)=A(I,J)-C*A(N,K)
0059           B(I)=B(I)-A(N,L)*B(N)
0060           A(N,L)=C
0061       520 CONTINUE
0062       510 CONTINUE
0063    C
0064    C**** WRITE BLOCK OF REDUCED EQUATIONS ON TAPE 2
0065    C
0066           IF (NUMBLK-NB) 30,35,30
0067        30 WRITE (1) (B(N),N=1,NN),((A(N,M),N=1,NN),M=2,MM)
0068           GO TO 15
0069    C
0070    C**** BACK-SUBSTITUTION
0071    C
0072        35 DO 540 M=1,NN
0073           N=NN+1-M
0074           DO 550 K=2,MM
0075           L=N+K-1
0076       550 B(N)=B(N)-A(N,K)*B(L)
0077           NM=N+NN
0078           B(NM)=B(N)
0079       540 A(NM,NB)=B(N)
0080           NB=NB-1
0081           IF (NB) 40,45,40
0082        40 BACKSPACE 1
0083           READ (1) (B(N),N=1,NN),((A(N,M),N=1,NN),M=2,MM)
0084           BACKSPACE 1
0085           GO TO 35
0086    C
0087    C**** ORDER UNKNOWNS IN B ARRAY
0088    C
0089        45 K=0
0090           DO 560 NB=1,NUMBLK
0091           DO 560 N=1,NN
0092           NM=N+NN
0093           K=K+1
0094           BBT(K)=A(NM,NB)+BBT(K)
0095       560 BB(K)=A(NM,NB)
0096           RETURN
0097           END


0001           SUBROUTINE CONSET(NUMEL)
0002    C                    (CALLED FROM "QUAD" - STRESS ANALYSIS - ISOTROPIC HARDENING)
0003    C                    (SEE 3-4-1)
0004    C****************************************************************************
0005    C
0006    C
0007    C      THIS SUBROUTINE EVALUATES THE DFDEDE AND DFDTDT TERMS OF THE
0008    C      CONSTITUTIVE EQUATION, AND (IF REQUIRED) SETS A NEW YIELD STRESS
0009    C      FOR AN  ELEMENT UNLOADING FROM A PLASTIC CONDITION.
0010    C
0011    C
0012    C****************************************************************************
0013    C
0014    C
0015           COMMON/STRAIN/EPSDI(378),EPSDOT(378),SIGAVE(378),DEVSIG(378,5),
0016          *       EPSDTC(378),EPSDTP(378)
```

```
0017          COMMON/STREZZ/RATS1(378),TSIG(378,5),TOTEPS(378,4),EPS(378),
0018      *        DLAM(378),RATST(378),DEPS(378,4)
0019          COMMON/PRPRTE/PROP1(5,11,5),PROP2(5,11,5),PROP3(5,11,5),
0020      *        PROP4(5,11,5),PROP5(5,11,5),PROP6(5,11,5),PROPS(5,11,5)
0021          COMMON/TEMP2/PROP(9),IPTRT,IPTRE,TEMD,RATIOT,RATIOE,
0022      *        TMPRNG,EPSRNG
0023          COMMON/TEMP3/EBR,EBRPRM,H1,DFDTDT,DFDEDE,HPR
0024          COMMON/HEATR/XA,DTI,TIME,TIMECR
0025          DIMENSION FSIGBR(4)
0026   C
0027   C**** INITIALIZE
0028   C
0029          DSGDT1=0.0
0030          DSGDT2=0.0
0031          DSGDE2=0.0
0032          DSGDE1=0.0
0033   C
0034   C**** FOR NO THERMAL OR STRAIN RATE DEPENDENCE OF MATERIAL
0035   C     PROPERTIES, OR FOR CREEP BEHAVIOUR, SKIP EVALUATION
0036   C
0037          IF (TIMECR.NE.0.0) GO TO 5
0038          IF (IPTRT.EQ.1 .AND. IPTRE.EQ.1) GO TO 5
0039   C
0040   C**** EVALUATE DSIGBAR/DEPS DSIGBAR/DT AS IN NTRPLT SUBROUTINE
0041   C
0042          DO 500 I=1,4
0043          GO TO (10,15,20,25),I
0044      10 MM=IPTRT
0045          LL=IPTRE
0046          GO TO 30
0047      15 MM=IPTRT-1
0048          LL=IPTRE
0049          IF (MM.EQ.0) MM=1
0050          GO TO 30
0051      20 MM=IPTRT
0052          LL=IPTRE-1
0053          IF (LL.EQ.0) LL=1
0054          GO TO 30
0055      25 MM=IPTRT-1
0056          IF (MM.EQ.0) MM=1
0057          LL=IPTRE-1
0058          IF (LL.EQ.0) LL=1
0059      30 PROPS(MM,8,LL)=3.*PROPS(MM,8,LL)/(3.-(1.-2.*PROPS(MM,3,LL))*
0060      *            (PROPS(MM,8,LL)/PROPS(MM,2,LL)))
0061          PROPS(MM,2,LL)=1.5*PROPS(MM,8,LL)/(1.+PROPS(MM,3,LL))
0062          T1=(1.-PROPS(MM,8,LL)/PROPS(MM,2,LL))*PROPS(MM,10,LL)+
0063      *       PROPS(MM,8,LL)*EPS(NUMEL)
0064          T2=(PROPS(MM,2,LL)*EPS(NUMEL)/T1)+1.0E-10
0065          T3=PROPS(MM,9,LL)*ALOG10(T2)
0066          IF (T3.GT.30.) GO TO 35
0067          IF (T3.LT.-20.) GO TO 40
0068          T4=T2**PROPS(MM,9,LL)
0069          FSIGBR(I)=PROPS(MM,2,LL)*EPS(NUMEL)/(1.+T4)**(1./PROPS(MM,9,LL))
0070          GO TO 500
0071      35 FSIGBR(I)=T1
0072          GO TO 500
0073      40 FSIGBR(I)=PROPS(MM,2,LL)*EPS(NUMEL)
0074     500 CONTINUE
0075          DSGDT2=FSIGBR(1)-FSIGBR(2)
0076          DSGDT1=FSIGBR(3)-FSIGBR(4)
0077          DSGDE2=FSIGBR(2)-FSIGBR(4)
0078          DSGDE1=FSIGBR(1)-FSIGBR(3)
0079          IF (TMPRNG.EQ.0.0) GO TO 45
0080          DSGDT2=DSGDT2/TMPRNG
0081          DSGDT1=DSGDT1/TMPRNG
0082      45 IF (EPSRNG.EQ.0.0) GO TO 5
0083          DSGDE2=DSGDE2/EPSRNG
0084          DSGDE1=DSGDE1/EPSRNG
0085       5 DSGDT=DSGDT1+RATIOT*(DSGDT2-DSGDT1)
0086          DSGDE=DSGDE1+RATIOE*(DSGDE2-DSGDE1)
```

340

```
0087          IF (DSGDT.GE.O.O) DSGDT=0.0
0088          IF (DSGDE.GE.O.O) DSGDE=0.0
0089          H11=PROP (7) /PROP (1)
0090          EBR=1.5*PROP (1) / (1.+PROP (2))
0091          EBRPRM=3.*PROP (7) / (3.- (1.-2.*PROP (2)) *H11)
0092          H1=EBRPRM/EBR
0093          S1= (1.-H1) *PROP (9) +EBRPRM*EPS (NUMEL)
0094          S2= (EBR*EPS (NUMEL) /S1) +1.0E-10
0095          S3=PROP (8) *ALOG10 (S2)
0096          IF (S3.GT.30.) GO TO 50
0097          IF (S3.LT.-20.) GO TO 55
0098          S4=S2**PROP (8)
0099          SGBR=EBR*EPS (NUMEL) / (1.+S4) ** (1./PROP (8))
0100          GO TO 60
0101       50 SGBR=S1
0102          GO TO 60
0103       55 SGBR=EBR*EPS (NUMEL)
0104       60 DFDTDT=- (2./3.) *SGBR*DSGDT*TEMD
0105          DFDEDE=- (2./3.) *SGBR*DSGDE*EPSDI (NUMEL)
0106          IF (DLAM (NUMEL) .GE.O.O) GO TO 65
0107          PROP (5) =PROP (9) + (EBR*EPS (NUMEL) -DEVSIG (NUMEL,5)) *H1/ (1.-H1)
0108       65 RETURN
0109          END




0001          SUBROUTINE CONTOT (IELNUM,NPP,KINHAR)
0002    C               (CALLED FROM "QUAD" - FOR STRESS ANALYSIS)
0003    C               (SEE 3-10)
0004    C************************************************************************
0005    C
0006    C
0007    C     THIS SUBROUTINE EVALUATES TERMS OF THE CONSTITUTIVE EQUATION
0008    C     FOR EACH ELEMENT IN THE STRUCTURE.
0009    C
0010    C
0011    C************************************************************************
0012    C
0013    C
0014          COMMON/STRAIN/EPSDI (378) ,EPSDOT (378) ,SIGAVE (378) ,DEVSIG (378,5) ,
0015        *       EPSDTC (378) ,EPSDTP (378)
0016          COMMON/STREZZ/RATS1 (378) ,TSIG (378,5) ,TOTEPS (378,4) ,EPS (378) ,
0017        *       DLAM (378) ,RATST (378) ,DEPS (378,4)
0018          COMMON/TEMP2/PROP (9) ,IPTRT,IPTRE,TEMD,RATIOT,RATIOE,
0019        *       TMPRNG,EPSRNG
0020          COMMON/TEMP3/EBR,EBRPRM,H1,DFDTDT,DFDEDE,hpr
0021          COMMON/TEMP4/CE (4,4) ,CEI (4,4) ,CEIPR (378,4,4) ,DCEIDE (4,4) ,
0022        *       DCEIDT (4,4) ,SIG (4) ,DCESIG (4) ,DCTSIG (4)
0023          COMMON/TEMP5/CEP (4,4) ,TOTTRM (4) ,SZERO
0024          COMMON/PLASTB/AVDEVI (378) ,TRDEVI (378,4)
0025          COMMON/ANEW/SPL (378) ,DLAMP (378) ,TOTEP (378,4)
0026          DIMENSION T2 (4) ,T3 (4)
0027    C
0028    C**** INITIALIZATION
0029    C
0030          DO 500 I=1,4
0031          T2 (I) =0.0
0032          T3 (I) =0.0
0033      500 TOTTRM (I) =0.0
0034    C
0035    C**** EVALUATION
0036    C
0037          DO 510 II=1,4
0038    C     T2 (II) = (CEP (II,1) *DCESIG (1) +CEP (II,2) *DCESIG (2) +CEP (II,3) *
0039    C   *       DCESIG (3) +CEP (II,4) *DCESIG (4)) *EPSDI (IELNUM)
0040          T2 (II) =T2 (II) + (CEP (II,1) * (PROP (4) +DCTSIG (1)) +CEP (II,2) * (PROP (4) +
0041        *       DCTSIG (2)) +CEP (II,3) * (PROP (4) +DCTSIG (3)) +CEP (II,4) *
0042        *       (DCTSIG (4))) *TEMD
0043      510 TOTTRM (II) =T2 (II)
0044          IF (RATS1 (IELNUM) .LT..01) GO TO 5
```

341

```
0045          DO 520 II=1,4
0046          T3(II)=0.0
0047          DO 530 JJ=1,4
0048          T4=DEVSIG(IELNUM,JJ)*(DFDTDT+DFDEDE)/(SZERO
0049     *       *2.*PROP(1)/(1.+PROP(2)))
0050          IF (KINHAR.EQ.1) T4=TRDEVI(IELNUM,JJ)*(DFDTDT+DFDEDE)
0051     *                            /SPL(IELNUM)
0052      530 T3(II)=T3(II)+CE(II,JJ)*T4
0053          IF (NPP.EQ.1) T3(3)=0.0
0054      520 TOTTRM(II)=TOTTRM(II)+T3(II)
0055        5 RETURN
0056          END



0001          SUBROUTINE CSUBE(IELNUM,NPP,ITIME,KINHAR)
0002  C                  (CALLED FROM "QUAD" - FOR STRESS ANALYSIS)
0003  C                  (SEE 3-3-4, 3-10)
0004  C*************************************************************************
0005  C
0006  C
0007  C    THIS SUBROUTINE CALCULATES CE, THE ELEMENT STRESS-STRAIN
0008  C    TRANSFORMATION MATRIX, ITS INVERSE, CEI, FOR AXISYMMETRIC,
0009  C    PLANE STRESS AND STRAIN GEOMETRIES.  IN ADDITION, THE DCEI/DE*SIG
0010  C    AND DCEI/DT*SIG PORTION OF THE DCEI/DE*SIG*DE AND DCEI/DT*SIG*DT
0011  C    TERMS OF THE CONSTITUTIVE EQUATION IS EVALUATED.
0012  C
0013  C
0014  C*************************************************************************
0015  C
0016  C
0017          COMMON/COOR/R(378),Z(378),T(378),IX(378,5),TK(378)
0018          COMMON/STRAIN/EPSDI(378),EPSDOT(378),SIGAVE(378),DEVSIG(378,5),
0019     *       EPSDTC(378),EPSDTP(378)
0020          COMMON/TEMP2/PROP(9),IPTRT,IPTRE,TEMD,RATIOT,RATIOE,
0021     *       TMPRNG,EPSRNG
0022          COMMON/TEMP4/CE(4,4),CEI(4,4),CEIPR(378,4,4),
0023     *       DCEIDE(4,4),DCEIDT(4,4),SIG(4),DCESIG(4),DCTSIG(4)
0024          COMMON/THICK/THK
0025          COMMON/HEATR/XA,DTI,TIME,TIMECR
0026          COMMON/PLASTB/AVDEVI(378),TRDEVI(378,4)
0027          COMMON/PASMAT/CECE(378,4,4),DCESGN(378,4),DCTSGN(378,4)
0028  C
0029  C**** INITIALIZATION
0030  C
0031          DO 500 II=1,4
0032          DO 500 JJ=1,4
0033      500 CE(II,JJ)=0.0
0034          I=IX(IELNUM,1)
0035          J=IX(IELNUM,2)
0036          K=IX(IELNUM,3)
0037          L=IX(IELNUM,4)
0038          IF (NPP-1) 5,10,15
0039  C
0040  C**** CE MATRIX FOR AXISYMMETRIC
0041  C
0042        5 COMM=PROP(1)*(1.-PROP(2))/((1.+PROP(2))*(1.-2.*
0043     *        PROP(2)))
0044          T1=PROP(2)/(1.-PROP(2))
0045          CE(1,1)=COMM
0046          CE(1,2)=COMM*T1
0047          CE(4,4)=COMM*(1.-2.*PROP(2))/(2.*(1.-PROP(2)))
0048          CE(1,3)=COMM*T1
0049          CE(2,1)=CE(1,2)
0050          CE(2,2)=COMM
0051          CE(2,3)=COMM*T1
0052          CE(3,1)=CE(1,3)
0053          CE(3,2)=CE(2,3)
0054          CE(3,3)=COMM
0055          GO TO 20
```

```
0056        10 COMM=PROP (1) / (1.-PROP (2) **2)
0057    C
0058    C**** CE MATRIX FOR PLANE STRESS
0059    C
0060           CE (1,1) =COMM
0061           CE (1,2) =COMM*PROP (2)
0062           CE (2,1) =CE (1,2)
0063           CE (2,2) =COMM
0064           CE (3,3) =1.E-30
0065           CE (4,4) =COMM* (1.-PROP (2) ) /2.
0066           IF (K-L) 25,30,25
0067        30 THK= (TK (I) +TK (J) +TK (K) ) /3.
0068           GO TO 20
0069        25 THK= (TK (I) +TK (J) +TK (K) +TK (L) ) /4.
0070           GO TO 20
0071    C
0072    C**** CE MATRIX FOR PLANE STRAIN
0073    C
0074        15 COMM=PROP (1) / (1.+PROP (2) )
0075           CE (1,1) =COMM* (1.-PROP (2) ) / (1.-2.*PROP (2) )
0076           CE (1,2) =COMM*PROP (2) / (1.-2.*PROP (2) )
0077           CE (2,1) =CE (1,2)
0078           CE (2,2) =CE (1,1)
0079           CE (3,3) =1.E-30
0080           CE (4,4) =.5*COMM
0081           IF (K-L) 35,40,35
0082        40 THK= (TK (I) +TK (J) +TK (K) ) /3.
0083           GO TO 20
0084        35 THK= (TK (I) +TK (J) +TK (K) +TK (L) ) /4.
0085        20 DO 510 IND1=1,4
0086           DO 510 IND2=1,4
0087       510 CEI (IND1,IND2) =CE (IND1,IND2)
0088    C
0089    C**** INVERT CE MATRIX, RETURN AS CE-1 MATRIX
0090    C
0091           CALL SYMINV (CEI,4)
0092    C
0093    C**** EVALUATE DCEIDEPS PORTION
0094    C
0095           DO 520 II=1,4
0096           DO 520 JJ=1,4
0097           IF (ABS (EPSDI (IELNUM) ) -1.E-05) 45,50,50
0098        45 DCEIDE (II,JJ) =0.0
0099           GO TO 520
0100        50 IF (ITIME.LT.2) GO TO 520
0101           IF (TIMECR.NE.0.0) GO TO 520
0102           DCEIDE (II,JJ) = (CEI (II,JJ) -CEIPR (IELNUM,II,JJ) ) /EPSDI (IELNUM)
0103       520 CEIPR (IELNUM,II,JJ) =CEI (II,JJ)
0104    C
0105    C**** EVALUATE DCEIDT PORTION
0106    C
0107           DO 530 II=1,4
0108           DO 530 JJ=1,4
0109           IF (ABS (TEMD) -1.E-05) 55,60,60
0110        55 DCEIDT (II,JJ) =0.0
0111           GO TO 530
0112        60 IF (ITIME.LT.2) GO TO 530
0113           IF (TIMECR.NE.0.0) GO TO 530
0114           DCEIDT (II,JJ) = (CEI (II,JJ) -CEIPR (IELNUM,II,JJ) ) /TEMD
0115       530 CEIPR (IELNUM,II,JJ) =CEI (II,JJ)
0116    C
0117    C**** EVALUATE DCESIG PORTION
0118    C
0119           DO 540 II=1,4
0120           DCESIG (II) =0.0
0121           DO 540 JJ=1,4
0122           IF (KINHAR.EQ.1) DEVSIG (IELNUM,JJ) =TRDEVI (IELNUM,JJ)
0123           IF (KINHAR.EQ.1) SIGAVE (IELNUM) =AVDEVI (IELNUM)
0124           IF (JJ-3) 65,65,70
0125        70 SIG (JJ) =DEVSIG (IELNUM,JJ)
0126           GO TO 75
```

```
0127        65 SIG(JJ)=DEVSIG(IELNUM,JJ)+SIGAVE(IELNUM)
0128        75 IF (NPP.EQ.1) SIG(3)=0.0
0129       540 DCESIG(II)=DCESIG(II)+DCEIDE(II,JJ)*SIG(JJ)
0130           IF (NPP.EQ.1) DCESIG(3)=0.0
0131           DO 150 I=1,4
0132       150 DCESGN(IELNUM,I)=DCESIG(I)
0133    C
0134    C**** EVALUATE DCTSIG PORTION
0135    C
0136           DO 550 II=1,4
0137           DCTSIG(II)=0.0
0138           DO 550 JJ=1,4
0139           IF (KINHAR.EQ.1) DEVSIG(IELNUM,JJ)=TRDEVI(IELNUM,JJ)
0140           IF (KINHAR.EQ.1) SIGAVE(IELNUM)=AVDEVI(IELNUM)
0141           IF (JJ-3) 80,80,85
0142        85 SIG(JJ)=DEVSIG(IELNUM,JJ)
0143           GO TO 90
0144        80 SIG(JJ)=DEVSIG(IELNUM,JJ)+SIGAVE(IELNUM)
0145        90 IF (NPP.EQ.1) SIG(3)=0.0
0146       550 DCTSIG(II)=DCTSIG(II)+DCEIDT(II,JJ)*SIG(JJ)
0147           IF (NPP.EQ.1) DCTSIG(3)=0.0
0148           DO 250 I=1,4
0149       250 DCTSGN(IELNUM,I)=DCTSIG(I)
0150           RETURN
0151           END



0001           SUBROUTINE CSUBEP(IELNUM,NPP,JOBTYP)
0002    C                     (CALLED FROM "QUAD" - STRESS ANALYSIS - ISOTROPIC HARDENING)
0003    C                     (SEE 3-15-3, 3-12)
0004    C**********************************************************************
0005    C
0006    C
0007    C      THIS SUBROUTINE CALCULATES THE ELEMENT CEP MATRIX (THE
0008    C      ELASTO-PLASTICITY MATRIX) FOR PLANE STRESS AND STRAIN, AND
0009    C      AXISYMMETRIC GEOMETRIES.
0010    C
0011    C
0012    C**********************************************************************
0013    C
0014    C
0015           COMMON/STRAIN/EPSDI(378),EPSDOT(378),SIGAVE(378),DEVSIG(378,5),
0016          *        EPSDTC(378),EPSDTP(378)
0017           COMMON/STREZZ/RATS1(378),TSIG(378,5),TOTEPS(378,4),
0018          *        EPS(378),DLAM(378),ratst(378),deps(378,4)
0019           COMMON/TEMP2/PROP(9),IPTRT,IPTRE,TEMD,RATIOT,RATIOE,
0020          *        TMPRNG,EPSRNG
0021           COMMON/TEMP3/EBR,EBRPRM,H1,DFDTDT,DFDEDE,HPR
0022           COMMON/TEMP4/CE(4,4),CEI(4,4),CEIPR(378,4,4),
0023          *        DCEIDE(4,4),DCEIDT(4,4),SIG(4),DCESIG(4),DCTSIG(4)
0024           COMMON/TEMP5/CEP(4,4),TOTTRM(4),SZERO
0025           DIMENSION S(4)
0026    C
0027    C**** INITIALIZE VARIABLES FOR EVALUATION
0028    C
0029           EPST=EPS(IELNUM)
0030           GNEW=PROP(1)*.5/(1.+PROP(2))
0031           T1=(1.-H1)*PROP(9)+EBRPRM*EPST
0032           T2=ALOG10(EBR*EPST/T1+1.0E-10)
0033           IF (PROP(8)*T2-20.0) 5,10,10
0034        10 HPR=EBRPRM+1.0E-15
0035           GO TO 15
0036         5 IF (PROP(8)*T2.GT.-20.0) GO TO 20
0037           HPR=10.0e+15
0038           GO TO 15
0039        20 HKP=EBR/(((1.0+((EBR*EPST/T1)**PROP(8)))**((PROP(8)+
0040          *    1.0)/PROP(8)))*(1.0+(EBR*EPST/T1)**(PROP(8)+1.)*H1)
0041           IF (ABS(1.0/HKP-1.0/EBR).LE. 1.0/EBR) GO TO 25
```

344

```
0042              HPR=1.0/(ABS(1.0/HKP-1.0/EBR))
0043              GO TO 15
0044          25 HPR=10.0E+15
0045          15 IF (NPP-1) 30,35,40
0046      C
0047      C**** CEP AND SZERO FOR AXISYMMETRIC
0048      C
0049          30 SZERO=(2./3.*DEVSIG(IELNUM,5)**2)*(1.+HPR/(3.*GNEW))
0050              COMM=2.*GNEW/SZERO
0051              DO 500 II=1,4
0052              DO 500 JJ=1,4
0053         500 CEP(II,JJ)=CE(II,JJ)-DEVSIG(IELNUM,II)*DEVSIG(IELNUM,JJ)*COMM
0054              GO TO 45
0055      C
0056      C**** CEP AND SZERO FOR PLANE STRESS
0057      C
0058          35 S(1)=PROP(1)/(1.-PROP(2)**2)*(DEVSIG(IELNUM,1)+PROP(2)*
0059         *          DEVSIG(IELNUM,2))
0060              S(2)=PROP(1)/(1.-PROP(2)**2)*(DEVSIG(IELNUM,2)+PROP(2)*
0061         *          DEVSIG(IELNUM,1))
0062              S(4)=PROP(1)/(1.+PROP(2))*DEVSIG(IELNUM,4)
0063              SZERO=4./9.*DEVSIG(IELNUM,5)**2*HPR+S(1)*DEVSIG(IELNUM,1)+
0064         *          S(2)*DEVSIG(IELNUM,2)+2.*S(4)*DEVSIG(IELNUM,4)
0065              DO 510 II=1,4
0066              DO 510 JJ=1,4
0067              IF (II.EQ.3.OR.JJ.EQ.3) GO TO 510
0068              CEP(II,JJ)=CE(II,JJ)-S(II)*S(JJ)/SZERO
0069         510 CONTINUE
0070              GO TO 45
0071      C
0072      C**** CEP AND SZERO FOR PLANE STRAIN
0073      C
0074          40 SZERO=2./3.*(1.+HPR/(3.*GNEW)-1.5*(1.+PROP(2))*DEVSIG(IELNUM,
0075         *          3)**2/DEVSIG(IELNUM,5)**2)*DEVSIG(IELNUM,5)**2
0076              COMM=2.*GNEW/SZERO
0077              DO 520 II=1,4
0078              DO 520 JJ=1,4
0079              IF (II.EQ.3.OR.JJ.EQ.3) GO TO 520
0080              CEP(II,JJ)=CE(II,JJ)-DEVSIG(IELNUM,II)*DEVSIG(IELNUM,JJ)*COMM
0081         520 CONTINUE
0082          45 RETURN
0083              END



0001              SUBROUTINE INTER
0002      C              (CALLED FROM "TRISTF" - FOR STRESS ANALYSIS)
0003      C              (NOT DISCUSSED)
0004      C*******************************************************************;
0005      C
0006      C
0007      C      THIS SUBROUTINE USES A NUMERICAL METHOD TO EVALUATE THE
0008      C      TERMS FOR THE ELEMENT STIFFNESS MATRIX.
0009      C      ALTHOUGH INTENDED PRIMARILY FOR AXISYMMETRIC ANALYSIS,
0010      C      INTEGRATIONS FOR PLANAR ANALYSES ARE DONE HERE ALSO.
0011      C
0012      C
0013      C*******************************************************************
0014      C
0015      C
0016              IMPLICIT REAL*8 (Q)
0017              DOUBLE PRECISION AI(3),AJ(3),AS(3),V(3),COJ,X1,X2,X3,Y1,Y2,DBLE
0018              COMMON/RRZZ/RR(4),ZZ(4),XI(6)
0019              COMMON/DUM/RRR(5),ZZZ(5),LM(4)
0020              COMMON/DBLP/QRR(4),QZZ(4),QRRR(5),QZZZ(5),QXI(6)
0021              COMMON/PLANE/NPP
0022              COMMON/THICK/THK
0023      C
0024      C**** GAUSS/RADAU INTEGRATION CONSTANTS FOR N=3
0025      C
```

345

```
0026          DATA AJ/0.112702D0,0.5D0,0.887298D0/
0027          DATA AI/0.088588D0,0.409467D0,0.787659D0/
0028          DATA AS/0.220462D0,0.388193D0,0.328844D0/
0029          DATA V/0.277778D0,0.444444D0,0.277778D0/
0030     C
0031     C**** INITIALIZATION
0032     C
0033          DO 500 I=1,6
0034      500 QXI(I)=0.D0
0035     C
0036     C**** CONVERT SINGLE (R*4) TO DOUBLE PRECISION (R*8)
0037     C
0038          QTHK=DBLE(THK)
0039          DO 510 K=1,5
0040          QRRR(K)=DBLE(RRR(K))
0041      510 QZZZ(K)=DBLE(ZZZ(K))
0042     C
0043     C**** DOUBLE PRECISION CALCULATIONS
0044     C
0045          II=LM(1)
0046          JJ=LM(2)
0047          KK=LM(3)
0048          QRR(1)=QRRR(II)
0049          QRR(2)=QRRR(JJ)
0050          QRR(3)=QRRR(KK)
0051          QZZ(1)=QZZZ(II)
0052          QZZ(2)=QZZZ(JJ)
0053          QZZ(3)=QZZZ(KK)
0054          N=3
0055          ID=0
0056          COJ=(QRR(1)-QRR(3))*(QZZ(2)-QZZ(3))-(QZZ(1)-QZZ(3))*
0057         1       (QRR(2)-QRR(3))
0058          IF (NPP.NE.0) COJ=COJ*QTHK
0059          DO 520 I=1,N
0060          DO 520 J=1,N
0061          X1=AI(I)
0062          X2=AJ(J)*(1.0D0-X1)
0063          X3=1.0D0-X1-X2
0064          Y1=X1*QRR(1)+X2*QRR(2)+X3*QRR(3)
0065          Y2=X1*QZZ(1)+X2*QZZ(2)+X3*QZZ(3)
0066          IF(NPP.EQ.0) GO TO 5
0067          QXI(1)=QXI(1)+AS(I)*V(J)*(1.0D0-X1)
0068          QXI(2)=QXI(2)+AS(I)*V(J)*(1.0D0-X1)/Y1
0069          QXI(3)=QXI(3)+AS(I)*V(J)*(1.0D0-X1)/Y1**2
0070          QXI(4)=QXI(4)+AS(I)*V(J)*(1.0D0-X1)*Y2/Y1
0071          QXI(5)=QXI(5)+AS(I)*V(J)*(1.0D0-X1)*Y2/Y1**2
0072          QXI(6)=QXI(6)+AS(I)*V(J)*(1.0D0-X1)*Y2**2/Y1**2
0073          GO TO 520
0074        5 QXI(1)=QXI(1)+AS(I)*V(J)*(1.0D0-X1)*Y1
0075          QXI(2)=QXI(2)+AS(I)*V(J)*(1.0D0-X1)
0076          QXI(3)=QXI(3)+AS(I)*V(J)*(1.0D0-X1)/Y1
0077          QXI(4)=QXI(4)+AS(I)*V(J)*(1.0D0-X1)*Y2
0078          QXI(5)=QXI(5)+AS(I)*V(J)*(1.0D0-X1)*Y2/Y1
0079          QXI(6)=QXI(6)+AS(I)*V(J)*(1.0D0-X1)*Y2**2/Y1
0080      520 CONTINUE
0081          DO 530 I=1,6
0082      530 QXI(I)=QXI(I)*COJ
0083     C
0084     C**** CONVERT DOUBLE PRECISION (R*8) TO SINGLE (R*4)
0085     C
0086          DO 540 I=1,6
0087      540 XI(I)=SNGL(QXI(I))
0088          DO 550 J=1,4
0089          RR(J)=SNGL(QRR(J))
0090      550 ZZ(J)=SNGL(QZZ(J))
0091          RETURN
0092          END
```

```
0001            SUBROUTINE KINSLO (MM,HPR)
0002      C                 (CALLED FROM "QUAD" - STRESS ANALYSIS - KINEMATIC HARDENING)
0003      C                 (SEE 3-12, 3-13, 3-16-(6))
0004      C*********************************************************************
0005      C
0006      C
0007      C       THIS SUBROUTINE EVALUATES THE CONSTITUTIVE EQUATION AND THE
0008      C       CEP MATRIX FOR A KINEMATIC HARDENING MATERIAL.
0009      C
0010      C
0011      C*********************************************************************
0012      C
0013      C
0014            COMMON/STREZZ/RATS1 (378) ,TSIG (378,5) ,TOTEPS (378,4) ,EPS (378),
0015          *        DLAM (378) ,RATST (378) ,DEPS (378,4)
0016            COMMON/PLASTA/D2 (378) ,EFSLOP (378) ,EPSPL (378,4)
0017            COMMON/PLASTB/AVDEVI (378) ,TRDEVI (378,4)
0018            COMMON/TEMP2/PROP (9) ,IPTRT,IPTRE,TEMD,RATIOT,RATIOE,
0019          *        TMPRNG,EPSRNG
0020            COMMON/TEMP5/C (4,4) ,TOTTRM (4) ,SZERO
0021            COMMON/TEMP4/CE (4,4) ,CEI (4,4) ,CEIPR (378,4,4) ,
0022          *        DCEIDE (4,4) ,DCEIDT (4,4) ,SIG (4) ,DCESIG (4) ,DCTSIG (4)
0023            COMMON/ANEW/SPL (378) ,DLAMP (378) ,TOTEP (378,4)
0024            COMMON/CHANGE/XEPS (5) ,YSIG (5) ,IUN
0025            COMMON/PLANE/NPP
0026            COMMON/TIMSTP/ITIME
0027            DIMENSION TRV (4) ,HPRM (4)
0028      C
0029      C**** INITIALIZATION
0030      C
0031            N=MM
0032            DEFS=10.E-30
0033            H1=PROP (7) /PROP (1)
0034            B1= (1.-H1) *PROP (9)
0035            ET=10.E-20
0036            IF (ITIME.EQ.1) GESH=10.E-20
0037            IF (RATST (N) .LE.0.01) GO TO 5
0038            IG=4
0039            E3G=PROP (1)
0040            DO 500 I=1,4
0041        500 IF (ABS (YSIG (I)) .LT.1.E-10) YSIG (I) =0.0
0042            SIGW=YSIG (1) **2+YSIG (2) **2+YSIG (3) **2+.5*YSIG (4) **2
0043            GO TO 10
0044          5 IG=1
0045            E3G=1.5*PROP (1)·/ (1.+PROP (2) )
0046         10 CONTINUE
0047      C
0048      C**** LOOP FOR WEIGHTED AVERAGE OF CONSTITUTIVE CURVE
0049      C
0050            DO 510 I=1,IG
0051            ET=ABS (XEPS (I) )
0052            IF (NPP.EQ.1 .AND. I.EQ.3) GO TO 510
0053            IF (ET.LT.10.E-20) ET=10.E-20
0054            BIL= (B1+PROP (7) *ET)
0055            AAA=ALOG10 (E3G*ET/BIL)
0056            IF (PROP (8) *AAA-30.) 15,20,20
0057         20 HPRM (I) =PROP (7) +1.0E-30
0058            GO TO 25
0059         15 IF (PROP (8) *AAA.GT.-20.) GO TO 30
0060            HPRM (I) =10.E+20
0061            GO TO 25
0062         30 ESUBT=E3G/ ((1.+ ((E3G*ET/BIL) **PROP (8) )) ** ((PROP (8) +1.0) /PROP (8) ))
0063          1* (1.+ (E3G*ET/BIL) ** (PROP (8) +1.) *H1)
0064            HPRM (I) =1.0/ (ABS ((1./ESUBT-1./E3G) +10.E-30) )
0065            IF (PROP (7) .LT.0.0) HPRM (I) =10.E-30
0066         25 WE=1.0
0067            IF (IG.NE.1) WE=YSIG (I) **2/SIGW
0068            DEFS=DEFS+WE/HPRM (I)
0069        510 CONTINUE
0070            DLAMP (N) =DLAM (N)
```

347

```
0071          IF (DEFS.LT.1.0E-20) DEFS=1.0E-20
0072          EFSLOP(N)=1.0/DEFS
0073          HPR=EFSLOP(N)
0074          IF (IUN.EQ.0) GO TO 35
0075     C
0076     C**** SET CEP=CE FOR UNLOADING FROM PLASTIC STATE
0077     C
0078          DO 520 I=1,4
0079          DO 520 J=1,4
0080      520 C(I,J)=CE(I,J)
0081          GO TO 40
0082     C
0083     C**** EVALUATE CEP MATRIX
0084     C
0085       35 GESH=0.
0086          DO 530 I=1,4
0087          GASH=0.
0088          DO 540 J=1,4
0089          GASH=GASH+CE(I,J)*TRDEVI(N,J)
0090      540 TRV(I)=GASH
0091      530 GESH=GESH+TRDEVI(N,I)*TRV(I)
0092          SPL(N)=6.*D2(N)*EFSLOP(N)+6.*GESH
0093          IF (D2(N).LE.10.E-6) SPL(N)=10.E+20
0094          DO 550 I=1,4
0095          DO 550 J=1,4
0096          IF (NPP.EQ.1.AND.((I.EQ.3).OR.(J.EQ.3))) GO TO 550
0097          CP=6.0*TRV(I)*TRV(J)/SPL(N)
0098          C(I,J)=CE(I,J)-CP
0099      550 CONTINUE
0100       40 RETURN
0101          END
```

```
0001          SUBROUTINE KITENS(MM,RATS,E13,DYDT,DYDE)
0002     C              (CALLED FROM "STRESS" - STRESS ANALYSIS - KINEMATIC HARDENIN
0003     C              (SEE 3-13, 3-15-8)
0004     C**************************************************************************
0005     C
0006     C
0007     C      THIS SUBROUTINE EVALUATES THE TRANSLATED STRESS DEVIATORS,
0008     C      STRAIN AND TRANSLATION TENSORS AND VARIOUS DERIVATIVES OF THE
0009     C      KINEMATIC YIELD FUNCTION
0010     C
0011     C
0012     C**************************************************************************
0013     C
0014     C
0015          COMMON/PLASTA/D2(378),EFSLOP(378),EPSPL(378,4)
0016          COMMON/PLASTB/AVDEVI(378),TRDEVI(378,4)
0017          COMMON/PLASTC/TRLSIG(378,4),PYS(378)
0018          COMMON/DSTRES/RATS2(378),DSIG(378,4),DIVEPS(378,4)
0019          COMMON/STREZZ/RATS1(378),TSIG(378,5),TOTEPS(378,4),EPS(378),
0020         *       DLAM(378),RATST(378),DEPS(378,4)
0021          COMMON/ALFA/ALF(378,4),EFFALF(378)
0022          COMMON/PLANE/NPP
0023          DIMENSION AVALF(438),WSIG(4)
0024     C
0025     C**** INITIALIZATION
0026     C
0027          N=MM
0028          DLAM(N)=.0
0029          D2(N)=10.E-6
0030          T1=.0
0031          D1=10.E-6
0032          DMU=.0
0033          AVDEVI(N)=.0
0034          AVALF(N)=(1./3.)*(ALF(N,1)+ALF(N,2)+ALF(N,3))
0035          IF (NPP.EQ.1)AVALF(N)=(ALF(N,1)+ALF(N,2))/3.
```

```
0036   C
0037   C**** DMU AND DLAM CORRECTION FACTOR (FOR RATSP<1.0 AND RATST>1.0)
0038   C
0039         GAB=1.0
0040         IF (RATS.LT.1.0 .AND. RATST(N).GE.1.0)   GAB=(RATST(N)-1.0)/
0041       *      (RATST(N)-RATS)
0042   C
0043   C**** EVALUATE DEVIATORS, TENSORS, AND DERIVATIVES
0044   C
0045         DO 500 I=1,4
0046         TRLSIG(N,I)=TSIG(N,I)-ALF(N,I)
0047   500 AVDEVI(N)=TRLSIG(N,I)+AVDEVI(N)
0048         AVDEVI(N)=(AVDEVI(N)-TRLSIG(N,4))/3.0
0049         DO 510 J=1,3
0050         IF (NPP.EQ.1.AND.J.EQ.3) GO TO 510
0051         TRDEVI(N,J)=TRLSIG(N,J)-AVDEVI(N)
0052         D2(N)=D2(N)+TRDEVI(N,J)**2
0053   510 CONTINUE
0054         TRDEVI(N,4)=TRLSIG(N,4)
0055         D2(N)=D2(N)+2.*TRDEVI(N,4)**2
0056         T1=TRDEVI(N,1)*DSIG(N,1)+TRDEVI(N,2)*DSIG(N,2)+TRDEVI(N,3)*
0057       *      DSIG(N,3)+2.*TRDEVI(N,4)*DSIG(N,4)
0058         D1=TRDEVI(N,1)*TRLSIG(N,1)+TRDEVI(N,2)*TRLSIG(N,2)+TRDEVI(N,3)*
0059       *      TRLSIG(N,3)+2.*TRDEVI(N,4)*TRLSIG(N,4)
0060         IF (D2(N).LT.10.E-5) GO TO 5
0061         DMU=(6.*T1+DYDT+DYDE)/(6.*D1)
0062         IF (ABS(D1).LE.10.E-6) DMU=0.0
0063         IF (EFSLOP(N).LT.10.E-6) EFSLOP(N)=10.E-6
0064         DLAM(N)=(1.0/EFSLOP(N))*(6.*T1+DYDT+DYDE)/(36.*D2(N))
0065     5 IF (RATST(N).LE.1.0 .OR. DLAM(N).LT.0.0) GO TO 10
0066         DO 520 I=1,4
0067         IF (NPP.EQ.1.AND.I.EQ.3) GO TO 520
0068         DEPSPL=6.*DLAM(N)*TRDEVI(N,I)*GAB
0069         EPSPL(N,I)=EPSPL(N,I)+DEPSPL
0070         DALF=DMU*(TSIG(N,I)-ALF(N,I))*GAB
0071         IF (RATST(N).LT.1.0 .AND. DLAM(N).LT.0.0) DALF=0.0
0072         ALF(N,I)=ALF(N,I)+DALF
0073         WSIG(I)=TSIG(N,I)-ALF(N,I)
0074   520 CONTINUE
0075         EFF=SQRT(0.5*((ALF(N,1)-ALF(N,2))**2+(ALF(N,2)-ALF(N,3))**2
0076       1+(ALF(N,3)-ALF(N,1))**2)+3.*ALF(N,4)**2)
0077         EFFALF(N)=EFF
0078         EFSIG=SQRT(.5*((WSIG(1)-WSIG(2))**2+(WSIG(2)-WSIG(3))**2+
0079       1(WSIG(3)-WSIG(1))**2)+3.*WSIG(4)**2)
0080         RATK=EFSIG/E13
0081         IF (GAB.LT.1.) PYS(N)=RATK
0082         IF (RATK.GE.PYS(N)) RATK=1.0
0083         RATST(N)=RATK
0084    10 RETURN
0085         END


0001         SUBROUTINE KNCONS(N)
0002   C                (CALLED FROM "QUAD" - STRESS ANALYSIS - KINEMATIC HARDENING)
0003   C                (SEE 3-4-1)
0004   C***********************************************************************
0005   C
0006   C
0007   C     THIS SUBROUTINE EVALUATES THE DFDTDT AND DFDEDE TERMS IN THE
0008   C     CONSTITUTIVE EQUATION FOR KINEMATIC HARDENING ANALYSIS.
0009   C
0010   C
0011   C***********************************************************************
0012   C
0013   C
0014         COMMON/STRAIN/EPSDI(378),EPSDOT(378),SIGAVE(378),DEVSIG(378,5),
0015       *      EPSDTC(378),EPSDTP(378)
0016         COMMON/STREZZ/RATS1(378),TSIG(378,5),TOTEPS(378,4),EPS(378),
0017       *      DLAM(378),RATST(378),DEPS(378,4)
```

```
0018          COMMON/PRPRTE/PROP1(5,11,5),PROP2(5,11,5),PROP3(5,11,5),
0019     *        PROP4(5,11,5),PROP5(5,11,5),PROP6(5,11,5),PROPS(5,11,5)
0020          COMMON/TEMP2/PROP(9),IPTRT,IPTRE,TEMD,RATIOT,RATIOE,
0021     *        TMPRNG,EPSRNG
0022          COMMON/TEMP3/EBR,EBRPRM,H1,DFDTDT,DFDEDE,hpr
0023          COMMON/HEATR/XA,DTI,TIME,TIMECR
0024          DYD1=0.0
0025          DYD2=0.0
0026          IF (IPTRT.EQ.1 .AND. IPTRE.EQ.1) GO TO 5
0027          IF (IPTRE.EQ.1) GO TO 10
0028          IF (IPTRT.EQ.1) GO TO 15
0029          DYDT2=PROPS(IPTRT,6,IPTRE)-PROPS(IPTRT-1,6,IPTRE)
0030          DTDT1=PROPS(IPTRT,6,IPTRE-1)-PROPS(IPTRT-1,6,IPTRE-1)
0031          DYDE1=PROPS(IPTRT-1,6,IPTRE)-PROPS(IPTRT-1,6,IPTRE-1)
0032          DYDE2=PROPS(IPTRT,6,IPTRE)-PROPS(IPTRT,6,IPTRE-1)
0033          DYDT1=DYDT1/TMPRNG
0034          DYDT2=DYDT2/TMPRNG
0035          DYDE1=DYDE1/EPSRNG
0036          DYDE2=DYDE2/EPSRNG
0037          DYD1=DYDT1+RATIOT*(DYDT2-DYDT1)
0038          DYD2=DYDE1+RATIOE*(DYDE2-DYDE1)
0039          IF (DYD1.GE.0.0) DYD1=0.0
0040          IF (DYD2.GE.0.0) DYD2=0.0
0041          GO TO 5
0042    10 DYDT=PROPS(IPTRT,6,IPTRE)-PROPS(IPTRT-1,6,IPTRE)
0043          DYD1=(DYDT/TMPRNG)*RATIOT
0044          IF (DYD1.GE.0.0) DYD1=0.0
0045          GO TO 5
0046    15 DYDE=PROPS(IPTRT,6,IPTRE)-PROPS(IPTRT,6,IPTRE-1)
0047          DYD2=(DYDE/EPSRNG)*RATIOE
0048          IF (DYD2.GE.0.0) DYD2=0.0
0049     5 DFDTDT=-2./3.*PROP(9)*DYD1*TEMD
0050          DFDEDE=-2./3.*PROP(9)*DYD2*EPSDI(N)
0051          RETURN
0052          END


0001          SUBROUTINE KNQUAD(N)
0002    C                (CALLED FROM "QUAD" - STRESS ANALYSIS - KINEMATIC HARDENING)
0003    C                (SEE 3-13)
0004    C****************************************************************************
0005    C
0006    C
0007    C     THIS SUBROUTINE IS THE EQUIVALENT QUAD SUBROUTINE FOR
0008    C     KINEMATIC HARDENING ANALYSIS
0009    C
0010    C
0011    C****************************************************************************
0012    C
0013    C
0014          COMMON/MECH1/NMSTRC,Q,QT(378)
0015          COMMON/STRAIN/EPSDI(378),EPSDOT(378),SIGAVE(378),DEVSIG(378,5),
0016     *        EPSDTC(378),EPSDTP(378)
0017          COMMON/DISP1/UR(378),UZ(378),CODE(378)
0018          COMMON/SYSMSH/NUMEL,NUMNP,NRMAX
0019          COMMON/COOR/R(378),Z(378),T(378),IX(378,5),TK(378)
0020          COMMON/TIMSTP/ITIME
0021          COMMON/MATRIX/C(4,4),EE(20)
0022          COMMON/RRZZ/RR(4),ZZ(4),XI(6)
0023          COMMON/DUM/RRR(5),ZZZ(5),LM(4)
0024          COMMON/TRIPLE/TT(4),TP(6),P(10),S(10,10)
0025          COMMON/DBLH/HH(6,10)
0026          COMMON/STREZZ/RATS1(378),TSIG(378,5),TOTEPS(378,4),EPS(378),
0027     *        DLAM(378),ratst(378),deps(378,4)
0028          COMMON/PLANE/NPP
0029          COMMON/TEMP1/NUMMAT,NUMTC,DENSTY(6)
0030          COMMON/TEMP2/PROP(9),IPTRT,IPTRE,TEMD,RATIOT,RATIOE,
0031     *        TMPRNG,EPSRNG
```

350

```
0032          COMMON/TEMP3/EBR,EBRPRM,H1,DFDTDT,DFDEDE,hpr
0033          COMMON/TEMP4/CENEW(4,4),CEI(4,4),CEIPR(378,4,4),
0034        *        DCEIDE(4,4),DCEIDT(4,4),SIG(4),DCESIG(4),DCTSIG(4)
0035          COMMON/TEMP5/CEPNEW(4,4),TOTTRM(4),SZERO
0036          COMMON/PLASTB/AVDEVI(378),TRDEVI(378,4)
0037          COMMON/ANEW/SPL(378),DLAMP(378),TOTEP(378,4)
0038          COMMON/MECH2/TSIP(378),AEPS(378),EEPS(378),ANP(378),ANG(378),
0039        *        STP(378),EFFCLF(378)
0040          COMMON/MECH3/BLF(378,4),BEPS(378,4),CLF(378,4),CEPS(378,4),NEU
0041          COMMON/KINH/IRE(378),ICK(378),LCY(378),EFFBLF(378),RATSP(378)
0042          COMMON/CHANGE/XEPS(5),YSIG(5),IUN
0043          COMMON/ALFA/ALF(378,4),EFFALF(378)
0044          ISTORE=0
0045          IF (DLAM(N).GE.0.0) GO TO 5
0046          IUN=1
0047          IRE(N)=0
0048          IF (DLAMP(N).LT.0.0) GO TO 10
0049          ISTORE=1
0050          GO TO 10
0051        5 IUN=0
0052          IF (DLAMP(N).GE.0.0) GO TO 10
0053          ICK(N)=1
0054          LCY(N)=LCY(N)+1
0055          ANP(N)=ANG(N)
0056          EFFBLF(N)=EFFCLF(N)
0057          AEPS(N)=EEPS(N)
0058          DO 500 I=1,4
0059          BLF(N,I)=CLF(N,I)
0060      500 BEPS(N,I)=CEPS(N,I)
0061       10 CONTINUE
0062          IF (ICK(N).NE.1) GO TO 6211
0063          ST=TSIG(N,105)/ABS(TSIG(N,105))
0064          IF (NEU-1) 6216,6217,6216
0065     6217 IF (Q.LT.75.0 .AND. Q.GE.65.0) ST=0.0
0066     6216 IF (ST*STP(N)) 6212,6211,6213
0067     6213 IF (ABS(TSIG(N,105))-ABS(TSIP(N))) 6211,6211,6214
0068     6214 IRE(N)=-1
0069          ANG(N)=0.0+ANP(N)
0070          GO TO 6215
0071     6212 IRE(N)=1
0072          ANG(N)=3.1416+ANP(N)
0073     6215 ICK(N)=0
0074          STP(N)=0.0
0075     6211 CONTINUE
0076          IF (NEU.GT.0 .AND. N.EQ.NUMEL) NEU=NEU-1
0077          IF (LCY(N).NE.1) GO TO 15
0078          YSIG(5)=DEVSIG(N,5)
0079          XEPS(5)=EPS(N)
0080          DO 510 I=1,4
0081          YSIG(I)=TSIG(N,I)
0082      510 XEPS(I)=TOTEPS(N,I)
0083          GO TO 20
0084       15 IQ=1
0085          IF (IRE(N).EQ.1 .AND. ICK(N).EQ.0) IQ=-1
0086          YSIG(5)=ABS(DEVSIG(N,5)-EFFBLF(N)*IQ)
0087          IZ=1
0088          IF (IRE(N).EQ.1 .AND. TSIG(N,105).LE.0.0 .AND.
0089        *     TOTEPS(N,105).LE.0.0) IZ=-1
0090          IF (IRE(N).EQ.1 .AND. TSIG(N,105).GT.0.0 .AND.
0091        *     TOTEPS(N,105).GT.0.0) IZ=-1
0092          XEPS(5)=ABS(EPS(N)-AEPS(N)*IZ)
0093          DO 520 I=1,4
0094          YSIG(I)=COS(ANG(N))*(TSIG(N,I)-BLF(N,I))
0095      520 XEPS(I)=COS(ANG(N))*(TOTEPS(N,I)-BEPS(N,I))
0096       20 IF (ISTORE.EQ.0) GO TO 25
0097          STP(N)=TSIG(N,105)/ABS(TSIG(N,105))
0098          EFFCLF(N)=EFFALF(N)
0099          DO 530 I=1,4
0100          CLF(N,I)=ALF(N,I)
0101          AEW=(ALF(N,I)-BLF(N,I))*COS(ANG(N))
0102          EP=XEPS(I)-(YSIG(I)-AEW)/prop(I)
```

351

```
0103          CEPS(N,I)=BEPS(N,I)+EP/COS(ANG(N))
0104      530 TOTEP(N,I)=AEW/prop(1)
0105          EEP=XEPS(5)-(DEVSIG(N,5)-EFFCLF(N)*COS(ANG(N)))*2.*
0106        *    (1.+PROP(2))/3./PROP(1)
0107          EEPS(N)=ABS(AEPS(N)+EEP*COS(ANG(N)))
0108       25 RETURN
0109          END
```

```
0001          SUBROUTINE MATDEL(EPSON,EE1,EE2,EE15,EE16,EE17,BBB)
0002      C             (CALLED FROM "STRESS" - FOR STRESS ANALYSIS)
0003      C             (SEE 3-5-4)
0004      C***********************************************************************
0005      C
0006      C
0007      C      THIS SUBROUTINE EVALUATES STRESS FROM THE POLYNOMIAL FUNCTION
0008      C      USING THE EFFECTIVE STRAIN.
0009      C
0010      C
0011      C***********************************************************************
0012      C
0013      C
0014          EM1=1.5*EE1/(1.+EE2)
0015          EM2=3.*EE15/(3.-(1.-2.*EE2)*EE15/EE1)
0016          H1=EM2/EM1
0017          AA1=(1.-EE15/EE1)*EE17+EE15*EPSON
0018          AA=(EE1*EPSON/AA1)+1.0E-10
0019          AAA=EE16*ALOG10(AA)
0020          IF (AAA.GT.30.) GO TO 5
0021          IF (AAA.LT.-30.) GO TO 10
0022          AA=AA**EE16
0023          BBB=EE1*EPSON/(1+AA)**(1./EE16)
0024          GO TO 15
0025        5 BBB=AA1
0026          GO TO 15
0027       10 BBB=EE1*EPSON
0028       15 RETURN
0029          END
```

```
0001          SUBROUTINE MODIFY(A,B,NEQ,MBAND,N,U)
0002      C             (CALLED FROM "STIFF" - FOR STRESS ANALYSIS)
0003      C             (SEE 3-16-(10) AND (11))
0004      C***********************************************************************
0005      C
0006      C
0007      C      THIS SUBROUTINE ALTERS THE STIFFNESS MATRIX (A) AND LOAD
0008      C      VECTOR (B) WHERE APPROPRIATE, FOR EITHER PRESCRIBED CONSTANT
0009      C      INCREMENT NODAL DISPLACEMENTS AND/OR SPECIFIED BOUNDARY
0010      C      CONDITIONS.
0011      C
0012      C
0013      C***********************************************************************
0014      C
0015      C
0016          DIMENSION A(197,54),B(108)
0017          DO 500 M=2,MBAND
0018          K=N-M+1
0019          IF (K) 5,5,10
0020       10 B(K)=B(K)-A(K,M)*U
0021          A(K,M)=0.0
0022        5 K=N+M-1
0023          IF (NEQ-K) 500,15,15
0024       15 B(K)=B(K)-A(N,M)*U
0025          A(N,M)=0.0
0026      500 CONTINUE
0027          A(N,1)=1.0
0028          B(N)=U
0029          RETURN
0030          END
```

```
0001          SUBROUTINE NTRPLT(NUMEL)
0002    C          (CALLED FROM "QUAD" - FOR STRESS ANALYSIS)
0003    C          (SEE 3-16-(3))
0004    C************************************************************************
0005    C
0006    C
0007    C     THIS SUBROUTINE DETERMINES MATERIAL PROPERTIES, BASED ON THE
0008    C     CURRENT STRAIN RATE AND TEMPERATURE VALUE OF THE  ELEMENT UNDER
0009    C     CONSIDERATION, BY INTERPOLATING BETWEEN USER-SPECIFIED STRAIN
0010    C     RATE AND TEMPERATURE DEPENDENT MATERIAL PROPERTIES.
0011    C
0012    C
0013    C************************************************************************
0014    C
0015    C
0016          COMMON/COOR/R(378),Z(378),T(378),IX(378,5),TK(378)
0017          COMMON/MECH1/NMSTRC,Q,QT(378)
0018          COMMON/STRAIN/EPSDI(378),EPSDOT(378),SIGAVE(378),DEVSIG(378,5),
0019         *      EPSDTC(378),EPSDTP(378)
0020          COMMON/TIMSTP/ITIME
0021          COMMON/PRPRTE/PROP1(5,11,5),PROP2(5,11,5),PROP3(5,11,5),
0022         *      PROP4(5,11,5),PROP5(5,11,5),PROP6(5,11,5),PROPS(5,11,5)
0023          COMMON/TEMP1/NUMMAT,NUMTC,DENSTY(6)
0024          COMMON/TEMP2/PROP(9),IPTRT,IPTRE,TEMD,RATIOT,RATIOE,
0025         *      TMPRNG,EPSRNG
0026          COMMON/HEATR/XA,DTI,TIME,TIMECR
0027          DIMENSION PRPTEL(9),PRPTEH(9)
0028    C
0029    C**** FIND CORRECT MATERIAL SET FOR THIS ELEMENT
0030    C
0031          MTYPE=IX(NUMEL,5)
0032          DO 500 I=1,NUMTC
0033          DO 500 J=1,11
0034          DO 500 K=1,NMSTRC
0035          GO TO (5,10,15,20,25,30),MTYPE
0036        5 PROPS(I,J,K)=PROP1(I,J,K)
0037          GO TO 500
0038       10 PROPS(I,J,K)=PROP2(I,J,K)
0039          GO TO 500
0040       15 PROPS(I,J,K)=PROP3(I,J,K)
0041          GO TO 500
0042       20 PROPS(I,J,K)=PROP4(I,J,K)
0043          GO TO 500
0044       25 PROPS(I,J,K)=PROP5(I,J,K)
0045          GO TO 500
0046       30 PROPS(I,J,K)=PROP6(I,J,K)
0047      500 CONTINUE
0048    C
0049    C**** INITIALIZE POINTER TO FIRST STRAIN RATE
0050    C     LOOP TO FIND UPPER AND LOWER STRAIN RATE BOUNDS
0051    C
0052          M=1
0053          L=1
0054          IPTRT=1
0055          IPTRE=1
0056          TEMD=0.0
0057          TMPRNG=0.0
0058          EPSRNG=0.0
0059          RATIOT=0.0
0060          RATIOE=0.0
0061          I=IX(NUMEL,1)
0062          J=IX(NUMEL,2)
0063          K=IX(NUMEL,3)
0064          LL=IX(NUMEL,4)
0065          IF (LL-K) 35,40,35
0066       35 TEMP=(T(I)+T(J)+T(K)+T(LL))/4.0
0067          GO TO 45
0068       40 TEMP=(T(I)+T(J)+T(K))/3.0
0069       45 IF (ITIME.LE.1) GO TO 50
0070          Q=QT(NUMEL)
0071       50 TEMD=TEMP-Q
```

353

```
0072          QT (NUMEL) =TEMP
0073          IF (NMSTRC.EQ.1) GO TO 55
0074          IF (NUMTC.EQ.1) GO TO 60
0075      55 IF (NUMTC.EQ.1) GO TO 65
0076          M=2
0077      75 IF (PROPS (M,1,1) -TEMP) 70,60,60
0078      70 M=M+1
0079          IF (M.GT.5) WRITE (6,2000)
0080          IF (M.GT.50) CALL EXIT
0081          GO TO 75
0082      60 IPTRT=M
0083          IF (NMSTRC.EQ.1) GO TO 80
0084          L=2
0085      95 IF (PROPS (M,11,L) -ABS (EPSDOT (NUMEL))) 85,90,90
0086      85 L=L+1
0087          IF (L.GT.5) WRITE (6,2001)
0088          IF (L.GT.50) CALL EXIT
0089          GO TO 95
0090   C
0091   C**** EVALUATE
0092   C
0093      90 IPTRE=L
0094          IF (NUMTC.EQ.1) GO TO 100
0095          TMPRNG=PROPS (M,1,L) -PROPS (M-1,1,L)
0096          RATIOT= (TEMP-PROPS (M-1,1,L)) /TMPRNG
0097          DO 510 I=1,9
0098          PRPTEL (I) =PROPS (M-1,I+1,L-1) +RATIOT* (PROPS (M,I+1,L-1) -
0099      *              PROPS (M-1,I+1,L-1))
0100     510 PRPTEH (I) =PROPS (M-1,I+1,L) +RATIOT* (PROPS (M,I+1,L) -
0101      *              PROPS (M-1,I+1,L))
0102          EPSRNG=PROPS (M,11,L) -PROPS (M,11,L-1)
0103          RATIOE= (ABS (EPSDOT (NUMEL) -PROPS (M,11,L-1))) /EPSRNG
0104          DO 520 I=1,9
0105     520 PROP (I) =PRPTEL (I) +RATIOE* (PRPTEH (I) -PRPTEL (I))
0106          GO TO 105
0107      80 TMPRNG=PROPS (M,1,L) -PROPS (M-1,1,L)
0108          RATIOT= (TEMP-PROPS (M-1,1,L)) /TMPRNG
0109          DO 530 I=1,9
0110     530 PROP (I) =PROPS (M-1,I+1,1) +RATIOT* (PROPS (M,I+1,1) -
0111      *              PROPS (M-1,I+1,1))
0112          GO TO 105
0113     100 EPSRNG=PROPS (M,11,L) -PROPS (M,11,L-1)
0114          RATIOE= (ABS (EPSDOT (NUMEL) -PROPS (M,11,L-1))) /EPSRNG
0115          DO 540 I=1,9
0116     540 PROP (I) =PROPS (M,I+1,L-1) +RATIOE* (PROPS (M,I+1,L) -PROPS (M,I+1,L-1))
0117          GO TO 105
0118      65 DO 550 I=1,9
0119     550 PROP (I) =PROPS (M,I+1,L)
0120     105 RETURN
0121   C
0122   C**** OUTPUT FORMATS FOR ERROR MESSAGES
0123   C
0124    2000 FORMAT(' EXCEEDS THE MAX. # (5) OF TEMPERATURE DEPENDENT PROPS.
0125      1    CHECK NUMTC VALUE')
0126    2001 FORMAT(' EXCEEDS THE MAX. # (5) OF STRAIN RATE DEPENDENT PROPS.
0127      1    CHECK NMSTRC VALUE')
0128          END
```

```
0001          SUBROUTINE PRPRTY
0002    C                   (CALLED FROM "MAIN" - FOR STRESS ANALYSIS)
0003    C                   (SEE USER MANUAL)
0004    C******************************************************************
0005    C
0006    C
0007    C       THIS SUBROUTINE READS AND WRITES THE USER-SPECIFIED STRAIN RATE
0008    C       AND TEMPERATURE DEPENDENT MATERIAL PROPERTIES TO BE USED FOR
0009    C       THE ANALYSIS.  IT IS POSSIBLE TO SPECIFY PROPERTIES FOR UP
0010    C       SIX MATERIALS, AND UP TO FIVE DIFFERENT STRAIN RATES AND
0011    C       TEMPERATURES FOR EACH MATERIAL.
0012    C
0013    C
0014    C******************************************************************
0015    C
0016    C
0017          COMMON/COOR/R(378),Z(378),T(378),IX(378,5),TK(378)
0018          COMMON/MECH1/NMSTRC,Q,QT(378)
0019          COMMON/PRPRTE/PROP1(5,11,5),PROP2(5,11,5),PROP3(5,11,5),
0020         *       PROP4(5,11,5),PROP5(5,11,1),PROP6(5,11,5),PROPS(5,11,5)
0021          COMMON/TEMP1/NUMMAT,NUMTC,DENSTY(6)
0022          DIMENSION RER(5),WRTPRP(5,11)
0023    C
0024    C**** BEGIN LOOP FOR EACH MATERIAL SPECIFIED
0025    C
0026          DO 500 M=1,NUMMAT
0027          READ (5,1000) MTYPE,NUMTC,NMSTRC,DENSTY(MTYPE)
0028          WRITE (6,2000) MTYPE,NUMTC,DENSTY(MTYPE),NMSTRC
0029          IF (NUMTC-5) 5,5,10
0030       10 WRITE (6,2001)
0031          CALL EXIT
0032        5 IF (NMSTRC-5) 15,15,20
0033       20 WRITE (6,2002)
0034          CALL EXIT
0035    C
0036    C**** READ IN 11 PROPERTIES
0037    C
0038       15 DO 510 L=1,NMSTRC
0039          DO 520 K=1,11
0040          IF (L.GT.1.AND.K.EQ.1) GO TO 25
0041          IF (L.GT.1.AND.K.EQ.5) GO TO 25
0042          READ (5,1001) (RER(I),I=1,NUMTC)
0043       25 DO 530 I=1,NUMTC
0044          IF (L.GT.1.AND.K.EQ.1) PROPS(I,K,L)=PROPS(I,K,1)
0045          IF (L.GT.1.AND.K.EQ.5) PROPS(I,K,L)=PROPS(I,K,1)
0046          IF (L.GT.1.AND.K.EQ.1) GO TO 530
0047          IF (L.GT.1.AND.K.EQ.5) GO TO 530
0048          PROPS(I,K,L)=RER(I)
0049      530 CONTINUE
0050      520 CONTINUE
0051      510 CONTINUE
0052    C
0053    C**** WRITE MATERIAL PROPERTIES
0054    C
0055          DO 540 I=NUMTC,5
0056          DO 540 L=1,NMSTRC
0057          DO 540 J=1,11
0058      540 PROPS(I,J,L)=PROPS(NUMTC,J,L)
0059          DO 550 L=1,NMSTRC
0060      550 WRITE (6,2003) ((PROPS(I,J,L),I=1,5),J=1,11)
0061    C
0062    C**** PLACE IN APPROPRIATE PROPERTY MATRIX
0063    C
0064          DO 560 I=1,NUMTC
0065          DO 560 J=1,11
0066          DO 560 L=1,NMSTRC
0067          GO TO (30,35,40,45,50,55),MTYPE
0068       30 PROP1(I,J,L)=PROPS(I,J,L)
0069          GO TO 560
0070       35 PROP2(I,J,L)=PROPS(I,J,L)
```

355

```
0071            GO TO 560
0072         40 PROP3(I,J,L)=PROPS(I,J,L)
0073            GO TO 560
0074         45 PROP4(I,J,L)=PROPS(I,J,L)
0075            GO TO 560
0076         50 PROP5(I,J,L)=PROPS(I,J,L)
0077            GO TO 560
0078         55 PROP6(I,J,L)=PROPS(I,J,L)
0079        560 CONTINUE
0080        500 CONTINUE
0081            RETURN
0082      C
0083      C**** INPUT FORMATS
0084      C
0085       1000 FORMAT(3I5,E12.5)
0086       1001 FORMAT(5E10.0)
0087      C
0088      C**** OUTPUT FORMATS
0089      C
0090       2000 FORMAT(1H1,' MATERIAL NO. = ',I2,',',' NO. OF TEMPERATURE',
0091          * ' INPUTS = ',I2,',',' MASS DENSITY = ',F7.4,',',' NO. OF',
0092          * ' STRAIN RATES = ',I2,////)
0093       2001 FORMAT(' MAX # (5) OF TEMP. DEP. PROPS. EXCEEDED --------------'
0094          * ,'-- CHECK NUMTC VALUE')
0095       2002 FORMAT(' MAX # (5) OF STRAIN RATE DEP. PROPS. EXCEEDED --------'
0096          * ,'-- CHECK NMSTRC VALUE')
0097       2003 FORMAT(1H0,'TEMPERATURE--------------------',5E15.5/'0',
0098          *'YOUNGS MODULUS (E) -----------',5E15.5/'0','POISSON RATIO (NU) --
0099          *----------',5E15.5/'0','SHEAR MODULUS (G) -----------',5E15.5/'0',
0100          *         'THERM. EXP. COEFF. (ALPHA)-----',5E15.5/'0',
0101          *         'YIELD STRENGTH (SIGYLD)--------',5E15.5/'0',
0102          *         'RUPTURE STRENGTH (SIGRUP)------',5E15.5/'0',
0103          *         'PLASTICITY MODULUS (EPRIME)----',5E15.5/'0',
0104          *         'STRESS-STRAIN POWER (EN)-------',5E15.5/'0',
0105          *         'BI-LI STRS/STRN KINK (SIGKNK)--',5E15.5/'0',
0106          *         'STRAIN RATE (EPSDOT)-----------',5E15.5/'0')
0107            END
```

```
0001            SUBROUTINE QUAD(N,JOBTYP,VOL,KINHAR)
0002      C             (CALLED FROM "STIFF" - FOR STRESS ANALYSIS)
0003      C             (SEE 3-10 TO 3-15 : 3-16-(3) TO (8))
0004      C*****************************************************************************
0005      C
0006      C
0007      C     THIS SUBROUTINE LINKS TOGETHER A NUMBER OF SUBROUTINES, AND
0008      C     IN SO DOING EVALUATES THE CONSTITUTIVE EQUATION, THE STIFFNESS
0009      C     MATRIX, THE STRAIN-DISPLACEMENT AND STRESS-STRAIN TRANSFORMATION
0010      C     MATRIX FOR EACH ELEMENT IN THE STRUCTURE.  PERTINENT VARIABLES
0011      C     FROM THIS SEQUENCE OF CALLS ARE STORED ON DISK FOR LATER
0012      C     PROCESSING IN STRESS SUBROUTINE.
0013      C
0014      C
0015      C*****************************************************************************
0016      C
0017      C
0018            COMMON/MECH1/NMSTRC,Q,QT(378)
0019            COMMON/STRAIN/EPSDI(378),EPSDOT(378),SIGAVE(378),DEVSIG(378,5),
0020          *      EPSDTC(378),EPSDTP(378)
0021            COMMON/DISP1/UR(378),UZ(378),CODE(378)
0022            COMMON/SYSMSH/NUMEL,NUMNP,NRMAX
0023            COMMON/COOR/R(378),Z(378),T(378),IX(378,5),TK(378)
0024            COMMON/TIMSTP/ITIME
0025            COMMON/MATRIX/C(4,4),EE(20)
0026            COMMON/RRZZ/RR(4),ZZ(4),XI(6)
0027            COMMON/DUM/RRR(5),ZZZ(5),LM(4)
0028            COMMON/TRIPLE/TT(4),TP(6),P(10),S(10,10)
0029            COMMON/DBLH/HH(6,10)
```

356

```
0030          COMMON/STREZZ/RATS1(378),TSIG(378,5),TOTEPS(378,4),EPS(378),
0031        *      DLAM(378),RATST(378),DEPS(378,4)
0032          COMMON/PLANE/NPP
0033          COMMON/TEMP1/NUMMAT,NUMTC,DENSTY(6)
0034          COMMON/TEMP2/PROP(9),IPTRT,IPTRE,TEMD,RATIOT,RATIOE,
0035        *      TMPRNG,EPSRNG
0036          COMMON/TEMP3/EBR,EBRPRM,H1,DFDTDT,DFDEDE,hpr
0037          COMMON/TEMP4/CENEW(4,4),CEI(4,4),CEIPR(378,4,4),
0038        *      DCEIDE(4,4),DCEIDT(4,4),SIG(4),DCESIG(4),DCTSIG(4)
0039          COMMON/TEMP5/CEPNEW(4,4),TOTTRM(4),SZERO
0040          COMMON/PLASTB/AVDEVI(378),TRDEVI(378,4)
0041          COMMON/ANEW/SPL(378),DLAMP(378),TOTEP(378,4)
0042          COMMON/MECH2/TSIP(378),AEPS(378),EEPS(378),ANP(378),ANG(378),
0043        *      STP(378),EFFCLF(378)
0044          COMMON/MECH3/BLF(378,4),BEPS(378,4),CLF(378,4),CEPS(378,4),NEW
0045          COMMON/KINH/IRE(378),ICK(378),LCY(378),EFFBLF(378),RATSP(378)
0046          COMMON/CHANGE/XEPS(5),YSIG(5),IUN
0047          COMMON/ALFA/ALF(378,4),EFFALF(378)
0048          COMMON/PASMAT/CECE(378,4,4),DCESGN(378,4),DCTSGN(378,4)
0049      C
0050      C**** INITIALIZE
0051      C
0052          HPR=10.0E+15
0053          SZERO=10.0E+15
0054          SPL(N)=10.0E+20
0055          SLOPLA=10.0E+30
0056          IF (ITIME.GT.1) GO TO 5
0057          AEPS(N)=0.0
0058          ANP(N)=0.0
0059          ANG(N)=0.0
0060          DLAMP(N)=0.0
0061          ICK(N)=0
0062          EPSDI(N)=1.0E-20
0063          DO 500 I=1,4
0064          BEPS(N,I)=0.0
0065      500 BLF(N,I)=0.0
0066          QT(N)=0.0
0067        5 DO 510 III=1,4
0068          DO 510 JJJ=1,4
0069          C(III,JJJ)=0.0
0070          CENEW(III,JJJ)=0.0
0071          CEPNEW(III,JJJ)=0.0
0072      510 CEI(III,JJJ)=0.0
0073      C
0074      C**** DETERMINE MATERIAL PROPERTIES
0075      C
0076          MTYPE=IX(N,5)
0077          CALL NTRPLT(N)
0078          IF (KINHAR.EQ.1) CALL KNQUAD(N)
0079      C
0080      C**** EVALUATE DFDEDE AND DFDTDT TERMS OF THE CONSTITUTIVE EQUATION
0081      C
0082          IF (KINHAR.EQ.0) CALL CONSET(N)
0083          IF (KINHAR.EQ.1) CALL KNCONS(N)
0084      C
0085      C**** FORM CE MATRIX
0086      C
0087          CALL CSUBE(N,NPP,ITIME,KINHAR)
0088          DO 520 II=1,4
0089          DO 520 JJ=1,4
0090          CECE(N,II,JJ)=CENEW(II,JJ)
0091      520 CEPNEW(II,JJ)=CENEW(II,JJ)
0092          IF (RATST(N).LT..05 .AND. ITIME.LE.1) GO TO 10
0093          IF (KINHAR.EQ.1) CALL KINSLO(N,HPR)
0094          IF (KINHAR.EQ.1) GO TO 10
0095      C
0096      C**** IF DLAM LESS THAN 0.0, ELASTIC UNLOADING - SKIP PLASTIC ANALYSIS.
0097      C      (ISOTROPIC HARDENING ONLY)
0098      C
0099          IF (DLAM(N).LT.0.0) GO TO 10
0100      C
```

357

APPENDIX 5

```
0101    C**** CALCULATE CEP MATRIX (ISOTROPIC HARDENING ONLY)
0102    C
0103          CALL CSUBEP (N,NPP,JOBTYP)
0104    C
0105    C**** EVALUATE TERMS OF CONSTITUTIVE EQUATION
0106    C
0107       10 CALL CONTOT (N,NPP,KINHAR)
0108          I=IX (N,1)
0109          J=IX (N,2)
0110          K=IX (N,3)
0111          L=IX (N,4)
0112          RRR (5) = (R (I) +R (J) +R (K) +R (L)) /4.0
0113          ZZZ (5) = (Z (I) +Z (J) +Z (K) +Z (L)) /4.0
0114          DO 530 M=1,4
0115          MM=IX (N,M)
0116          RRR (M) =R (MM)
0117      530 ZZZ (M) =Z (MM)
0118    C
0119    C**** INITIALIZE ELEMENT DISPLACEMENT TRANSFORMATION AND
0120    C     STIFFNESS MATRIX
0121    C
0122          DO 540 II=1,10
0123          P (II)=0.0
0124          DO 550 JJ=1,6
0125      550 HH (JJ,II)=0.0
0126          DO 540 JJ=1,10
0127      540 S (II,JJ)=0.0
0128          IF (K-L) 15,20,15
0129    C
0130    C**** EVALUATE STIFFNESS AND DISPLACEMENT FOR TRIANGLE ELEMENT
0131    C
0132       20 CALL TRISTF (1,2,3,N,MTYPE,JOBTYP)
0133          RRR (5) = (RRR (1) +RRR (2) +RRR (3)) /3.0
0134          ZZZ (5) = (ZZZ (1) +ZZZ (2) +ZZZ (3)) /3.0
0135          VOL=XI (1)
0136          GO TO 25
0137       15 VOL=0.0
0138    C
0139    C**** EVALUATE STIFFNESS AND DISPLACEMENT FOR QUADRILATERAL ELEMENT
0140    C
0141          CALL TRISTF (4,1,5,N,MTYPE,JOBTYP)
0142          VOL=VOL+XI (1)
0143          CALL TRISTF (1,2,5,N,MTYPE,JOBTYP)
0144          VOL=VOL+XI (1)
0145          CALL TRISTF (2,3,5,N,MTYPE,JOBTYP)
0146          VOL=VOL+XI (1)
0147          CALL TRISTF (3,4,5,N,MTYPE,JOBTYP)
0148          VOL=VOL+XI (1)
0149          DO 560 II=1,6
0150          DO 560 JJ=1,10
0151      560 HH (II,JJ)=HH (II,JJ) /4.0
0152       25 IND8=N
0153    C
0154    C**** STORE PARTICULAR VALUES FOR PROCESSING IN STRESS SUBROUTINE
0155    C
0156          WRITE (8'IND8) RRR (5) ,ZZZ (5) ,PROP (1) ,PROP (2) ,PROP (5) ,PROP (6) ,
0157        *      PROP (7) ,PROP (8) ,PROP (9) ,DFDEDE,DFDTDT, (TOTTRM (IA) ,IA=1,4) ,
0158        *      (S (9,IC) ,IC=1,10) , (P (IB) ,IB=9,10) , (S (10,IC) ,IC=1,10) ,
0159        * ((CEPNEW (ID,IE) ,IE=1,4) ,ID=1,4) , ((HH (IF,IG) ,IG=1,10) ,IF=1,6) ,HPR,
0160        * PROP (4) ,TEMD,QT (N) ,PROP (3)
0161          RETURN
0162          END
```

```
0001            SUBROUTINE STIFF(MBAND,JOBTYP,KINHAR)
0002      C            (CALLED FROM "MAIN" - FOR STRESS ANALYSIS)
0003      C            (SEE 3-16-(9))
0004      C***********************************************************************
0005      C
0006      C
0007      C      THIS SUBROUTINE ASSEMBLES THE STRUCTURAL STIFFNESS MATRIX (K)
0008      C      (FROM THE INDIVIDUAL ELEMENT STIFFNESS MATRICES, AS EVALUATED
0009      C      IN THE CALL TO "QUAD"), AND THE STRUCTURAL LOAD VECTOR F (FROM
0010      C      USER-SUPPLIED CONCENTRATED NODAL LOADS FROM "DATAI", AND/OR
0011      C      PRESSURE BOUNDARY LOADS FROM "MAIN").
0012      C
0013      C      NOTE THAT THE STRUCTURAL STIFFNESS MATRIX AND LOAD VECTOR ARE
0014      C      STORED IN BLOCKS OF A SIZE APPROPRIATE TO THE MAXIMUM
0015      C      BANDWIDTH (HERE 26) OF THE SYSTEM.  WHERE REQUIRED, THE
0016      C      STRUCTURAL STIFFNESS MATRIX AND LOAD VECTOR ARE MODIFIED FOR
0017      C      THE GIVEN BOUNDARY CONDITIONS BY A CALL TO "MODIFY".
0018      C
0019      C
0020      C***********************************************************************
0021      C
0022      C
0023            COMMON/DISP1/UR(378),UZ(378),CODE(378)
0024            COMMON/BOUND/IBC(200),JBC(200),NPBCR(10),NUMPC,PR(200)
0025            COMMON/SYSMSH/NUMEL,NUMNP,NRMAX
0026            COMMON/COOR/R(378),Z(378),T(378),IX(378,5),TK(378)
0027            COMMON/DUM/RRR(5),ZZZ(5),LM(4)
0028            COMMON/TRIPLE/TT(4),TP(6),P(10),S(10,10)
0029            COMMON/PLANE/NPP
0030            COMMON/DISPL/BB(756),BBT(756)
0031            COMMON/BANARG/MBANN,NUMBLK,B(108),A(197,54)
0032            DIMENSION IGAB(378)
0033            MBANN=MBAND
0034      C
0035      C**** INITIALIZATION OF BLOCK NUMBERS, STIFFNESS MATRIX, AND LOAD VECTOR
0036      C
0037            REWIND 2
0038            NB=27
0039            ND=2*NB
0040            STOP=0.0
0041            ND2=2*ND
0042            NUMBLK=0
0043            DO 500 N=1,ND2
0044            B(N)=0.0
0045            DO 500 M=1,ND
0046        500 A(N,M)=0.0
0047            DO 510 IG=1,NUMEL
0048        510 IGAB(IG)=0
0049      C
0050      C**** FORM STIFFNESS MATRIX IN BLOCKS
0051      C
0052        190 NUMBLK=NUMBLK+1
0053            NH=NB*(NUMBLK+1)
0054            NM=NH-NB
0055            NL=NM-NB+1
0056            KSHIFT=2*NL-2
0057      C
0058      C**** BEGIN LOOP FOR ASSEMBLING ELEMENT STIFFNESS FOR THIS BLOCK
0059      C
0060            DO 520 N=1,NUMEL
0061            IF (IX(N,5)) 520,520,5
0062          5 IF (IGAB(N).EQ.1) GO TO 520
0063            DO 530 I=1,4
0064            IF (IX(N,I)-NL) 530,10,10
0065         10 IF (IX(N,I)-NM) 15,15,530
0066        530 CONTINUE
0067            GO TO 520
0068         15 CALL QUAD(N,JOBTYP,VOL,kinhar)
0069            IF (VOL) 20,20,25
0070         20 WRITE (6,2000) N
0071            STOP=1.0
```

359

```
0072        25 IF  (IX(N,3)-IX(N,4))  30,35,30
0073     C
0074     C**** CONDENSATION FOR QUADRILATERAL (TRIANGULAR COMPOSITE) ELEMENT
0075     C
0076        30 DO 540 II=1,9
0077           IF  (S(II,10).LT.1.0E-10 .AND. S(II,10).GT.-1.0E-10)  S(II,10)=0.0
0078           CC=S(II,10)/S(10,10)
0079           P(II)=P(II)-CC*P(10)
0080           DO 540 JJ=1,9
0081       540 S(II,JJ)=S(II,JJ)-CC*S(10,JJ)
0082           DO 550 II=1,8
0083           CC=S(II,9)/S(9,9)
0084           P(II)=P(II)-CC*P(9)
0085           DO 550 JJ=1,8
0086       550 S(II,JJ)=S(II,JJ)-CC*S(9,JJ)
0087     C
0088     C**** ADD ELEMENT STIFFNESS TO TOTAL STIFFNESS
0089     C
0090        35 DO 560 I=1,4
0091       560 LM(I)=2*IX(N,I)-2
0092           DO 570 I=1,4
0093           DO 570 K=1,2
0094           II=LM(I)+K-KSHIFT
0095           KK=2*I-2+K
0096           B(II)=B(II)+P(KK)
0097           DO 570 J=1,4
0098           DO 570 L=1,2
0099           JJ=LM(J)+L-II+1-KSHIFT
0100           LL=2*J-2+L
0101           IF (JJ) 570,570,40
0102        40 IF (ND-JJ) 45,50,50
0103        45 WRITE (6,2001) N
0104           STOP=1.0
0105           GO TO 520
0106        50 A(II,JJ)=A(II,JJ)+S(KK,LL)
0107       570 CONTINUE
0108           IGAB(N)=1
0109       520 CONTINUE
0110     C
0111     C**** ADD CONCENTRATED FORCES WITHIN BLOCK
0112     C
0113           DO 580 N=NL,NM
0114           K=2*N-KSHIFT
0115           IF (CODE(N)) 80,60,65
0116        60 B(K)=B(K)+UZ(N)
0117           B(K-1)=B(K-1)+UR(N)
0118           GO TO 580
0119        65 IF (CODE(N)-1.0) 580,70,75
0120        70 B(K)=B(K)+UZ(N)
0121           GO TO 580
0122        75 IF (CODE(N)-2.0) 580,80,580
0123        80 B(K-1)=B(K-1)+UR(N)
0124       580 CONTINUE
0125     C
0126     C**** BOUNDARY CONDITIONS EVALUATIONS
0127     C
0128     C
0129     C**** PRESSURE BOUNDARY CONDITIONS
0130     C
0131           IF (NUMPC) 85,90,85
0132        85 DO 590 L=1,NUMPC
0133           I=IBC(L)
0134           J=JBC(L)
0135           PP=PR(L)/6.
0136           DZ=(Z(I)-Z(J))*PP
0137           DR=(R(J)-R(I))*PP
0138           RX=2.0*R(I)+R(J)
0139           ZX=R(I)+2.0*R(J)
0140           IF (NPP.NE.1) GO TO 95
0141     C
0142     C**** FOR PLANE STRESS
```

```
0143      C
0144            THI=3.*TK(I)+TK(J)
0145            THJ=3.*TK(J)+TK(I)
0146       95 II=2*I-KSHIFT
0147            JJ=2*J-KSHIFT
0148            IF (II) 100,100,105
0149      105 IF (II-ND) 110,110,100
0150      C
0151      C**** DO R COMPONENT FIRST
0152      C
0153      110 IF (NPP-1) 115,120,125
0154      C
0155      C**** PLANE STRAIN
0156      C
0157      125 B(II-1)=B(II-1)+DZ*3.
0158            B(II)=B(II)+DR*3.
0159            GO TO 100
0160      C
0161      C**** PLANE STRESS
0162      C
0163      120 B(II-1)=B(II-1)+DZ*THI*.75
0164            B(II)=B(II)+DR*THI*.75
0165            GO TO 100
0166      C
0167      C**** AXISYMMETRIC
0168      C
0169      115 B(II-1)=B(II-1)+RX*DZ
0170            B(II)=B(II)+RX*DR
0171      100 IF (JJ) 590,590,130
0172      C
0173      C**** NOW DO Z COMPONENT
0174      C
0175      130 IF (JJ-ND) 135,135,590
0176      135 IF (NPP-1) 140,145,150
0177      C
0178      C**** PLANE STRAIN
0179      C
0180      150 B(JJ-1)=B(JJ-1)+DZ*3.
0181            B(JJ)=B(JJ)+DR*3.
0182            GO TO 590
0183      C
0184      C**** PLANE STRESS
0185      C
0186      145 B(JJ-1)=B(JJ-1)+DZ*THJ*.75
0187            B(JJ)=B(JJ)+DR*THJ*.75
0188            GO TO 590
0189      C
0190      C**** AXISYMMETRIC
0191      C
0192      140 B(JJ-1)=B(JJ-1)+ZX*DZ
0193            B(JJ)=B(JJ)+ZX*DR
0194      590 CONTINUE
0195       90 DO 600 M=NL,NH
0196            IF (M-NUMNP) 155,155,600
0197      155 U=UR(M)
0198            N=2*M-1-KSHIFT
0199            IF (CODE(M)) 160,600,165
0200      165 IF (CODE(M)-1.) 170,175,170
0201      170 IF (CODE(M)-2.) 180,160,180
0202      180 IF (CODE(M)-3.) 160,185,160
0203      175 CALL MODIFY(A,B,ND2,MBAND,N,U)
0204            GO TO 600
0205      185 CALL MODIFY(A,B,ND2,MBAND,N,U)
0206      160 U=UZ(M)
0207            N=N+1
0208            CALL MODIFY(A,B,ND2,MBAND,N,U)
0209      600 CONTINUE
0210      C
0211      C**** WRITE BLOCK OF EQUATIONS ON TAPE AND SHIFT UP LOWER BLOCK
```

361

```
0212      C
0213            WRITE (2) (B(N),N=1,ND),((A(N,M),N=1,ND),M=1,MBAND)
0214            DO 610 N=1,ND
0215            K=N+ND
0216            B(N)=B(K)
0217            B(K)=0.0
0218            DO 610 M=1,ND
0219            A(N,M)=A(K,M)
0220        610 A(K,M)=0.0
0221      C
0222      C**** CHECK FOR LAST BLOCK
0223      C
0224            IF (NM-NUMNP) 190,195,195
0225        195 IF (STOP) 200,205,200
0226        200 CALL EXIT
0227        205 RETURN
0228      C
0229      C**** FORMATS FOR ERROR STATEMENTS
0230      C
0231       2000 FORMAT(26HONEGATIVE AREA ELEMENT NO. I4)
0232       2001 FORMAT(29HOBAND WIDTH EXCEEDS ALLOWABLE I4)
0233            END
```

```
0001            SUBROUTINE STRESS(DT1,TIME1,JOBTYP,KINHAR)
0002      C           (CALLED FROM "MAIN" - FOR STRESS ANALYSIS)
0003      C           (SEE 3-10, 3-13, 3-16-(13) TO (15), 4-6)
0004      C************************************************************************
0005      C
0006      C
0007      C     THIS SUBROUTINE RECOVERS ELEMENT INFORMATION (ASSEMBLED IN
0008      C     "QUAD") FOR CALCULATING ELEMENT STRAINS AND STRESSES (BOTH
0009      C     INCREMENTAL AND TOTAL). EACH ELEMENT IS CHECKED FOR YIELDING
0010      C     AND THE LOADING/UNLOADING PARAMETER IS EVALUATED.
0011      C
0012      C
0013      C************************************************************************
0014      C
0015      C
0016            COMMON/PRINT1/IPRINT,NELPR,NODPR,IOPRNT,IDELPR(378),IDNDPR(378)
0017            COMMON/PRINT2/IPSNDD,IPSELR,IPSELS
0018            COMMON/STRAIN/EPSDI(378),EPSDOT(378),SIGAVE(378),DEVSIG(378,5),
0019          *       EPSDTC(378),EPSDTP(378)
0020            COMMON/SYSMSH/NUMEL,NUMNP,NRMAX
0021            COMMON/COOR/R(378),Z(378),T(378),IX(378,5),TK(378)
0022            COMMON/TIMSTP/ITIME
0023            COMMON/DSTRES/RATS2(378),DSIG(378,4),DIVEPS(378,4)
0024            COMMON/STREZZ/RATS1(378),TSIG(378,5),TOTEPS(378,4),EPS(378),
0025          *       DLAM(378),RATST(378),DEPS(378,4)
0026            COMMON/DISPL/BB(756),BBT(756)
0027            COMMON/PLANE/NPP
0028            DIMENSION SIG(10),P(10),TP(6),HH(6,10),S(10,10),
0029          *       RR(4),GOSH(4),TOTTRM(4)
0030            COMMON/RMTRIX/C(4,4),EE(20)
0031            COMMON/HEATR/XA,DTI,TIME,TIMECR
0032            COMMON/CCRP/SCP(378,14),SSCP(378,2)
0033            COMMON/ATEM/ACTQ,RCONT,COEF,ENC
0034            COMMON/PMET/TEMP
0035            COMMON/DTMEN/TIMAX,TIMAXO
0036            COMMON/PPPP/FEPC,BBC,ACC
0037            COMMON/RMATR/SMATRX(4,4)
0038            COMMON/CREEP/DEPCJ(4),EPCJ(4),QDEPCJ(4),
0039          *       QP(10),CDSIG(4),QDSIG(4),QH(6,10)
0040            COMMON/RESTR/MBANX,DTIME(1000)
0041            COMMON/MATRI/III
0042            COMMON/RITBE/T11,T12
0043            COMMON/DOT/DTOO
```

```
0044            COMMON/DDDD/FEPCDT,FECDDT,DEPCD
0045            COMMON/EFBB/BBBB
0046            COMMON/TIMKEP/TIMEK
0047            COMMON/FLAGS/PLAST
0048            COMMON/KINH/IRE(378),ICK(378),LCY(378),EFFBLF(378),RATSP(378)
0049            COMMON/PASMAT/CECE(378,4,4),DCESGN(378,4),DCTSGN(378,4)
0050            LOGICAL PLAST(378)
0051            DOUBLE PRECISION FEPCDT,FECDDT,DEPCD
0052            DOUBLE PRECISION FEPC,BBC,AAC,TIMAX,TIMAXO
0053            DIMENSION TEMP1(4),TEMP2(4)
0054            MPRINT=0
0055     C
0056     C**** INITIALIZATION OF MATRIX FOR CREEP CALCULATION
0057     C
0058            IF (NPP.EQ.0) GO TO 5
0059            SMATRX(1,3)=0.0
0060            SMATRX(2,3)=0.0
0061            SMATRX(3,3)=0.0
0062            SMATRX(3,1)=0.0
0063            SMATRX(3,2)=0.0
0064            GO TO 10
0065          5 SMATRX(1,3)=-0.5
0066            SMATRX(2,3)=-0.5
0067            SMATRX(3,3)=1.0
0068            SMATRX(3,1)=-0.5
0069            SMATRX(3,2)=-0.5
0070         10 SMATRX(1,1)=1.0
0071            SMATRX(2,2)=1.0
0072            SMATRX(1,2)=-0.5
0073            SMATRX(2,1)=-0.5
0074            SMATRX(4,4)=1.5
0075            SMATRX(1,4)=0.0
0076            SMATRX(2,4)=0.0
0077            SMATRX(3,4)=0.0
0078            SMATRX(4,1)=0.0
0079            SMATRX(4,2)=0.0
0080            SMATRX(4,3)=0.0
0081     C
0082     C**** BEGIN LOOP FOR EACH ELEMENT
0083     C
0084            DO 500 M=1,NUMEL
0085            N=M
0086            IELPRT=1
0087            IF (NELPR.EQ.0) GO TO 15
0088            DO 510 I=1,NELPR
0089            IELPRT=0
0090            IF ((N-IDELPR(I)).EQ.0) GO TO 20
0091            GO TO 510
0092         20 IELPRT=1
0093            GO TO 15
0094        510 CONTINUE
0095         15 MTYPE=IX(N,5)
0096     C
0097     C**** RECOVER ELEMENT INFORMATION ASSEMBLED IN QUAD
0098     C
0099            IND8=N
0100            READ (8'IND8) RC,ZC,EE(1),EE(2),EE(13),EE(14),EE(15),EE(16),
0101          *    EE(17),DYDE,DYDT,(TOTTRM(IA),IA=1,4),(S(9,IC),IC=1,10),
0102          *    (RR(IB),IB=1,2),(S(10,IC),IC=1,10),((C(ID,IE),IE=1,4),ID=1,4),
0103          *    ((HH(IF,IG),IG=1,10),IF=1,6),HPR,EE(10),TEMD,TEMP,EE(7)
0104            EPSOFF=EE(13)/EE(1)
0105            IX(N,5)=MTYPE
0106            DO 520 I=1,4
0107            JJ=2*IX(N,I)
0108            II=2*I
0109            P(II-1)=BB(JJ-1)
0110        520 P(II)=BB(JJ)
0111     C
0112     C**** RECOVER REACTION FORCES FOR ELEMENT CENTROID
0113     C
0114            DO 530 I=1,2
```

363

```
0115              DO 530 K=1,8
0116        530 RR(I)=RR(I)-S(I+8,K)*P(K)
0117              COMM=S(9,9)*S(10,10)-S(9,10)*S(10,9)
0118              IF (COMM) 25,30,25
0119         25 P(9)=(S(10,10)*RR(1)-S(9,10)*RR(2))/COMM
0120              P(10)=(-S(10,9)*RR(1)+S(9,9)*RR(2))/COMM
0121      C
0122      C**** CALCULATE ELEMENT STRAINS
0123      C
0124         30 DO 540 I=1,6
0125              TP(I)=0.0
0126              DO 540 K=1,10
0127        540 TP(I)=TP(I)+HH(I,K)*P(K)
0128              RR(1)=TP(2)
0129              RR(2)=TP(6)
0130              RR(3)=(TP(1)+TP(2)*RC+TP(3)*ZC)/RC
0131              IF (NPP.NE.0) RR(3)=0.0
0132              RR(4)=TP(3)+TP(5)
0133      C
0134      C**** DETERMINE ELEMENT STRESSES FROM THE CONSTITUTIVE EQUATION
0135      C
0136              DO 550 I=1,4
0137              SIG(I)=-TOTTRM(I)
0138              I9=I+9
0139              IF (JOBTYP.EQ.0 .OR. JOBTYP.EQ.3 .OR. JOBTYP.EQ.4) GO TO 35
0140              TT11=T11+1.0E-04
0141              IF (TIME.GE.TT11) GO TO 40
0142         35 DO 560 J=1,4
0143              J9=J+9
0144        560 SCP(N,J9)=0.0
0145         40 GOSH(I)=0.0
0146              DO 570 K=1,4
0147              K9=K+9
0148        570 GOSH(I)=GOSH(I)+C(I,K)*(RR(K)-SCP(N,K9))
0149        550 SIG(I)=SIG(I)+GOSH(I)
0150              IF (NPP.NE.2) GO TO 45
0151      C
0152      C**** RECOVER OUT-OF-PLANE STRESSES FOR PLANE STRAIN
0153      C
0154              SIG(3)=EE(2)*(SIG(1)+SIG(2))-EE(1)*EE(10)*TEMD
0155              IF (.NOT.PLAST(N)) GO TO 45
0156      C
0157      C**** (HERE, FOR PLASTIC REGION)
0158      C
0159              AVSIG=(SIG(1)+SIG(2)+SIG(3))/3.
0160              DO 580 I=1,3
0161        580 DEVSIG(N,I)=SIG(I)-AVSIG
0162              SIG(3)=SIG(3)-(DEVSIG(N,1)+EE(2)*DEVSIG(N,3))*SIG(1)
0163           1 -(DEVSIG(N,2)+EE(2)*DEVSIG(N,3))*SIG(2)
0164           2 -2.0*TSIG(N,4)*SIG(4)
0165              SIG(3)=SIG(3)*DEVSIG(N,3)/(DEVSIG(N,3)**2-(4.0*HPR
0166           1 *DEVSIG(N,5)**2/(9.0*EE(1))))
0167      C
0168      C**** OUT-OF-PLANE STRAIN FOR PLANE STRESS
0169      C
0170         45 IF (NPP.EQ.1) RR(3)=-EE(2)/EE(1)*(SIG(1)+SIG(2))+EE(10)*TEMD-RR(1)
0171           1-RR(2)+(SIG(1)+SIG(2))*(1.-EE(2))/EE(1)
0172      C
0173      C**** CALCULATE ELEMENT TOTAL STRAINS
0174      C
0175              DO 590 I=1,4
0176              DEPS(N,I)=RR(I)
0177              TOTEPS(N,I)=TOTEPS(N,I)+RR(I)
0178        590 RR(I)=TOTEPS(N,I)
0179      C
0180      C**** CALCULATE EFFECTIVE STRAIN
0181      C
0182              EPS1N1=SQRT((RR(1)-RR(2))**2+(RR(1)-RR(3))**2+(RR(2)-RR(3))**2+6.*
0183           *    RR(4)**2)*1.414213562/3.
0184              EPSDOT(N)=(EPS1N1-EPS(N))/DT1
0185              EPSDTC(N)=EPSDOT(N)
```

364

```
0186          DEPSDT=EPSDTC(N)-EPSDTP(N)
0187          EPSDTP(N)=EPSDTC(N)
0188          EPSDI(N)=DEPSDT
0189          IF (ITIME.LT.3) EPSDI(N)=0.0
0190          EPS(N)=EPS1N1
0191    C
0192    C**** OUTPUT STRESSES
0193    C        CALCULATE PRINCIPAL STRESSES
0194    C        CALCULATE TOTAL AND DEVIATORIC STRESS
0195    C
0196          DO 600 IIS=1,4
0197          DSIG(N,IIS)=SIG(IIS)
0198          TSIG(N,IIS)=TSIG(N,IIS)+SIG(IIS)
0199      600 SIG(IIS)=TSIG(N,IIS)
0200          AVSIG=.3333333*(SIG(1)+SIG(2)+SIG(3))
0201          IF (NPP.EQ.1) AVSIG=(SIG(1)+SIG(2))/3.
0202          SIGAVE(N)=AVSIG
0203          DO 610 IIS=1,3
0204      610 DEVSIG(N,IIS)=SIG(IIS)-AVSIG
0205          CC=(SIG(1)+SIG(2))/2.0
0206          BBBB=1.0E-20+(SIG(1)-SIG(2))/2.0
0207          CR1=SQRT(BBBB**2+SIG(4)**2)
0208          SIG(5)=CC+CR1
0209          SIG(6)=CC-CR1
0210          SIG(7)=28.648*ATAN2(SIG(4),BBBB)
0211          SIG(5)=AMAX1(SIG(3),SIG(5))
0212          SIG(6)=AMIN1(SIG(3),SIG(6))
0213    C
0214    C        STRESSES PARALLEL TO LINE I-J
0215    C
0216          I=IX(N,1)
0217          J=IX(N,2)
0218          ANG=2.*ATAN2(Z(J)-Z(I),R(J)-R(I))
0219          SIN2A=SIN(ANG)
0220          COS2A=COS(ANG)
0221          CX=.5*(SIG(1)-SIG(2))
0222          SIG(9)=CX*COS2A+SIG(4)*SIN2A+CC
0223          SIG(10)=2.*CC-SIG(8)
0224          SIG(8)=-CX*SIN2A+SIG(4)*COS2A
0225          IF (IPRINT-IOPRNT) 55,60,55
0226       60 IF (NELPR.NE.0 .AND. MPRINT.NE.0) GO TO 55
0227          IF (MPRINT) 65,70,65
0228       70 IF (IPSELS.EQ.1) GO TO 75
0229          WRITE (6,2000)
0230       75 MPRINT=55
0231       65 MPRINT=MPRINT-1
0232    C
0233    C**** CALCULATE EFFECTIVE STRESS
0234    C
0235       55 BBBB=SQRT(0.5*((SIG(1)-SIG(2))**2+(SIG(2)-SIG(3))**2+(SIG(3)-
0236        *     SIG(1))**2)+3.0*SIG(4)**2)
0237          DTOO=0.0
0238          IF (JOBTYP.EQ.0 .OR. TIME.LT.T11) GO TO 80
0239          IF (JOBTYP.EQ.3 .OR. TIME.LT.T11) GO TO 80
0240          IF (JOBTYP.EQ.4 .OR. TIME.LT.T11) GO TO 80
0241          DTOO=1.0
0242          CALL CREEPP(N,SIG)
0243       80 DEVSIG(N,4)=SIG(4)
0244          DEVSIG(N,5)=BBBB
0245          RATS1(N)=BBBB/EE(13)
0246    C
0247    C*** CHECK FOR ELEMENT YIELDING
0248    C
0249          IF (EPS(N).GE.EPSOFF.OR.BBBB.GT.EE(13)) PLAST(N)=.TRUE.
0250          IF (.NOT.PLAST(N)) GO TO 85
0251          CALL MATDEL (EPS(N),EE(1),EE(2),EE(15),EE(16),EE(17),BBB)
0252          RATS2(N)=BBBB/BBB
0253          GO TO 90
0254       85 RATS2(N)=0.
0255    C
0256    C**** CALCULATE DLAM (LOADING OR UNLOADING PARAMETER)
```

365

```
0257    C
0258      90 IF (KINHAR.EQ.0) GO TO 95
0259    C
0260    C**** FOR KINEMATIC HARDENING ONLY
0261    C
0262         RATST (N) =RATS1 (N)
0263         IF (LCY (N).EQ.1) GO TO 100
0264         IR=1
0265         IF (IRE (N).EQ.1 .AND. ICK (N).EQ.0) IR=-1
0266         RATST (N) =ABS (DEVSIG (N,5) -EFFBLF (N) *IR) /EE (13)
0267     100 CALL KITENS (N,RATSP (N) ,EE (13) ,DYDT,DYDE)
0268         IF (EPS (N).GE.EPSOFF .OR. BBBB.GT.EE (13)) PLAST (N) =.TRUE.
0269         IF (.NOT.PLAST (N)) GO TO 105
0270         CALL MATDEL (EPS (N) ,EE (1) ,EE (2) ,EE (15) ,EE (16) ,EE (17) ,BBB)
0271         RATS2 (N) =BBBB/BBB
0272         GO TO 110
0273     105 RATS2 (N) =0.0
0274         GO TO 110
0275      95 TA=DEVSIG (N,1) *DSIG (N,1) +DEVSIG (N,2) *DSIG (N,2) +DEVSIG (N,3) *
0276       *    DSIG (N,3) +2.*DEVSIG (N,4) *DSIG (N,4)
0277         TB=DEVSIG (N,1) **2+DEVSIG (N,2) **2+DEVSIG (N,3) **2+2.*DEVSIG (N,4) **2
0278         DO 1530 IND=1,4
0279    1530 TEMP2 (IND) =0.0
0280         ES= (4./9.) *BBBB**2* (1.5*EE (1) / (1.+EE (2)) +HPR)
0281         DO 1510 I=1,4
0282    1510 TEMP1 (I) = (EE (10) +DCTSGN (N,I)) *TEMD+DCESGN (N,I) *EPSDI (N)
0283         DO 1520 I=1,4
0284         DO 1520 J=1,4
0285    1520 TEMP2 (I) =TEMP2 (I) +CECE (N,I,J) * (DEPS (N,J) -TEMP1 (J))
0286         T1=0.0
0287         DO 1540 I=1,4
0288    1540 T1=T1+DEVSIG (N,I) *TEMP2 (I)
0289         DLAM (N) = (T1+DYDT+DYDE) /ES
0290     110 IF (IPRINT-IOPRNT) 115,120,115
0291     120 IF (IELPRT.EQ.0) GO TO 115
0292         IF (IPSELS.EQ.1) GO TO 115
0293    C
0294    C**** WRITE OUT ELEMENT STRESSES
0295    C
0296         WRITE (6,2001) N,RC,ZC, (SIG (I) ,I=1,8) ,BBBB
0297     115 IF (JOBTYP.EQ.0 .OR. TIME.LT.T11) GO TO 500
0298         IF (JOBTYP.EQ.3 .OR. TIME.LT.T11) GO TO 500
0299         IF (JOBTYP.EQ.4 .OR. TIME.LT.T11) GO TO 500
0300    C
0301    C**** ASSIGNMENTS FOR CREEP ANALYSIS
0302    C
0303         FEPCC=SNGL (FEPC)
0304         BBCC=SNGL (BBC)
0305         AACC=SNGL (AAC)
0306         SSCP (N,2) =SCP (N,8)
0307         SCP (N,8) =FEPCC
0308         SCP (N,9) =BBCC
0309         SCP (N,14) =BBBB
0310         SCP (N,1) =SNGL (FEPCDT)
0311         SCP (N,2) =SNGL (FECDDT)
0312         SCP (N,3) =SNGL (DEPCD)
0313         DO 620 ICC=1,4
0314         IICC=ICC+3
0315     620 SCP (N,IICC) =SCP (N,IICC) +DEPCJ (ICC)
0316         DO 630 ICC=1,4
0317         IICC=ICC+9
0318     630 SCP (N,IICC) =DEPCJ (ICC)
0319     500 CONTINUE
0320         IF (JOBTYP.EQ.0 .OR. JOBTYP.EQ.3 .OR. JOBTYP.EQ.4) GO TO 125
0321    C
0322    C**** FOR CREEP ANALYSIS
0323    C
0324         TAXO=SNGL (TIMAX)
0325         IF (TAXO.GT.1.5*DTIME (ITIME)) TAXO=1.5*DTIME (ITIME)
0326         IF (TIME.LT.T11) GO TO 130
```

```
0327              III=III+1
0328              GO TO 135
0329        130  III=1
0330              TAXO=0.
0331              DTOO=0.0
0332        135  IF (TIME.EQ.T11) TIMEK=DTIME(ITIME)
0333              IF (TIME.GT.T11.AND.ITIME.GT.2) TIMEK=DTIME(ITIME-2)
0334              IF (DTOO.NE.O..AND.III.GT.1) DTIME(ITIME)=TAXO
0335        125  RETURN
0336      C
0337      C**** OUTPUT FORMATS
0338      C
0339       2000 FORMAT(4H1EL.,2X,11HCOORDINATES,3X,98H * * * * * * * * * * * * *
0340             1 * * * * *  STRESSES          * * * * * * * * * * * * * * * *
0341             2* * '    '/' NO.',5X,'R',9X,'Z',10X,'R',11X,'Z',11X,'T',
0342             310X,'RZ',10X,'MAX',9X,'MIN',7X,'ANGLE',8X,'SHEAR',3X,'EFFECTIVE')
0343       2001 FORMAT(I4,2E10.3,6E12.4,E11.3,2E12.4)
0344             END


0001             SUBROUTINE SYMINV(A,NMAX)
0002      C              (CALLED FROM "CSUBE" - FOR STRESS ANALYSIS)
0003      C              (SEE 3-10, 3-13 TO 3-15)
0004      C***********************************************************************
0005      C
0006      C
0007      C     THIS SUBROUTINE INVERTS THE CE MATRIX (EVALUATED IN CSUBE
0008      C     SUBROUTINE) TO YIELD CE-1 FOR THE CONSTITUTIVE EQUATION.
0009      C
0010      C
0011      C***********************************************************************
0012      C
0013      C
0014             DIMENSION A(4,4)
0015             DO 500 N=1,NMAX
0016             D=A(N,N)
0017             DO 510 J=1,NMAX
0018        510  A(N,J)=-A(N,J)/D
0019             DO 520 I=1,NMAX
0020             IF (N-I) 5,520,5
0021          5  DO 530 J=1,NMAX
0022             IF (N-J) 10,530,10
0023         10  A(I,J)=A(I,J)+A(I,J)*A(N,J)
0024        530  CONTINUE
0025        520  A(I,N)=A(I,N)/D
0026        500  A(N,N)=1./D
0027             RETURN
0028             END


0001             SUBROUTINE TRISTF(II,JJ,KK,N,MTYPE,JOBTYP)
0002      C              (CALLED FROM "QUAD" - FOR STRESS ANALYSIS)
0003      C              (SEE 3-15-2, 3-15-4, 3-15-5)
0004      C***********************************************************************
0005      C
0006      C
0007      C     THIS SUBROUTINE CALCULATES H AS USED IN THE STRAIN-DISPLACEMENT
0008      C     TRANSFORMATION MATRIX AND IN THE ELEMENT STIFFNESS MATRIX K.  THE
0009      C     INTEGRAND IS EVALUATED FOR BOTH AXISYMMETRIC AND PLANAR GEOMETRIES.
0010      C
0011      C
0012      C***********************************************************************
0013      C
C0014      C
0015             IMPLICIT REAL*8 (Q)
0016             DOUBLE PRECISION COMM,DBLE
0017             COMMON/MATRIX/C(4,4),EE(20)
0018             COMMON/TEMP5/CEPNEW(4,4),TOTTRM(4),SZERO
0019             COMMON/DUM/RRR(5),ZZZ(5),LM(4)
```

367

```
0020          COMMON/TRIPLE/TT(4),TP(6),P(10),S(10,10)
0021          COMMON/DBLH/HH(6,10)
0022          COMMON/DBLP/QRR(4),QZZ(4),QRRR(5),QZZZ(5),QXI(6)
0023          COMMON/PLANE/NPP
0024          COMMON/THICK/THK
0025          COMMON/COOR/R(378),Z(378),T(378),IX(378,5),TK(378)
0026          COMMON/RITBE/T11,T12
0027          COMMON/HEATR/XA,DTI,TIME,TIMECR
0028          COMMON/CREEP/DEPCJ(4),EPCJ(4),QDEPCJ(4),
0029       *        QP(10),CDSIG(4),QDSIG(4),QH(6,10)
0030          DIMENSION QF(6,10),QD(6,6),QC(4,4),QDD(3,3),
0031       *        QS(10,10),QHH(6,10),QTT(4),QTP(6)
0032    C
0033    C**** INITIALIZATION
0034    C
0035          DO 500 IND1=1,4
0036          TT(IND1)=TOTTRM(IND1)
0037          DO 500 IND2=1,4
0038      500 C(IND1,IND2)=CEPNEW(IND1,IND2)
0039          LM(1)=II
0040          LM(2)=JJ
0041          LM(3)=KK
0042          DO 510 I=1,6
0043          DO 520 J=1,10
0044          QF(I,J)=0.D0
0045      520 QH(I,J)=0.D0
0046          DO 510 J=1,6
0047      510 QD(I,J)=0.D0
0048    C
0049    C**** CONVERT SINGLE PRECISION (R*4) TO DOUBLE PRECISION (R*)
0050    C
0051          DO 530 I=1,6
0052          DO 540 K=1,10
0053          QP(K)=DBLE(P(K))
0054          DO 550 J=1,10
0055          QHH(I,J)=DBLE(HH(I,J))
0056      550 QS(K,J)=DBLE(S(K,J))
0057      540 CONTINUE
0058      530 CONTINUE
0059          DO 560 I=1,4
0060          QTT(I)=DBLE(TT(I))
0061          DO 570 J=1,4
0062      570 QC(I,J)=DBLE(C(I,J))
0063      560 CONTINUE
0064    C
0065    C**** FORM INTEGRAL (G) TRANSPOSE*(C)*(G)
0066    C
0067          CALL INTER
0068          IF (NPP-1) 5,10,10
0069    C
0070    C**** FOR PLANAR ANALYSIS
0071    C
0072       10 QD(2,2)=QXI(1)*QC(1,1)
0073          QD(2,3)=QXI(1)*QC(1,4)
0074          QD(2,5)=QD(2,3)
0075          QD(2,6)=QXI(1)*QC(1,2)
0076          QD(3,3)=QXI(1)*QC(4,4)
0077          QD(3,5)=QD(3,3)
0078          QD(3,6)=QXI(1)*QC(2,4)
0079          QD(5,5)=QD(3,3)
0080          QD(5,6)=QD(3,6)
0081          QD(6,6)=QXI(1)*QC(2,2)
0082          GO TO 15
0083    C
0084    C**** FOR AXISYMMETRIC ANALYSIS
0085    C
0086        5 QD(1,1)=QXI(3)*QC(3,3)
0087          QD(1,2)=QXI(2)*(QC(1,3)+QC(3,3))
0088          QD(1,3)=QXI(5)*QC(3,3)+QXI(2)*QC(3,4)
0089          QD(1,5)=QXI(2)*QC(3,4)
0090          QD(1,6)=QXI(2)*QC(2,3)
```

368

```
0091            QD(2,2)=QXI(1)*(QC(1,1)+2.0D0*QC(1,3)+QC(3,3))
0092            QD(2,3)=QXI(4)*(QC(1,3)+QC(3,3))+QXI(1)*(QC(1,4)+QC(3,4))
0093            QD(2,5)=QXI(1)*(QC(1,4)+QC(3,4))
0094            QD(2,6)=QXI(1)*(QC(1,2)+QC(2,3))
0095            QD(3,3)=QXI(6)*QC(3,3)+2.0D0*QXI(4)*QC(3,4)+QXI(1)*QC(4,4)*1.00D0
0096            QD(3,5)=QXI(4)*QC(3,4)+QXI(1)*QC(4,4)*1.00D0
0097            QD(3,6)=QXI(4)*QC(2,3)+QXI(1)*QC(2,4)
0098            QD(5,5)=QXI(1)*QC(4,4)*1.00D0
0099            QD(5,6)=QXI(1)*QC(2,4)
0100            QD(6,6)=QXI(1)*QC(2,2)
0101      C
0102      C**** ESTABLISH SYMMETRIC MATRIX
0103      C
0104       15 DO 580 J=2,6
0105            K=J-1
0106            DO 580 I=1,K
0107      580 QD(J,I)=QD(I,J)
0108      C
0109      C**** FORM COEFFICIENT-DISPLACEMENT TRANSFORMATION,(H),MATRIX
0110      C
0111            COMM=QRR(2)*(QZZ(3)-QZZ(1))+QRR(1)*(QZZ(2)-QZZ(3))+
0112        1         QRR(3)*(QZZ(1)-QZZ(2))
0113            QDD(1,1)=(QRR(2)*QZZ(3)-QRR(3)*QZZ(2))/COMM
0114            QDD(1,2)=(QRR(3)*QZZ(1)-QRR(1)*QZZ(3))/COMM
0115            QDD(1,3)=(QRR(1)*QZZ(2)-QRR(2)*QZZ(1))/COMM
0116            QDD(2,1)=(QZZ(2)-QZZ(3))/COMM
0117            QDD(2,2)=(QZZ(3)-QZZ(1))/COMM
0118            QDD(2,3)=(QZZ(1)-QZZ(2))/COMM
0119            QDD(3,1)=(QRR(3)-QRR(2))/COMM
0120            QDD(3,2)=(QRR(1)-QRR(3))/COMM
0121            QDD(3,3)=(QRR(2)-QRR(1))/COMM
0122      C
0123      C**** FORM (H) MATRIX
0124      C
0125            DO 590 I=1,3
0126            J=2*LM(I)-1
0127            QH(1,J)=QDD(1,I)
0128            QH(2,J)=QDD(2,I)
0129            QH(3,J)=QDD(3,I)
0130            QH(4,J+1)=QDD(1,I)
0131            QH(5,J+1)=QDD(2,I)
0132      590 QH(6,J+1)=QDD(3,I)
0133      C
0134      C**** FORM K-MATRIX=(H) TRANSPOSE*INTEGRAL((G)T*(C)*(G))*(H)
0135      C
0136            DO 600 J=1,10
0137            DO 600 K=1,6
0138            IF (QH(K,J)) 20,600,20
0139       20 DO 610 I=1,6
0140      610 QF(I,J)=QF(I,J)+QD(I,K)*QH(K,J)
0141      600 CONTINUE
0142            DO 620 I=1,10
0143            DO 620 K=1,6
0144            IF (QH(K,I)) 25,620,25
0145       25 DO 630 J=1,10
0146      630 QS(I,J)=QS(I,J)+QH(K,I)*QF(K,J)
0147      620 CONTINUE
0148      C
0149      C**** FORM THERMAL LOAD VECTOR
0150      C
0151            IF (NPP-1) 30,35,35
0152      C
0153      C**** (AXISYMMETRIC)
0154      C
0155       30 QTP(1)=QXI(2)*QTT(3)
0156            QTP(2)=QXI(1)*(QTT(1)+QTT(3))
0157            QTP(3)=QXI(4)*QTT(3)
0158            QTP(4)=0.D0
0159            QTP(5)=0.D0
0160            QTP(6)=QXI(1)*QTT(2)
0161            GO TO 40
```

```
0162      C
0163      C**** (PLANAR)
0164      C
0165          35 QTP(1)=0.DO
0166             QTP(2)=QXI(1)*QTT(1)
0167             QTP(3)=0.DO
0168             QTP(4)=0.DO
0169             QTP(5)=0.DO
0170             QTP(6)=QXI(1)*QTT(2)
0171          40 DO 640 I=1,10
0172             DO 640 K=1,6
0173         640 QP(I)=QP(I)+QH(K,I)*QTP(K)
0174             IF (JOBTYP.EQ.0 .OR. JOBTYP.EQ.3 .OR. JOBTYP.EQ.4) GO TO 45
0175      C
0176      C**** CREEP ANALYSIS ONLY
0177      C
0178             TT11=T11+1.0E-04
0179             IF (TIME.GE.TT11) CALL PCREEP(N,QC)
0180      C
0181      C**** FORM STRAIN TRANSFORMATION MATRIX, (HH)=SUM(H)
0182      C
0183          45 DO 650 I=1,6
0184             DO 650 J=1,10
0185         650 QHH(I,J)=QHH(I,J)+QH(I,J)
0186      C
0187      C**** CONVERT DOUBLE PRECISION (R*8) TO SINGLE PRECISION (R*4)
0188      C
0189             DO 660 L=1,6
0190             TP(L)=SNGL(QTP(L))
0191             DO 670 I=1,10
0192             P(I)=SNGL(QP(I))
0193             DO 680 M=1,10
0194             HH(L,M)=SNGL(QHH(L,M))
0195         680 S(I,M)=SNGL(QS(I,M))
0196         670 CONTINUE
0197         660 CONTINUE
0198             RETURN
0199             END




0001             SUBROUTINE ADCON(DE,A,C,N,COND,SH)
0002      C                  (CALLED FROM "SOLVE" - FOR THERMAL ANALYSIS)
0003      C                  (SEE 2-10-(2))
0004      C**************************************************************************
0005      C
0006      C
0007      C      THIS SUBROUTINE ASSEMBLES EACH ELEMENT CONDUCTIVITY
0008      C      AND HEAT CAPACITY MATRIX INTO THE SYSTEM MATRICES
0009      C
0010      C
0011      C**************************************************************************
0012      C
0013      C
0014             COMMON/SWITCH/IX(5,378)
0015             COMMON/SYSMSH/NUMELX,NUMNP,NRMAX
0016             COMMON/HEAT1/NMATX,NBCX,NTEMX,NTIMX,NTFX,MBANT,NMBX
0017             COMMON/LAM/MTYPE,CONN,H(5,5)
0018             COMMON/COOR/XXX(378),Y(378),TX(378),IXXX(378,5),TKX(378)
0019             COMMON/BIGR/X(378)
0020             COMMON/LIM/S(5,5)
0021             DIMENSION IXX(5),A(NUMNP,MBANT),C(NUMNP,MBANT)
0022             IF (DE.EQ.0.) GO TO 5
0023             IXX(1)=IX(1,N)
0024             IXX(2)=IX(2,N)
0025             IXX(3)=IX(3,N)
0026             IXX(4)=IX(4,N)
0027             IXX(5)=IX(5,N)
0028             CALL CONDUC(IXX,DE,SH)
```

```
0029          IF (S(1,1).NE.O.) GO TO 5
0030          WRITE (6,2000) N
0031          STOP
0032        5 DO 500 L=1,4
0033          I=IX(L,N)
0034          DO 500 M=1,4
0035          J=IX(M,N)-I+1
0036          IF (J.LE.O) GO TO 500
0037          A(I,J)=A(I,J)+S(L,M)
0038          C(I,J)=C(I,J)+H(L,M)
0039      500 CONTINUE
0040          RETURN
0041    C
0042    C**** ERROR MESSAGE FORMAT
0043    C
0044     2000 FORMAT(1X,' ELEMENT WITH NEGATIVE AREA IN ADCON ',I5)
0045          END
```

```
0001          SUBROUTINE BACKS(A,B)
0002    C              (CALLED FROM "SOLVE" - FOR THERMAL ANALYSIS)
0003    C              (SEE 2-8-1, 2-10-(8))
0004    C***********************************************************************:
0005    C
0006    C
0007    C     THIS SUBROUTINE PERFORMS BACK SUBSTITUTION TO SOLVE FOR
0008    C     THE NODAL TEMPERATURE INCREMENTS
0009    C
0010    C
0011    C***********************************************************************
0012    C
0013    C
0014          COMMON/SYSMSH/NUMELX,NUMNP,NRMAX
0015          COMMON/HEAT1/NMATX,NBCX,NTEMX,NTIMX,NTFX,MBANT,NMB
0016          DIMENSION A(NMB),B(NUMNP)
0017    C
0018    C**** INITIALIZE
0019    C
0020          MMM=MBANT-1
0021          N=0
0022       10 N=N+1
0023          C=B(N)
0024          IF (A(N).NE.O.O) B(N)=B(N)/A(N)
0025          IF (N.EQ.NUMNP) GO TO 5
0026          IL=N+1
0027          IH=AMINO(NUMNP,N+MMM)
0028          M=N
0029          DO 500 I=IL,IH
0030          M=M+NUMNP
0031      500 B(I)=B(I)-A(M)*C
0032          GO TO 10
0033        5 IL=N
0034          N=N-1
0035          IF (N.EQ.O) RETURN
0036          IH=AMINO(NUMNP,N+MMM)
0037          M=N
0038          DO 510 I=IL,IH
0039          M=M+NUMNP
0040      510 B(N)=B(N)-A(M)*B(I)
0041          GO TO 5
0042          END
```

371

```
0001          SUBROUTINE BANTIM(A,Q1,Q2,IK)
0002    C                (CALLED FROM "SOLVE" - FOR THERMAL ANALYSIS)
0003    C                (SEE 2-10)
0004    C***********************************************************************
0005    C
0006    C
0007    C     THIS SUBROUTINE MULTIPLIES A BANDED MATRIX WITH A VECTOR
0008    C
0009    C
0010    C***********************************************************************
0011    C
0012    C
0013          COMMON/SYSMSH/NUMELX,NUMNP,NRMAX
0014          COMMON/HEAT1/NMATX,NBCX,NTEMX,NTIMX,NTFX,MBANT,NMBX
0015          COMMON/DELTA/DELT(100)
0016          DIMENSION A(NUMNP,MBANT),Q1(NUMNP),Q2(NUMNP)
0017          DO 500 I=1,NUMNP
0018          NKK=I
0019          NKK1=NKK-1
0020          SUM=0.
0021          KP=0
0022          KPP=NKK+1
0023          IF (NKK1.LT.1) GO TO 5
0024          DO 510 NK=1,NKK1
0025          KP=KP+1
0026          KPP=KPP-1
0027          IF (KPP.GT.MBANT) GO TO 510
0028          SUM=SUM+A(KP,KPP)*Q2(KP)
0029      510 CONTINUE
0030        5 CONTINUE
0031          JP=NKK1
0032          DO 520 J=1,MBANT
0033          JP=JP+1
0034          IF (JP.GT.NUMNP) GO TO 520
0035          SUM=SUM+A(I,J)*Q2(JP)
0036      520 CONTINUE
0037          Q1(I)=Q1(I)-SUM/DELT(IK)
0038      500 CONTINUE
0039          RETURN
0040          END


0001          SUBROUTINE CONDUC(IX,DE,SH)
0002    C                (CALLED FROM "ADCON" - FOR THERMAL ANALYSIS)
0003    C                (SEE 2-5,2-6)
0004    C***********************************************************************
0005    C
0006    C
0007    C     THIS SUBROUTINE CALCULATES THE ELEMENT CONDUCTIVITIES
0008    C     AND HEAT CAPACITIES
0009    C
0010    C
0011    C***********************************************************************
0012    C
0013    C
0014          COMMON/COOR/XXX(378),Y(378),TX(378),IXX(378,5),TK(378)
0015          COMMON/BIGR/X(378)
0016          COMMON/LIM/S(5,5)
0017          COMMON/LAM/MTYPE,COND,H(5,5)
0018          DIMENSION E(3,3),KX(6),IX(5),F(3,3),LN(6)
0019    C
0020    C**** INITIALIZATION
0021    C
0022          DO 500 I=1,5
0023          LN(I)=IX(I)
0024          DO 500 J=1,5
0025          H(I,J)=0.0
0026      500 S(I,J)=0.
0027          LN(5)=IX(1)
0028          LN(6)=IX(2)
```

```
0029          IDK=4
0030          IF (LN(4).EQ.LN(3)) IDK=1
0031          DO 510 KD=1,IDK
0032          I=LN(KD)
0033          J=LN(KD+1)
0034          K=LN(KD+2)
0035          AJ=X(J)-X(I)
0036          AK=X(K)-X(I)
0037          BJ=Y(I)-Y(J)
0038          BK=Y(K)-Y(I)
0039          CJ=Y(J)-Y(K)
0040          CK=X(K)-X(J)
0041          XLAM=0.5*(X(I)*CJ+X(J)*BK+X(K)*BJ)
0042          IF (XLAM.LE.O.) RETURN
0043          COM=COND*(X(I)+X(J)+X(K))/(24.*XLAM)
0044          COMC=DE*SH*XLAM/60.
0045          E(1,1)=CJ**2+CK**2
0046          E(1,2)=BK*CJ-AK*CK
0047          E(1,3)=BJ*CJ+AJ*CK
0048          E(2,1)=E(1,2)
0049          E(2,2)=BK**2+AK**2
0050          E(2,3)=BJ*BK-AJ*AK
0051          E(3,1)=E(1,3)
0052          E(3,2)=E(2,3)
0053          E(3,3)=BJ**2+AJ**2
0054          F(1,1)=3.*X(I)+X(J)+X(K)
0055          F(1,3)=X(I)+X(J)/2.+X(K)
0056          F(1,2)=X(I)+X(J)+X(K)/2.
0057          F(2,1)=F(1,2)
0058          F(2,2)=X(I)+3.*X(J)+X(K)
0059          F(2,3)=X(I)/2.+X(J)+X(K)
0060          F(3,1)=F(1,3)
0061          F(3,2)=F(2,3)
0062          F(3,3)=X(I)+X(J)+3.*X(K)
0063          KX(1)=1
0064          KX(2)=2
0065          KX(3)=3
0066          KX(4)=4
0067          KX(5)=1
0068          KX(6)=2
0069          DO 520 IH=1,3
0070          II=KX(IH+KD-1)
0071          DO 520 JH=1,3
0072          JJ=KX(JH+KD-1)
0073          H(II,JJ)=H(II,JJ)+F(IH,JH)*COMC
0074          S(II,JJ)=S(II,JJ)+E(IH,JH)*COM
0075    520 CONTINUE
0076    510 CONTINUE
0077          RETURN
0078          END


0001          SUBROUTINE FLOTEM(TAVE,SH,DE,COND,MTYPE)
0002    C                (CALLED FROM "SOLVE" - FOR THERMAL ANALYSIS)
0003    C                (SEE USER MANUAL)
0004    C*********************************************************************
0005    C
0006    C
0007    C     THIS SUBROUTINE CALCULATES THE TEMPERATURE-DEPENDENT
0008    C     THERMOPHYSICAL MATERIAL PROPERTIES
0009    C
0010    C
0011    C*********************************************************************
0012    C
0013    C
0014          COMMON/HEAT3/CFX(378),XCONX(6),CPX(6),ROX(6),
0015         *      TFUN(9,6,3),TIFX(21,99),XLX(378)
0016          M=MTYPE+1
0017          IK=1
```

373

```
0018          DO 500 IC=1,3
0019       10 IF (TAVE.GE.TFUN(IK,1,IC).AND.TAVE.LE.TFUN(IK+1,1,IC)) GO TO 5
0020          IF (TAVE.GT.TFUN(IK+1,1,IC)) IK=IK+1
0021          IF (TAVE.LT.TFUN(IK,1,IC)) IK=IK-1
0022          GO TO 10
0023        5 DT=TFUN(IK+1,1,IC)-TFUN(IK,1,IC)
0024          DC=TFUN(IK+1,M,IC)-TFUN(IK,M,IC)
0025          DTEM= (TAVE-TFUN(IK,1,IC))/DT
0026          IF (IC.EQ.1) COND=TFUN(IK,M,IC)+DTEM*DC
0027          IF (IC.EQ.2) SH=TFUN(IK,M,IC)+DTEM*DC
0028      500 IF (IC.EQ.3) DE=TFUN(IK,M,IC)+DTEM*DC
0029          RETURN
0030          END




0001          SUBROUTINE FLOW(Q,A,XA)
0002    C               (CALLED FROM "SOLVE" - FOR THERMAL ANALYSIS)
0003    C               (SEE 2-9-1, 2-9-2)
0004    C*****************************************************************
0005    C
0006    C
0007    C     THIS SUBROUTINE CALCULATES FLOW BOUNDARY CONDITION VALUES
0008    C     AND CHANGES THE CONDUCTIVITY MATRIX ACCORDINGLY
0009    C
0010    C
0011    C*****************************************************************
0012    C
0013    C
0014          COMMON/SYSMSH/NUMELX,NUMNP,NRMAX
0015          COMMON/HEAT1/NMATX,NBC,NTEMX,NTIM,NTFX,MBANT,NMBX
0016          COMMON/COOR/XXX(378),Y(378),T(378),IXX(378,5),TK(378)
0017          COMMON/BIGR/X(378)
0018          COMMON/HEAT3/CF(378),XCONX(6),CPX(6),ROX(6),
0019        *      TFUNX(9,6,3),TIF(21,99),XL(378)
0020          COMMON/GRID/IK,KKK,DT,TM
0021          COMMON/PAR/IDTEMP(378)
0022          COMMON/HTBOND/IBCT(75),JBCT(75),IDB(6,75)
0023          COMMON/DELTA/DELT(100)
0024          DIMENSION A(NUMNP,MBANT),Q(NUMNP),DF(378),F(6)
0025    C
0026    C**** INITIALIZATION
0027    C
0028          DO 500 I=1,NUMNP
0029      500 Q(I)=0.
0030          IKP=0
0031          TAU=TM
0032       20 TAU1=TAU-1.0E-06
0033          IF (TAU.GE.TIF(1,IK).AND.TAU1.LE.TIF(1,IK+1)) GO TO 5
0034          IF (TAU.GT.TIF(1,IK+1)) GO TO 10
0035          GO TO 15
0036       10 TM=TM-DELT(IK)
0037          IK=IK+1
0038          IKP=1
0039          TAU=TM+DELT(IK)
0040          TM=TAU
0041       15 CONTINUE
0042          IF (TAU.LT.TIF(1,IK)) IK=IK-1
0043          GO TO 20
0044        5 D=TIF(1,IK+1)-TIF(1,IK)
0045          DT=(TAU-TIF(1,IK))/D
0046          IF (NBC.EQ.0) GO TO 25
0047    C
0048    C**** FIND FLOW VECTOR Q AND CHANGE THE CONDUCTIVITY MATRIX A
0049    C
0050          MM=0
0051          DO 510 N=1,NBC
0052          DO 520 M=1,6
```

374

```
0053            F (M) =0.
0054            K=IDB (M,N)
0055            IF (K.LE.O) GO TO 520
0056            K=K+1
0057            DH=TIF (K, IK+1) -TIF (K, IK)
0058            F (M) =TIF (K, IK) +DT*DH
0059        520 CONTINUE
0060            I=IBCT (N)
0061            II=-IDB (2,N)
0062            J=JBCT (N)
0063            XAV= (X (I) +X (J) ) /2.0-XA
0064            IF (XAV.EQ.0.) XAV=0.0001
0065            YAV= (Y (I) +Y (J) ) /2.0
0066            THETA=ATAN (YAV/XAV)
0067            TAVE= (T (I) +T (J) ) /2.
0068            S2=0.
0069            IF ( (F (1) .NE.TAVE) .AND. (F (2) .NE.0.) ) GO TO 30
0070            TC=F (2)
0071            GO TO 35
0072         30 TC=ABS (F (1) -TAVE) **F (5) *F (2)
0073         35 IF (F (3) .GT.0.) GO TO 40
0074            TC2=0.
0075            GO TO 45
0076         40 TC2=F (3) **F (5)
0077         45 CONTINUE
0078            S3=TC*F (1)
0079            S2=TC2*F (1)
0080            F (1) =F (1) +460.
0081            TAVE=TAVE+460.
0082            SIG= (F (1) **2+TAVE**2) * (F (1) +TAVE) *F (4)
0083            S1=SIG* (F (1) -460.0)
0084            TC3=F (6)
0085            S4=TC3
0086            R= (S1+S2+S3+S4) *XL (N)
0087            IF (KKK) 50,50,55
0088         55 Q (I) =Q (I) +R
0089            Q (J) =Q (J) +R
0090            GO TO 510
0091         50 TC= (TC+SIG+TC2) *XL (N) /2.0
0092            A (I, 1) =TC+A (I, 1)
0093            A (J, 1) =TC+A (J, 1)
0094            K=J-I+1
0095            IF (K) 60,60,65
0096         65 A (I, K) =A (I, K) +TC
0097            GO TO 510
0098         60 K=I-J+1
0099            A (J, K) =A (J, K) +TC
0100        510 CONTINUE
0101         25 CONTINUE
0102      C
0103      C**** CHANGE OF THE TEMPERATURE TO FLOW BOUNDARY CONDITION
0104      C
0105            IF (KKK.EQ.0) GO TO 70
0106            DO 530 I=1,NUMNP
0107            DF (I) =.0
0108            K=IDTEMP (I) +1
0109            IF (K.LE.1) GO TO 530
0110            DH=TIF (K, IK) +DT* (TIF (K, IK+1) -TIF (K, IK) )
0111            DF (I) =DH
0112            Q (I) =Q (I) +CF (I) *DH
0113        530 CONTINUE
0114         70 RETURN
0115            END
```

APPENDIX 5

```
0001          SUBROUTINE HETDAT
0002     C              (CALLED FROM "THERM" - FOR THERMAL ANALYSIS)
0003     C              (SEE USER MANUAL)
0004     C*******************************************************************
0005     C
0006     C
0007     C    THIS SUBROUTINE READS IN HEAT TRANSFER BOUNDARY INPUT DATA
0008     C
0009     C
0010     C*******************************************************************
0011     C
0012     C
0013          COMMON/COOR/XXX(378),Y(378),T(378),IXX(378,5),TK(378)
0014          COMMON/BIGR/X(378)
0015          COMMON/HEAT1/NMAT,NBC,NTEM,NTIM,NTF,MBANTX,NMBX
0016          COMMON/PAR/IDTEMP(378)
0017          COMMON/HEAT3/CF(378),XCON(6),CP(6),RO(6),
0018         *       TFUN(9,6,3),TIF(21,99),XL(378)
0019          COMMON/HTBOND/IBCT(75),JBCT(75),IDB(6,75)
0020          COMMON/SYSMSH/NMLTES,NMNPTS,NRMAX
0021          NUMNP=NMNPTS
0022     C
0023     C**** READ AND PRINT MATERIAL PROPERTIES.
0024     C
0025          DO 500 I=1,NUMNP
0026      500 CF(I)=1.0
0027          DO 510 I=1,NMAT
0028      510 READ (5,1000) MTYPE,XCON(MTYPE),CP(MTYPE),RO(MTYPE)
0029          WRITE (6,2000) (I,XCON(I),CP(I),RO(I),I=1,NMAT)
0030     C
0031     C**** READ AND PRINT ALL TIME FUNCTIONS.
0032     C
0033          IF (NTIM.EQ.0) GO TO 5
0034          DO 520 I=1,NTIM
0035      520 READ (5,1001) (TIF(J,I),J=1,NTF)
0036          N=NTF+9
0037          NF=10
0038          NA=2
0039       10 IF (NF.GT.NTF) NF=NTF
0040          NN=NA-1
0041          NK=NF-1
0042          WRITE (6,2001) (I,I=NN,NK)
0043          DO 530 I=1,NTIM
0044      530 WRITE (6,2002) TIF(1,I),(TIF(J,I),J=NA,NF)
0045          NA=NF+1
0046          NF=NF+9
0047          IF (N.NE.NF) GO TO 10
0048          NTSUM=0
0049          NTIM1=NTIM-1
0050        5 IF (NBC.EQ. 0) GO TO 15
0051     C
0052     C**** READ AND PRINT BOUNDARY CONDITIONS
0053     C
0054          N=0
0055          DO 540 I=1,NBC
0056          READ (5,1002) IBCT(I),JBCT(I),(IDB(J,I),J=1,6)
0057          IF (NUMNP.EQ.276) IBCT(I)=IBCT(I)+10
0058          IF (NUMNP.EQ.276) JBCT(I)=JBCT(I)+10
0059          II=IBCT(I)
0060          IJ=JBCT(I)
0061          IF (NUMNP.EQ.276) IDB(3,I)=0
0062          IF (IDB(2,I).GE.0) GO TO 540
0063          IF (II.GT.0) GO TO 20
0064          II=-II
0065          N=N+1
0066          K=IDB(1,I)+1
0067          LEIGH=NUMNP+N
0068          T(LEIGH)=TIF(K,I)
0069       20 N=N+1
0070          LEIGH=NUMNP+N
```

376

```
0071        T(LEIGH)=T(IJ)
0072    540 XL(I)=SQRT((X(II)-X(IJ))**2+(Y(II)-Y(IJ))**2)*(X(II)+X(IJ))*0.25
0073        WRITE (6,2003) (IBCT(I),JBCT(I),(IDB(J,I),J=1,6),I=1,NBCT)
0074  C
0075  C**** READ AND PRINT CONDUCTIVITY OF MATERIALS V.S. TEMPERATURE.
0076  C
0077     15 IF (NTEM.EQ.0) GO TO 25
0078        M=NMAT+1
0079        DO 550 IC=1,3
0080        DO 550 I=1,NTEM
0081        READ (5,1003) (TFUN(I,N,IC),N=1,M)
0082    550 IF (M.EQ.6) TFUN(I,6,IC)=TFUN(I,2,IC)
0083        DO 560 IC=1,3
0084        WRITE (6,2004)
0085        N=M+9
0086        NF=10
0087        NA=2
0088        IF (NF.GT.M) NF=M
0089        NN=NA-1
0090        NK=NF-1
0091        WRITE (6,2005) (I,I=NN,NK)
0092        DO 560 I=1,NTEM
0093    560 WRITE (6,2002) TFUN(I,1,IC),(TFUN(I,J,IC),J=NA,NF)
0094        NA=NF+1
0095        NF=NF+9
0096     25 RETURN
0097  C
0098  C**** INPUT FORMATS
0099  C
0100   1000 FORMAT(I5,3E10.0)
0101   1001 FORMAT(8E10.0)
0102   1002 FORMAT(8I5)
0103   1003 FORMAT(6F10.0)
0104  C
0105  C**** OUTPUT FORMATS
0106  C
0107   2000 FORMAT (1H1,(10H MATERIAL=,I3/7X,
0108       * 'CONDUCTIVITY=',E12.5//7X,
0109       * 'SPECIFIC HEAT=',E12.5//7X,
0110       * 'DENSITY     =',E12.5//))
0111   2001 FORMAT(1H1,3X,'TIME        ',I6,8I12)
0112   2002 FORMAT(10E12.5)
0113   2003 FORMAT (1H1,32X,31H FUNCTION IDENTIFICATION NO.   ,/1X,
0114       * '   I        J  EXTER. TEMP.  FREE CONVEC.  FORCED CONVEC.
0115       * RAD. FACTOR  CONVECTION EXP.'/(2I7,I9,5I15))
0116   2004 FORMAT(///)
0117   2005 FORMAT(1H1,2X,'TEMPERATURE    ',I6,8I12)
0118        END


0001        SUBROUTINE SOLVE (DTI,TIME,XA,Q,E,C,VOL,A,JOBTYP)
0002  C          (CALLED FROM "THERM" - FOR THERMAL ANALYSIS)
0003  C          (SEE 2-10)
0004  C*************************************************************************
0005  C
0006  C
0007  C     THIS SUBROUTINE ASSEMBLES AND SOLVES THE
0008  C     SYSTEM OF EQUATIONS FOR THERMAL ANALYSIS
0009  C
0010  C
0011  C*************************************************************************
0012  C
0013  C
0014        COMMON/COOR/XXX(378),Y(378),T(378),IXX(378,5),TK(378)
0015        COMMON/BIGR/X(378)
0016        COMMON/SYSMSH/NUMEL,NUMNP,NRMAX
0017        COMMON/PAR/IDTEMP(378)
0018        COMMON/SWITCH/IX(5,378)
0019        COMMON/HTBOND/IBCTX(75),JBCTX(75),IDBX(6,75)
```

377

```
0020        COMMON/HEAT1/NMAT,NBC,NTEM,NTIM,NTF,MBANT,NMB
0021        COMMON/HEAT3/CF(378),XCON(6),CP(6),RO(6),
0022     *       TFUN(9,6,3),TIF(21,99),XL(378)
0023        COMMON/TIMSTP/ITIME
0024        COMMON/GRID/IK,KKK,DTX,TM
0025        COMMON/DELTA/DELT(100)
0026        COMMON/LAM/MTYPE,COND,HX(5,5)
0027        COMMON/LOCAL2/REFV(378)
0028        DIMENSION VOL(378),TFAKE(378)
0029        DIMENSION A(NUMNP,MBANT),Q(NUMNP),E(NUMNP),C(NUMNP,MBANT)
0030    C
0031    C**** INITIALIZE
0032    C
0033        IK=0
0034     5  IK=IK+1
0035        IF ((TIME-TIF(1,IK)).GE.0.0) GO TO 5
0036        IK=IK-1
0037        TM=TIF(1,IK)
0038        DELT(IK)=DTI
0039        DO 500 I=1,NUMNP
0040        Q(I)=0.
0041        E(I)=0.
0042        DO 500 J=1,MBANT
0043        C(I,J)=0.
0044    500 A(I,J)=0.
0045        DO 510 N=1,NUMEL
0046        I=IX(1,N)
0047        J=IX(2,N)
0048        K=IX(3,N)
0049        L=IX(4,N)
0050        XX=(X(I)+X(J)+X(K)+X(L))/4.
0051        IF (K.EQ.L) XX=(X(I)+X(J)+X(K))/3.
0052        VOL(N)=((X(I)-X(K))*(Y(J)-Y(L))-(X(J)-X(L))*(Y(I)-Y(K)))/2.*XX
0053        IF (ITIME.EQ.1) REFV(N)=VOL(N)
0054        MTYPE=IX(5,N)
0055        SH=CP(MTYPE)
0056        DE=RO(MTYPE)
0057        TAVE=(T(I)+T(J)+T(K)+T(L))/4.0
0058        COND=XCON(MTYPE)
0059    C
0060    C**** DETERMINE ELEMENT CONDUCTANCE
0061    C
0062        IF (COND.EQ.0.) CALL FLOTEM(TAVE,SH,DE,COND,MTYPE)
0063    510 CALL ADCON (DE,A,C,N,COND,SH)
0064    C
0065    C**** REVISE CONDUCTANCE MATRIX
0066    C
0067        DO 520 IFAKE=1,378
0068    520 TFAKE(IFAKE)=T(IFAKE)
0069        IF (JOBTYP.EQ.1 .OR. JOBTYP.EQ.3 .OR. JOBTYP.EQ.52) GO TO 10
0070        CALL BANTIM(C,E,TFAKE,IK)
0071        DO 530 I=1,NUMNP
0072        DO 530 J=1,MBANT
0073    530 A(I,J)=A(I,J)+C(I,J)/DELT(IK)
0074    C
0075    C**** MODIFY CONDUCTANCE MATRIX FOR BOUNDARY CONDITIONS
0076    C
0077     10 KKK=0
0078        CALL FLOW(Q,A,XA)
0079        ENORM=0.
0080        CNORM=0.
0081        DO 540 J=1,NUMNP
0082        CNORM=CNORM+C(J,1)
0083    540 ENORM=ENORM+A(J,1)
0084        ENORM=ENORM*1000000.
0085        DELTRC=CNORM*1000000./(ENORM+ENORM)
0086        DO 550 I=1,NUMNP
0087        IF (IDTEMP(I).EQ.0) GO TO 550
0088        A(I,1)=ENORM
0089        CF(I)=ENORM
```

```
0090      550 CONTINUE
0091      C
0092      C**** TRIANGULARIZATION OF THE CONDUCTANCE MATRIX
0093      C
0094          CALL TRIA(A)
0095      C
0096      C**** START THE STEP BY STEP PROCEDURE
0097      C
0098          KKK=1
0099          TM=TIME
0100      C
0101      C**** EVALUATE FLOW VECTOR AND REVISE THE CONDUCTIVITY MATRIX
0102      C
0103          CALL FLOW(Q,A,XA)
0104      C
0105      C**** CALCULATE EFFECTIVE FLOW VECTOR Q
0106      C
0107          DO 560 I=1,NUMNP
0108          TEMP=Q(I)
0109          Q(I)=Q(I)-E(I)
0110      560 E(I)=TEMP
0111          CALL BACKS(A,Q)
0112      C
0113      C**** COMPUTE THE TEMPERATURE
0114      C
0115          DO 570 I=1,NUMNP
0116          T(I)=Q(I)
0117      570 Q(I)=Q(I)-T(I)
0118          CALL BANTIM(C,E,Q,IK)
0119          RETURN
0120          END



0001          SUBROUTINE THERM(DTI,TIME,XA,JOBTYP)
0002      C              (CALLED FROM "MAIN" - FOR THERMAL ANALYSIS)
0003      C              (SEE USER MANUAL)
0004      C*****************************************************************
0005      C
0006      C
0007      C     THIS SUBROUTINE FUNCTIONS AS THE "MAIN" PROGRAM
0008      C     FOR THERMAL ANALYSIS
0009      C
0010      C
0011      C*****************************************************************
0012      C
0013      C
0014          COMMON/PRINT1/IPRINT,NELPR,NODPR,IOPRNT,IDELPR(378),IDNDPR(378)
0015          COMMON/PAR/IDTEMP(378)
0016          COMMON/COOR/R(378),Z(378),T(378),IX(378,5),TK(378)
0017          COMMON/BIGR/RBIGR(378)
0018          COMMON/SYSMSH/NUMEL,NUMNP,NRMAX
0019          COMMON/SWITCH/IXI(5,378)
0020          COMMON/HEAT1/NMAT,NBC,NTEM,NTIM,NTF,MBANT,NMB
0021          COMMON/HEAT3/CFX(378),XCONX(6),CPX(6),ROX(6),
0022         *        TFUNX(9,6,3),TIFX(21,99)
0023          COMMON/HTBOND/IBCTX(75),JBCTX(75),IDBX(6,75)
0024          COMMON/TIMSTP/ITIME
0025          INTEGER HED(12)
0026          DIMENSION A(378,28),Q(378),E(378),C(378,28),VOL(378)
0027          IF (ITIME.GT.1) GO TO 5
0028      C
0029      C**** INPUT AND OUTPUT THERMAL ANALYSIS INFORMATION (ONCE ONLY)
0030      C
0031          READ (5,1000) HED,NMAT,NBC,NTEM,NTIM,NTF
0032          WRITE (6,2000) HED,NUMNP,NUMEL,NMAT,NBC,NTEM,NTIM,NTF
0033      C
0034      C**** INITIALIZE (ONCE ONLY)
0035      C
0036          DO 500 I=1,NUMNP
```

379

```
0037      500 RBIGR(I)=R(I)+XA
0038          DO 510 I=1,5
0039          DO 510 J=1,NUMEL
0040      510 IXI(I,J)=IX(J,I)
0041          MBANT=0
0042          DO 520 N=1,NUMEL
0043          MB=0
0044          DO 530 I=1,4
0045          DO 530 J=1,4
0046          MM=IABS(IXI(I,N)-IXI(J,N))
0047      530 IF (MM.GT.MB) MB=MM
0048          MB=MB+1
0049      520 IF (MB.GT.MBANT) MBANT=MB
0050          NTF=NTF+1
0051      C
0052      C**** INPUT HEAT TRANSFER INFORMATION
0053      C
0054          CALL HETDAT
0055          NMB=NUMNP*MBANT
0056      C
0057      C**** SOLVE THE HEAT TRANSFER PROBLEM
0058      C
0059        5 CALL SOLVE(DTI,TIME,XA,Q,E,C,VOL,A,JOBTYP)
0060          IF (JOBTYP.NE.1 .AND. JOBTYP.NE.2) GO TO 10
0061          IF (IPRINT-IOPRNT) 10,15,10
0062      C
0063      C**** OUTPUT NODAL TEMPERATURES
0064      C
0065       15 WRITE (6,2001) ITIME,DTI,TIME
0066          IF (NODPR.NE.0) GO TO 20
0067          WRITE (6,2002) (N,R(N),Z(N),T(N),N=1,NUMNP)
0068          GO TO 10
0069       20 WRITE (6,2002) (IDNDPR(J),R(IDNDPR(J)),Z(IDNDPR(J)),T(IDNDPR(J)),
0070        *     J=1,NODPR)
0071       10 RETURN
0072      C
0073      C**** INPUT FORMAT
0074      C
0075     1000 FORMAT(12A4/5I5)
0076      C
0077      C**** OUTPUT FORMATS
0078      C
0079     2000 FORMAT (1H1,12A4//5X,
0080        *    'NUMBER OF NODAL POINTS-------------',I5//5X,'NUMBER OF ELEMENT
0081        *S-----------------',I5//5X,'NUMBER OF MATERIALS----------------',
0082        * I5//5X,'NUMBER OF BOUNDARY SIDES-----------',I5//5X,'NUMBER OF TE
0083        *MPERATURE POINTS--------',I5//5X,'NUMBER OF TIME POINTS----------
0084        *---',I5//5X,'NUMBER OF TIME FUNCTIONS-----------',I5/)
0085     2001 FORMAT(1H1,' TIME STEP=',I5,5X,'TIME INCREMENT=',E15.6,5X,'TIME=',
0086        * E15.6,///)
0087     2002 FORMAT(1X,' NODE NUMBER    R COOR.    Z COOR.        TEMPERATURE'
0088        *//(4X,I4,7X,F8.3,7X,F8.3,8X,F13.4))
0089          END


0001          SUBROUTINE TRIA(A)
0002      C            (CALLED FROM "SOLVE" - FOR THERMAL ANALYSIS)
0003      C            (SEE 2-10- (4))
0004      C*******************************************************************
0005      C
0006      C
0007      C    THIS SUBROUTINE TRIANGULARIZES THE OVERALL CONDUCTIVITY MATRIX
0008      C
0009      C
0010      C*******************************************************************
0011      C
0012      C
0013          COMMON/SYSMSH/NUMELX,NUMNP,NRMAX
0014          COMMON/HEAT1/NMATX,NBCX,NTEMX,NTIMX,NTFX,MBANT,NMB
0015          DIMENSION A(NMB)
```

```
0016      C
0017      C**** INITIALIZE
0018      C
0019            NE=NUMNP-1
0020            MN=MBANT-1
0021            MM=MN*NUMNP
0022            MK=NUMNP-MN
0023            DO 500 N=1,NE
0024            NT=N-MK
0025            IF (NT.GT.0) MM=MM-NUMNP
0026            IF (A(N).EQ.0.0) GO TO 500
0027            L=N
0028            IL=N+NUMNP
0029            IH=N+MM
0030            DO 510 I=IL,IH,NUMNP
0031            L=L+1
0032            J=L
0033            C=A(I)/A(N)
0034            DO 520 K=I,IH,NUMNP
0035            A(J)=A(J)-C*A(K)
0036        520 J=J+NUMNP
0037            A(I)=C
0038        510 CONTINUE
0039        500 CONTINUE
0040            RETURN
0041            END


0001            SUBROUTINE CREEPP(N,SIG)
0002      C                 (CALLED FROM "STRESS" - FOR CREEP ANALYSIS)
0003      C                 (SEE 4-6-(3) AND (4))
0004      C*************************************************************************
0005      C
0006      C
0007      C     THIS SUBROUTINE FUNCTIONS AS THE "MAIN"
0008      C     PROGRAM FOR CREEP ANALYSIS
0009      C
0010      C
0011      C*************************************************************************
0012      C
0013      C
0014
0015            IMPLICIT REAL*8 (Q)
0016            COMMON /CREEP/DEPCJ(4),EPCJ(4),QDEPCJ(4),
0017           *         QP(10),CDSIG(4),QDSIG(4),QH(6,10)
0018            COMMON/TIMSTP/ITIME
0019            COMMON/CCRP/SCP(378,14),SSCP(378,2)
0020            COMMON/RESTR/MBANX,DTIME(1000)
0021            COMMON/EFBB/BBBB
0022            COMMON/DOT/DTOO
0023            COMMON/ATEM/ACTQ,RCONT,COEF,ENC
0024            COMMON/STREZZ/RATS1(378),TSIG(378,5),TOTEPS(378,4),EPS(378),
0025           *         DLAM(378),ratst(378),deps(378,4)
0026            COMMON/PMET/TEMP
0027            COMMON/DTMEN/TIMAX,TIMAXO
0028            COMMON/DDDD/FEPCDT,FECDDT,DEPCD
0029            COMMON/PPPP/FEPC,BBC,AAC
0030            COMMON/RMATR/SMATRX(4,4)
0031            COMMON/MATRI/III
0032            COMMON/RMTRIX/C(4,4),EE(20)
0033            COMMON/TIMKEP/TIMEK
0034            COMMON/RITBE/T11,T12
0035            COMMON/HEATR/XA,DTI,TIME,TIMECR
0036            DIMENSION QMATRX(4,4),QEPC(4),QSIG(10),SIG(10)
0037            DIMENSION EFALF1(438),QDTIME(1000)
0038            DOUBLE PRECISION DBBB,BBC,AAC,FEPC,FEPCDT,FECDDT,DEPCD,
0039           *               FEPC1,DEPC,DDEPCJ(4),TMAX1,TIMAX,TIMAXX,QFEPCO
0040      C
0041      C**** INITIALIZE
0042      C
```

381

```
0043        QBBBB=DBLE (BBBB)
0044        ENC1=ENC-1
0045        DO 500 I=1,4
0046        DEPCJ(I)=0.0
0047        DDEPCJ(I)=DBLE(DEPCJ(I))
0048        QSIG(I)=DBLE(SIG(I))
0049        DO 500 J=1,4
0050    500 QMATRX(I,J)=DBLE(SMATRX(I,J))
0051        IF (DTOO.LE.0.0) GO TO 5
0052        TMAX1=1.0E+05
0053        TIMAXX=1.5*(1+EE(2))/(3*EE(1)*COEF*FUCR(TEMP)*ENC*QBBBB**(ENC-1))
0054        IF (TMAX1.LT.TIMAXX) TIMAXX=TMAX1
0055        IF (N.EQ.1) TIMAX=TIMAXX
0056        IF (N.EQ.1.AND.TIME.EQ.T11.AND.TIMAX.GT.1.5*DTIME(ITIME))
0057      *    TIMAX=1.5*DTIME(ITIME)
0058        IF (N.EQ.1.AND.TIME.EQ.T11) DTIME(ITIME)=SNGL(TIMAX)*0.01
0059        IF (TIMAXX.LT.TIMAX) TIMAX=TIMAXX
0060      5 QDTIME(ITIME)=DBLE(DTIME(ITIME))
0061        IF (III.GT.2) GO TO 10
0062        IF (III.EQ.1) GO TO 15
0063        QBBBB1=DBLE(SCP(N,14))
0064        FEPC1=DBLE(SCP(N,8))
0065        DBBB=QBBBB-QBBBB1
0066        BBC=DBBB/FEPC1
0067        AAC=QBBBB1
0068        CALL DEPC1(COEF,ENC,QBBBB1,TEMP,N)
0069        GO TO 20
0070     15 FEPC=FUCR(TEMP)*COEF*QBBBB**ENC*QDTIME(ITIME)
0071        FEPC1=0.
0072        SSCP(N,1)=BBBB
0073        GO TO 20
0074     10 FEPCDT=COEF*QBBBB**ENC*FUCR(TEMP)
0075        QBBBB1=DBLE(SCP(N,14))
0076        FEPC1=DBLE(SCP(N,8))
0077        QFEPCO=DBLE(SSCP(N,2))
0078        IF (III.EQ.3) QBBBB1=DBLE(SSCP(N,1))
0079        IF (III.EQ.3) QFEPCO=0.DO
0080        QFEPCO=0.DO
0081        DBBB=QBBBB-QBBBB1
0082        BBC=DBBB/(FEPC1-QFEPCO)
0083        AAC=QBBBB-BBC*FEPC1
0084  C
0085  C**** CALCULATE THE CREEP STRAIN
0086  C
0087        CALL DEPC1(COEF,ENC,QBBBB,TEMP,N)
0088     20 DEPC=FEPC-FEPC1
0089        QFB=DEPC/QBBBB
0090        DO 510 I=1,4
0091        DDEPCJ(I)=0.DO
0092        DO 520 J=1,4
0093    520 DDEPCJ(I)=QFB*QMATRX(I,J)*QSIG(J)+DDEPCJ(I)
0094    510 DEPCJ(I)=SNGL(DDEPCJ(I))
0095        RETURN
0096        END


0001        SUBROUTINE DEPC1(COEF,ENC,QBBBB,TEMP,N)
0002  C              (CALLED FROM "CREEPP" - FOR CREEP ANALYSIS)
0003  C              (SEE 4-2-3, 4-5)
0004  C*************************************************************************
0005  C
0006  C
0007  C    THIS SUBROUTINE CALCULATES THE CREEP STRAIN
0008  C
0009  C
0010  C*************************************************************************
0011  C
0012  C
0013        COMMON/RESTR/MBANX,DTIME(1000)
0014        COMMON/DDDD/FEPCDT,FECDDT,DEPCD
```

```
0015          COMMON/TIMSTP/ITIME
0016          COMMON/PPPP/FEPC,BBC,AAC
0017          COMMON/CCRP/SCP(378,14),SSCP(378,2)
0018          DOUBLE PRECISION QDTIME(1000),QBBBB,FEPC,FEPCDT,FECDDT,BBC,AAC
0019          DOUBLE PRECISION DEPCD,SEPC
0020          QDTIME(ITIME)=DBLE(DTIME(ITIME))
0021          FEPCDT=COEF*QBBBB**ENC*FUCR(TEMP)
0022          FECDDT=ENC*COEF*QBBBB**(ENC-1)*BBC*FUCR(TEMP)
0023          DEPCD=FEPCDT*QDTIME(ITIME)
0024          SEPC=DBLE(SCP(N,8))
0025          FEPC=SEPC+DEPCD
0026          RETURN
0027          END


0001          FUNCTION FUCR(TEMP)
0002    C              (USED IN "CREEPP" - FOR CREEP ANALYSIS)
0003    C              (SEE 4-2-2)
0004    C************************************************************************
0005    C
0006    C
0007    C     CREEP EXPONENTIAL FUNCTION
0008    C
0009    C
0010    C************************************************************************
0011    C
0012    C
0013          COMMON /ATEM/ACTQ,RCONT,COEF,ENC
0014          QRT=ACTQ/TEMP/RCONT
0015          EEXPL=QRT*ALOG10(2.71828)
0016          IF(EEXPL.GT.30.0) GO TO 5
0017          FUCR=1/EXP(QRT)
0018          GO TO 10
0019       5 EEXXP=1.0E30
0020          FUCR=1/EEXXP
0021      10 RETURN
0022          END


0001          SUBROUTINE PCREEP(N,QC)
0002    C              (CALLED FROM "TRISTF" - FOR CREEP ANALYSIS)
0003    C              (SEE 4-4)
0004    C************************************************************************
0005    C
0006    C
0007    C     THIS SUBROUTINE CALCULATES THE ASSOCIATED
0008    C     CREEP STRAIN NODAL LOAD
0009    C
0010    C
0011    C************************************************************************
0012    C
0013    C
0014          IMPLICIT REAL*8 (Q)
0015          COMMON/CCRP/SCP(378,14),SSCP(378,2)
0016          COMMON/CREEP/DEPCJ(4),EPCJ(4),QDEPCJ(4),
0017         *        QP(10),CDSIG(4),QDSIG(4),QH(6,10)
0018          COMMON/RMATR/SMATRX(4,4)
0019          COMMON/MATRI/III
0020          COMMON/RMTRIX/C(4,4),EE(20)
0021          COMMON/DBLP/QRR(4),QZZ(4),QRRR(5),QZZZ(5),QXI(11)
0022          DIMENSION QF(6,10),QD(6,6),QC(4,4),QDD(3,3),QANGLE(4),
0023         *        QS(10,10),QRO(6),QTT(4),QHH(6,10),QTP(6)
0024          DIMENSION QDPC(10),QPC(10)
0025          SSG=DBLE(EE(7))
0026          IF (III.EQ.1) GO TO 5
0027          K9=9
```

```
0028         DO 500 ICC=1,4
0029         IICC=ICC+K9
0030         DEPCJ(ICC)=SCP(N,IICC)
0031     500 QDEPCJ(ICC)=DBLE(DEPCJ(ICC))
0032         DO 510 I=1,4
0033         QDSIG(I)=0.DO
0034         DO 510 J=1,4
0035     510 QDSIG(I)=QDSIG(I)+QC(I,J)*QDEPCJ(J)
0036         QDPC(1)=QDSIG(3)*QXI(2)
0037         QDPC(2)=QXI(1)*(QDSIG(1)+QDSIG(3))
0038         QDPC(3)=QXI(4)*QDSIG(3)+QXI(1)*QDSIG(4)
0039         QDPC(4)=0.DO
0040         QDPC(5)=QXI(1)*QDSIG(4)
0041         QDPC(6)=QXI(1)*QDSIG(2)
0042         DO 520 I=1,10
0043         QPC(I)=0.DO
0044         DO 530 K=1,6
0045     530 QPC(I)=QPC(I)+QH(K,I)*QDPC(K)
0046     520 QP(I)=QP(I)+QPC(I)
0047       5 RETURN
0048         END
```

```
COMMAND QUALIFIERS

  FORTRAN/SHOW=(NOMAP)/LIST TEPSAC

  /CHECK=(NOBOUNDS,OVERFLOW,NOUNDERFLOW)
  /DEBUG=(NOSYMBOLS,TRACEBACK)
  /STANDARD=(NOSYNTAX,NOSOURCE_FORM)
  /SHOW=(NOPREPROCESSOR,NOINCLUDE,NOMAP,NODICTIONARY,SINGLE)
  /WARNINGS=(GENERAL,NODECLARATIONS)
  /CONTINUATIONS=19  /NOCROSS_REFERENCE  /NOD_LINES  /NOEXTEND_SOURCE  /F77
  /NOG_FLOATING  /I4  /NOMACHINE_CODE  /OPTIMIZE
```

```
COMPILATION STATISTICS

  Run Time:        172.66 seconds
  Elapsed Time:    408.69 seconds
  Page Faults:     2386
  Dynamic Memory:  727 pages
```

AUTHOR INDEX

385

387

SUBJECT INDEX